Sustainable Aviation

The Sustainable Aviation book series focuses on sustainability in aviation considering all aspects of the field. The books are developed in partnership with the International Sustainable Aviation Research Society (SARES), and include contributed volumes consisting of select contributions to SARES international symposiums and conferences, as well as cutting edge monographs and professional books focused on all aspects of sustainable aviation. The series aims at publishing state-of-the-art research and development in areas including, but not limited to:

- Green and renewable energy resources and aviation technologies
- Aircraft engine, control systems, production, storage, efficiency and planning
- Exploring the potential of integrating renewables within airports
- Sustainable infrastructure development under a changing climate
- Training and awareness facilities with aviation sector and social levels
- Teaching and professional development in renewable energy technologies and sustainability

Sergii Boichenko • Anna Yakovlieva
Oleksandr Zaporozhets • T. Hikmet Karakoc
Iryna Shkilniuk
Editors

Chemmotological Aspects of Sustainable Development of Transport

Springer

Editors
Sergii Boichenko
Institute of Energy Safety and Energy Management, Research (experimental) Interactive Laboratory for Diagnosing Operational Materials in Energy and Transport
National Technical University of Ukraine "Igor Sikorsky Kyiv Polytechnic Insitute"
Scientific-Technical Union of Chemmotologists
Kyiv, Ukraine

Oleksandr Zaporozhets
Institute of Aviation, Warsaw, Poland
Center of Ecological Problems of Airports
Kyiv, Ukraine

Iryna Shkilniuk
Institute of Energy Safety and Energy Management, Research (experimental) Interactive Laboratory for Diagnosing Operational Materials in Energy and Transport, Ukrainian Science Research and Educational Center of Chemmotology and Certification of Fuel, Lubricants, and Technical Liquids
National Technical University of Ukraine "Igor Sikorsky Kyiv Polytechnic Institute", Scientific-Technical Union of Chemmotologists
Kyiv, Ukraine

Anna Yakovlieva
Ukrainian Science Research and Educational Center of Chemmotology and Certification of Fuel, Lubricants, and Technical Liquids
Scientific-Technical Union of Chemmotologists, National Aviation University
Kyiv, Ukraine

T. Hikmet Karakoc
Faculty of Aeronautics and Astronautics
Eskisehir Technical University
Information Technology Research and Application Center, Istanbul Ticaret University
Eskisehir, Turkey

ISSN 2730-7778 ISSN 2730-7786 (electronic)
Sustainable Aviation
ISBN 978-3-031-06576-7 ISBN 978-3-031-06577-4 (eBook)
https://doi.org/10.1007/978-3-031-06577-4

This Springer imprint is published by the registered company Springer Nature Switzerland AG
The registered company address is: Gewerbestrasse 11, 6330 Cham, Switzerland

Preface

The book selectively represents the chemmotological aspects of providing sustainable development of modern transport sector. It is devoted to the modern problems of rational use of traditional and alternative fuels, lubricants, and other operational materials during exploitation of road and air transport.

This book, *Chemmotological Aspects of Providing Sustainable Development of Transport*, comprises selected outstanding papers presented at the VIII International Scientific-Technical Conference, "Problems of Chemmotology: Theory and Practice of Rational Use of Traditional and Alternative Fuels and Lubricants." This event was held in Kamianets-Pidilskyi, Ukraine, from June 21, 2022, to June 25, 2022, with the participation of researchers, scientists, practicians, and academics from all over the world.

The book is composed of 14 chapters in total. All chapters presented by the authors (co-authors) are published in the author's edition and aims to present an issue on how to achieve more sustainable development of modern transport. The first part of the book represents the current state of development of theoretical fundamentals of aviation chemmotology. The second part includes recent studies in the fields of chemical engineering and biotechnology. The third part of the book includes results of the studies on interrelations of quality of fuels, lubricants, and technical liquids with chemmotological reliability, operation efficiency, and economy of road and air transport. The fourth part presents results of the researches devoted to practical aspects of jet fuel supply. Modern trends in development and construction of aviation engines, environmental safety, and fuel efficiency of power units are presented in chapters of the fifth part. Finally, the sixth part of the book comprises studies on impact of aviation on the environment, and development of ecologistics, system of utilization, and recycling of transport operation materials.

The contributions of the authors and reviewers and the assistance of conference organizers in the preparation of this book are sincerely appreciated.

Kyiv, Ukraine	Sergii Boichenko
Kyiv, Ukraine	Anna Yakovlieva
Kyiv, Ukraine	Oleksandr Zaporozhets
Eskisehir, Turkey	T. Hikmet Karakoc
Kyiv, Ukraine	Iryna Shkilniuk

Introduction

The transport sector is an important component of the economy that has an impact on the development and prosperity of the population.

Rational use of fuels and lubricants, energy efficiency, and environmental safety are included in the list of the most important problems of the modern world. Solving these problems determines in a great manner the sustainable development of the world economy and keeping comfort conditions for human beings.

Efficiency, reliability of operation of vehicles, and rational use of operational materials depend on their correct selection. According to their quality, operational materials must conform to both the model and operating conditions of vehicles. The use of poor quality materials leads to a decrease in the durability and reliability of machinery and machine parts; the use of materials of higher quality than required causes unreasonable increase in costs.

The knowledge of machinery suggests not only the knowledge of construction, kinematic, dynamic, and temperature characteristics but also physico-chemical properties of constituent materials that are necessary for analyzing and forecasting of physico-chemical processes during use of a fuel or a lubricant. Thus, the efficiency and reliability of vehicle operation depend not only on their structural characteristics but also on the optimal selection of fuels and lubricants, technical liquids, and other operational materials.

Professional activity of specialists dedicated to petroleum refinery, organizing of storage, transportation and distribution of products, assurance of correspondence between the properties of Fuels, Lubricants, Technical liquids and the conditions of operation of technology and engines aimed at obtaining maximum technical, economical, ecological and social effects is called usage of Fuels, Lubricants and Technical liquids.

To *know* fuels, lubricants, and technical liquids is to clearly understand the interconnection of quality parameters with physico-chemical and energy processes, occurring in the process of their use under specific conditions, and also the connection with their chemical and group composition.

The *knowledge* of technology suggests not only the knowledge of construction, kinematic, dynamic, and temperature characteristics but also physico-chemical properties of constituent materials that are necessary for analyzing and forecasting of physico-chemical processes during use of a fuel or a lubricant.

The study of the essence, regularity (tendens), and connections of phenomena and the processes of use of fuels, lubricants, and technical liquids in aviation technology with the help of special methodological tools is the base of *aviation chemmotology*.

Aviation chemmotology is a part of *chemmotology* that studies and solves the problems of ensuring the necessary quality and application requirements of fuels and lubricants used in aviation technology.

Chemmotological reliability is reliability of technology depending on the quality of fuels and lubricants (the ability of technology to maintain good reliability when operated with fuel and lubricant grades that are of economically reasonable quality level).

This book as an integrative scientific work of many scholars is a striking example of the representation of these aspects and really illustrates the modern consolidated work of scientists and practitioners, trends in the development of scientific schools at different universities and in different countries, and science in general. Because, as is known, science does not have borders. Scientific achievements are global civilizational heritage.

Kyiv, Ukraine Sergii Boichenko
Kyiv, Ukraine Anna Yakovlieva
Kyiv, Ukraine Oleksandr Zaporozhets
Eskisehir, Turkey T. Hikmet Karakoc
Kyiv, Ukraine Iryna Shkilniuk

Contents

1 **Advance in Pathways to Sustainable Aviation Fuels** 1
Anna Yakovlieva and Sergii Boichenko

2 **Effect of Polymer Additives on the Rheological Properties of Heavy High-Viscosity Oil** 19
Tetiana Yarmola, Viktoria Romanchuk, Volodymyr Skorokhoda, and Petro Topilnytskyy

3 **Phenomenological Probabilistic Model of Friction Pair Wear Taking into Account Thermal Mechanical Stability of Boundary Layers** 31
Oksana Mikosianchyk, Rudolf Mnatsakanov, Vitalii Tokaruk, and Olena Kharchenko

4 **Revisiting the Synthesis of Fatty Acid Alkyl Esters of Lower Monohydric Alcohols by Homogeneous Base-Catalyzed Transesterification of Vegetable Oils** 49
Serhii Konovalov, Stepan Zubenko, Lyubov Patrylak, Anjela Yakovenko, Volodymyr Povazhnyi, and Kateryna Burlachenko

5 **Cultivating Microalgae in Wastewaters for Biofuel and Fertilizer Production** 81
S. Shamanskyi, S. Boichenko, I. Nezbrytska, and L. Pavliukh

6 **Development of New Structured Honeycomb Fiber Catalysts for Hydrocarbon Conversion to Carbon-Free Fuel** 101
Hennady Soloviov, Andryi Halstian, and Andryi Bushuev

7 **Providing Wasteless Manufacturing of Aviation Biofuels by Using Camelina Seed Residues for Producing Functional Bread** 127
Iryna Korniienko, Andrii Anatskyi, and Mykhailo Baranovskyi

8 Technologies for Restoring Soil from Oil Pollution 153
Nadiia Adamchuk-Chala, Sergii Ponomarenko,
Liubov Yankiv-Vitkovska, and Yelyzaveta Chala

9 Investigation for a Sustainable Use of Fossil Coal Through the Dynamics of Interaction of Smokeless Solid Fuel with Oxygen and the Possibilities of Its Practical Application 167
Yevgen Zbykovskyy and Iryna Shvets

10 Improvement of Diesel Engine Parameters by Using of Alcohol Conversion 187
Sviatoslav Kryshtopa, Liudmyla Kryshtopa, Myroslav Panchuk,
Volodymyr Korohodskyi, Igor Prunko, and Ivan Mykytii

11 Influence of Microbiological Pollution on Properties of Motor Fuels 209
Olena Shevchenko and Daryna Popytailenko

12 System for Monitoring Microbiological Contamination of Jet Fuels and Fuel Systems 231
Iryna Shkilniuk, Sergii Boichenko, Tetyana Kondratiuk,
and Kazimierz Lejda

13 Kinetics of Dissolution of Oil Deposits 247
Olena Tertyshna, Kostiantyn Zamikula, Oleh Tertyshnyi,
and Viacheslav Polishchuk

14 Cavitation Treatment of Gas Condensate Gasoline, Modified with Monohydric Alcohols 267
Sergii Kudryavtsev, Oleksii Tselishchev, Sergii Boichenko,
and Marina Loriia

Index 291

Chapter 1
Advance in Pathways to Sustainable Aviation Fuels

Anna Yakovlieva and Sergii Boichenko

Nomenclature

APU	Auxiliary Power Unit
ASTM	American Society for Testing and Materials
ATJ-SKA	Alcohol to Jet-Synthesized Kerosene with Aromatics
ATJ-SPK	Alcohol to Jet-Synthesized Paraffinic Kerosene
CH	Catalytic Hydrothermolysis
CAF	Conventional Aviation Fuels
DSHC	Direct Sugar to Hydrocarbons
FT	Fischer-Tropsch
HDCJ	Hydroprocessed Depolymerized Cellulosic Jet
HDO-SK	Hydrodeoxygenated Synthesized Kerosene
HDO-SKA	Hydrodeoxygenated Synthesized Aromatic Kerosene
HEFA	Hydrotreated Esters and Fatty Acids
HVO	Hydrotreated Vegetable Oils
SAF	Sustainable Aviation Fuels
SIP	Synthesized Isoparaffins
WSD	Wear Scar Diameter

A. Yakovlieva (✉)
Scientific-research department, National Aviation University, Kyiv, Ukraine
e-mail: anna.yakovlieva@nau.edu.ua

S. Boichenko
Institute of energy safety and energy management, National technical university of Ukraine
"Igor Sikorsky Kyiv Polytechnic Institute", Kyiv, Ukraine

S. Boichenko et al. (eds.), *Chemmotological Aspects of Sustainable Development of Transport*, Sustainable Aviation, https://doi.org/10.1007/978-3-031-06577-4_1

1.1 Introduction

Modern civil aviation produces about 2% of all global emissions of carbon. It is also the source of about 12% of all CO_2 emissions, which are formed by the transport sector (IATA 2018). According to forecasts, emissions of CO_2 from civil aviation will increase in about three times by 2050 (Ekici et al. 2016). This prediction is explained by the fact that passenger and cargo air transportation develops exceptionally quickly (Ekici et al. 2020; Maniatis 2018). And in the nearest future air transport may become a leader in global transportations. There is an opinion that today decarbonization of air transport is one of the most complicated tasks compared to other means of transport. Mainly, this relates to reduction of carbon dioxide emissions (IATA 2018). One of the possible ways to reduce CO_2 emissions from aviation is the implementation and use of low-carbon or sustainable aviation fuels (SAFs). Thus, this work aims to review and analyze the recent advancements and future challenges in progress and commercial use of alternative aviation fuels (IATA 2018; Ekici et al. 2016). The positive and negative aspects of utilizing alternative fuels have to be considered on the way toward decarbonization of modern air transport (Ekici et al. 2020).

Even considering significant progress in that has been made in SAF development, there are still significant problems and challenges to be solved (Maniatis 2018). Today we still can observe a number of problematic issues connected to technological processes, production, and supply of feedstock, transportation, storage, and delivery to airlines (Ekici et al. 2016; Hui et al. 2012).

The air transport sector is considered as one of the most significant contributors to global CO_2 emissions (IATA 2018; Hui et al. 2012). Moreover, these emissions, which are released at high altitudes, are determined as more harmful compared to low-level emissions (Ekici et al. 2020; Allen et al. 2013). The industry and policymakers are trying to reduce such adverse impact on environment and see advanced SAFs as one of the most reliable solutions. Liquid biofuels (sustainable fuels), which may be used as a "drop-in" fuel, can provide compatibility with existing infrastructure and aircraft design. To estimate the current state and possible near-term progress, it is necessary to analyze the existing and developing technologies for SAF manufacturing, its typical properties, and production potential to provide commercial aviation needs.

1.2 Pathways to Sustainable Aviation Fuels

The overall pathway to develop and introduce SAF to market on an industrial scale is complicated and requires a long time. Industrial-scale demonstration plants, which can produce sufficient amounts of SAFs, cannot be launched without substantial financial investments. Typically, these investments are estimated from tens to hundreds of million euros. Usually, it is pretty hard to attract such finances, as it is associated with numerous risks and uncertainties (Maniatis 2018).

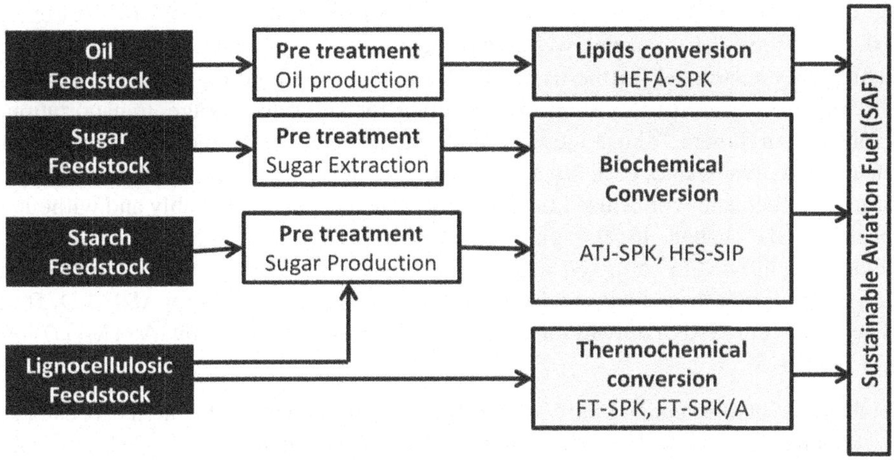

Fig. 1.1 Available aviation biofuel process pathways

When we speak about the rising production of SAF on a global scale, this is not an easy task, as this process requires strong industrial solid. It may be explained by the fact that today SAF cannot compete with traditional jet fuels produced from crude oil. The most popular pathways for alternative jet fuel production are shown in Fig. 1.1.

Decreasing CO_2 and other pollutant emissions in the air transport sector by using new SAFs on commercial flight requires passing long and complicated procedures of fuels' testing, certification, and approval. All these procedures and requirements for new fuel quality are determined in the following specification (IATA 2018; Maniatis 2018):

- ASTM D1655 Standard Specification for Aviation Turbine Fuels.
- ASTM D7566 Standard Specification for Aviation Turbine Fuel Containing Synthesized Hydrocarbons.
- ASTM D4054 Standard Practice for Qualification and Approval of New Aviation Turbine Fuels and Fuel Additives.

Fulfillment of the complex of activities, which are necessary to certify and approve new jet fuels, needs strong efforts and resources, with the involvement of different stakeholders. These activities should be done according to the specification ASTM D4054, except that passing from steps 1–2 to steps 3–4, as it is denoted in this specification, will also require production of a considerable amounts of jet fuel (about several tens of ton). Consequently, it is evident that this requires building small-scale demonstration plants. It needs more financial investments from producers. Thus, we may see a set of reasonable barriers on the way of introducing of new SAF to the commercial market (Allen et al. 2013).

Traditionally aircraft of civil aviation uses conventional jet fuel of grade Jet-A1. Due to a number of reasons, there is a need to develop a "drop-in" SAFs, which will be compatible with conventional oil-derived jet fuels at least till 2050 (Ekici et al. 2020). Among these reasons we should mention the high cost of existing aircraft, which are constructed for the use of kerosene, the long time for replacing existing aircraft fleet, as well as existing infrastructure for kerosene storage, transportation, fueling, and others. Thus, industry and scientists look for the development of sustainable, renewable, or alternative jet fuels, which will be produced from renewable feedstock and will be used in existing jet engines interchangeably and without a need to make changes in jet engine construction. Today there is a number of SAFs, which are allowed to be mixed with conventional jet fuel up to 50%. Requirements for the quality of these fuels are determined by standard specification ASTM D7566. The following SAF pathways are approved by the specification ASTM D7566 (Moses 2015):

- Hydrotreated esters and fatty acids (HEFA) – fuels received from oily feedstock: cooking oil, animal fats, algae, and plant oils (like camelina, rapeseed, jatropha, etc.) (HEFA-SPK).
- Fischer-Tropsch (FT) synthesis – fuels derived from solid biomass resources (wood or agricultural residues, municipal waste, etc.) (FT-SKA).
- Synthetic isoparaffin (SIP) – fuels derived from fermented hydroprocessed sugars, previously known as direct-sugar-to-hydrocarbon fuels. This fuel may be blended up to 10% with conventional jet fuel (SIP-SPK).
- Alcohol-to-jet (ATJ) – fuels derived from ethanol or iso-butanol. This fuel may be blended up to 30% with conventional jet fuel (ATJ-SPK).

Since the technology of SAF production has been developed, the new fuels should pass the procedure of certification and approval by the ASTM. At first, the new fuels should be tested for compliance with standard requirements; next, it is blended with conventional jet fuels (up to 50% vol.) and passes the same testing procedure. The maximal amount of SAF that can be blended with conventional one is determined by compliance to standard specification requirements and level of safety, which should be kept during its use. At the same time, probably in the following years, higher limits of SAF will be approved for use in jet fuel blends. In the case when SAF blended with conventional jet fuels meets standard requirements, it should pass certification procedure as conventional jet fuel of grade Jet A-1. It means that no changes to aircraft design or airport infrastructure should be done, and new fuel can be used in the same way as conventional jet fuel (IATA 2018; Yanovskii et al. 2020; Yakovlieva et al. 2017).

1.3 FT Kerosene Pathway

The Fischer-Tropsch (FT) process is already a well-known technology for alternative aviation fuel production and was the first technology approved by the ASTM for alternative fuel production. This technology is realized in four stages. At the first stage, the feedstock is transformed into syngas (CO + H2). This gas then enters the next stage, where long-chain hydrocarbons (paraffin or olefins) are received, and then it comes through the process of hydrocracking and isomerization. There is a possibility to vary product output. For example, paraffins are obtained at low-temperature FT with cobalt catalyst and olefins are obtained at high-temperature FT with iron catalyst. At the final stage, the obtained products pass through distillation stage, where specific fractions of hydrocarbons are received. One of these fractions is the kerosene fraction. In the case when natural gas is used as a feedstock, the syngas may be obtained by steam reforming or by partial oxidation of the gas (using oxygen). Simultaneously they apply conditioning; due to this, it is possible to delete particles and admixture. If solid matter (coal or wood residues) is used as a feedstock, syngas production requires partial oxidation and steam gasification, with the further step of conditioning. In the case when biomass is used as a feedstock, it is necessary to pass through the stage of initial feedstock treatment (Maniatis 2018; Starck et al. 2016).

FT fuel was firstly approved for blending with conventional jet fuel with maximum content of 50% vol. by ASTM in 2009. It was the reason for creating a new specification, ASTM D7566, for setting quality requirements to synthetic aviation fuel. The approval was given to FT fuel produced from coal feedstock; however now, the ASTM has certified aviation fuel produced by FT synthesis without relation to feedstock. Fuels produced by FT synthesis do not contain aromatic compounds. Because of this, they may be used only as a blend with conventional jet fuel up to 50% vol. Aromatic compounds are considered as essential for jet fuel properties, because they provide a high level of compatibility of fuel with construction materials of the aircraft fuel systems. According to ASTM specification the content of aromatics should be not less than 8% vol. Today commercial-scale production of FT fuel is organized by Sasol company in South Africa (coal is used as a feedstock), Shell company in Qatar (natural gas is used as a feedstock), and Solena company (municipal waste is used as a feedstock). For today, more than 1600 flights have been fulfilled successfully using aviation fuel blends, which contained 20–50% vol. of FT component (Maniatis 2018; Allen et al. 2013; Wormslev et al. 2017).

FT synthesis is a pretty well-developed process that has been used for tens of years in the industry. It allows the production of a mixture of saturated hydrocarbons through syngas conversion with an application of different metal catalysts at high temperatures and pressures. There is a possibility to regulate production process conditions to produce kerosene range hydrocarbons. Small amounts of oxygenated compounds are also produced during the FT process, but they are quickly eliminated from the process (IATA 2018; Yanovskii et al. 2020; Wormslev et al. 2017). The process of FT synthesis allows the production of a mixture of hydrocarbons that does not contain sulfur-containing and nitrogen-containing compounds, which are typical for oil-derived aviation fuels. Typically, the density of fuels, produced by FT synthesis, is lower compared to oil-derived aviation fuels (Table 1.1). It means that density may be considered as a parameter that limits the use of FT fuels in blends with jet fuel. FT jet fuel possesses improved energy properties: its heat of combustion is higher compared to conventional aviation fuel (Coppola 2014; Pavlenko and Kharina 2018).

Table 1.1 Properties of FT fuels and fossil jet fuel

Property	Units of measurement	ASTM D1655/ D7566 requirement	Results			
			Jet A-1	FT 1	FT 2	FT 3
Density at 15 °C	kg/m^3	775.0–840.0	795.0	761.2	761	756
Heat of combustion	MJ/kg	>42.80	43.39	n/a	44.0	43.9
Aromatic compounds	%, vol.	<25.0	13.7	0	0	0
Smoke point	mm	<25.0	26	>45	n/a	n/a

Fuel received by FT synthesis does not contain aromatic compounds; due to this, it is characterized by higher smoke points than conventional jet fuel. It means that blending FT fuel with conventional Jet A-1 may raise the completeness of fuel combustion and decrease the formation of soot and ash. However, some portion of aromatics is necessary for the aviation fuel according to ASTM specification. It means that FT fuel may be used only as a component for blending with conventional aviation fuel until standard requirements are met. The distillation profile of FT fuel is very similar to conventional one. This fuel may be used if the fractional composition of conventional jet fuels is too high. Alternative jet fuels produced by FT synthesis have similar or higher values of flash point that can positively affect fuel flammability and fire safety. FT fuel mainly contains stable saturated hydrocarbons, and no unsaturated hydrocarbons, and other compounds; this positively influences the chemical stability of aviation fuel. The absence of sulfur compounds is among other positive effects of FT jet fuel. It decreases corrosiveness of fuels and toxicity of emissions, except that FT aviation fuel is characterized by a very shallow freezing point, compared to conventional jet fuels.

1.4 HEFA-Kerosene Pathway

HEFA is an abbreviation for hydrotreated esters and fatty acids. The manufacturing route is quite close to the processing of conventional crude oil. At the first stage, oil-containing feedstock is pre-treated before the conversion itself. Then feedstock is treated with hydrogen: during this process, oxygen is removed from molecules, and various hydrocarbons are formed. Next, hydrocarbons are cracked into smaller molecules and isomerized to obtain n-paraffins and isoparaffins, which are necessary to provide good operation properties of future fuel. After that, a mixture of hydrocarbons is fractionated and divided into gasoline, kerosene, and diesel fractions. The feedstock for HEFA fuel is different from that of FT synthesis fuel. Usually, feedstock has oil-containing biomass composed of triglycerides or esters of fatty acids. The most popular kinds of biomass are rapeseed, palm, jatropha, camelina, soy oils, used cooking oils, animal fats, waste from the food industry, etc. Today we observe the tendency to shift from edible to waste or non-edible oil feedstock (IATA 2018; Hui et al. 2012; Wormslev et al. 2017).

HEFA is the second SAF fuel certified by ASTM in 2011 and included into specification ASTM D7566. Similar to FT synthesis fuel, HEFA does not contain aromatics in its content, and it is allowed to use it only as a blend up to 50% (vol.) with conventional fossil jet fuel. Today HEFA fuel is manufactured at several plants around the world, and mainly, this process is used to produce fuels for road vehicles. The most well-known producers of the HEFA fuel are Neste Oil company and UOP company (Moses 2015; Yakovlieva et al. 2017).

Fuels, produced by HEFA pathway, mainly contain paraffinic hydrocarbons and a very low amount of aromatic compounds. From one side, it is known that the presence of aromatics negatively influences emissions of soot in exhaust gases. From the other side, aromatic compounds are necessary to provide a good level of compatibility with operation materials of the fuel aircraft systems. Due to this reason, HEFA should be blended with conventional jet fuel until the required level of aromatics is provided (in a range 15–23% vol.). The average hydrocarbon content of HEFA fuels is given in Table 1.2 (Yanovskii et al. 2020; Pavlenko and Kharina 2018).

The absence of aromatic compounds and high content of isoparaffins results in some differences in properties of HEFA fuel compared to fossil aviation fuel. HEFA fuel possesses lower density, similar or higher heat of combustion, but lower energy value (Table. 1.3). Adding HEFA fuel improves the energy characteristics of blended jet fuel. Studies on test engines also showed lower consumption of fuel (about 10%). It may be explained by the difference in density and energy properties

Table 1.2 Hydrocarbon composition of HEFA fuels

Fuel sample	n-Paraffins	Isoparaffins	Olefin	Cycloparaffins	Aromatic compounds	Total
HEFA from camelina 1	11.7	87.3	0.1	0.9	–	100
HEFA from camelina 2	9.1	89.4	0.1	0.7	–	99.3
HEFA from tallow	12.8	86.9	0.1	0.3	–	100.1

Table 1.3 Characteristics of HEFA fuel and conventional aviation fuel

Property	Units of measurement	ASTM D1655/D7566 requirement	Results			
			Jet A-1	HEFA 1	HEFA 2	HEFA 3
Density at 15 °C	kg/m3	775.0–840.0	788.0	779.9	773.5	765.9
Heat of combustion	MJ/kg	> 42.80	43.45	43.70	44.07	44.11
Aromatic compounds	%, vol.	< 25.0	15.8	9.4	0	0

of HEFA fuel. The blending of HEFA fuel with conventional one allows reducing the content of aromatic compounds (till allowable limit), thus improving the quality of exhaust gas emissions (soot content). It is also proved by higher values of smoke point of HEFA fuels (the reason for this is the high content of n- and isoparaffins and absence of aromatic compounds) (Prussi et al. 2019; Stephen and Periyasamy 2018).

Due to the peculiarities of feedstock properties and production process, the main fraction of HEFA products are mainly the diesel fraction. Regulation of the production process parameters leads to stronger isomerization and break-down of diesel fraction into kerosene and gasoline. Because of this reason, HEFA fuel is usually characterized by a heavier distillation profile; it has higher initial and final boiling points. However, the distillation profile is within the limit, allowed by the specification.

Another essential property of HEFA fuel that should be considered is the flashpoint. This property determines fire safety of fuel during technological operations (transportation, fueling, refueling, etc.) and depends on chemical composition. Due to the reasons described above, HEFA fuel is characterized by values of flashpoint, which are higher compared to conventional aviation fuel and, thus, may lead to improvement of fuels' fire safety.

1.5 SIP-Kerosene Pathway

SIP is an abbreviation for synthesized isoparaffins, which are usually obtained from hydroprocessed fermented sugars. SIP is the abbreviation approved by the ASTM to designate the above-mentioned type of synthetic fuel. Earlier, this fuel production route was known as DSHC (direct sugar to hydrocarbons) (Hui et al. 2012; Allen et al. 2013; Yakovlieva et al. 2017).

SIP fuel manufacturing is usually composed of two main stages. At the first stage, sugars are fermented by microorganisms; this step results in farnesane unsaturated hydrocarbon with 15 carbon atoms of branched structure with four double bonds. Next, during the reaction of hydrogenation with an application of catalysts, this hydrocarbon is converted into saturated one. It is known as farnesane. Then farnesane comes through the distillation process to eliminate possible admixtures. The product is almost wholly composed of one farnesane compound; it may contain small amounts of unsaturated farnesane and other unsaturated hydrocarbons. As producers of SIP state, this product may be blended with conventional fuel in quantity up to 10% (vol.). It may significantly reduce emissions of particulates, sulfur, and nitrogen oxides during such fuel combustion (Prussi et al. 2019; Stephen and Periyasamy 2018).

Today SIP production process is developed for using sugar cane. In reality, SIP fuel may be produced from any sugar-containing feedstock, even cellulosic ones including wood residues or straw (Maniatis 2018; Van der Westhuizen et al. 2011).

SIP fuel was approved by ASTM for use as a component of conventional jet fuel in 2014 and included in ASTM D 7566 as *Annex A3*. Its maximal allowable content in jet fuel is only 10% (vol.) that is explained by its chemical composition – it contains only one type of molecule. However, testing procedures showed that even content of 20% (vol.) did not show a negative impact on fuel quality. Today this kind of SAF is produced by Total/Amyris company and used by Airbus for organizing delivery flights.

As it was described above, SIP fuel is a synthetic fuel composed of one type of hydrocarbon compound, farnesane. This compound is obtained by processing sugars into farnesane and then into farnesane. This chemical route is also called direct sugars to hydrocarbons (DSHC). The process of sugar conversion into hydrocarbon is done by microbial organisms, mainly yeast cells (Hui et al. 2012; Yakovlieva et al. 2017; Ratner et al. 2019).

According to studies, the life-cycle of greenhouse gases emissions is about 15 gCO_2e/MJ. It means that it is about 82% lower compared to traditional crude oil-derived aviation fuel. However, this value is quite controversial, as the study did not take into account emissions related to production and transportation of feedstock, mainly sugarcane (Wormslev et al. 2017).

The SIP fuel is composed almost solely of one hydrocarbon – 2,6,10-trimethyl dodecane or farnesane – and negligible amounts of cycloalkanes (naphthenes). Such chemical composition is a principal difference of SIP fuel from other sustainable fuels, which contains a wide range of hydrocarbons of different structures. The typical chemical composition of SIP fuel may be seen from Table 1.4.

As it is seen from Table 1.5, SIP fuel has a comparatively low value of density that is slightly lower than is required by the specification for jet fuels. It means that if the density of conventional jet fuel is typically around 790 kg/m^3, SIP fuel may be easily blended with no adverse effect (Allen et al. 2013; Starck et al. 2016; Prussi et al. 2019).

Table 1.4 Typical hydrocarbon composition of SIP fuel

Fuel sample	n-Paraffins	Isoparaffins	Olefin	Cycloparaffins	Aromatic compounds	Total
SIP (Farnesane)	–	96.4	0.2	1.3	–	97.9

Table 1.5 Typical properties of SIP fuel and conventional aviation fuel

Property	Units of measurement	ASTM D1655/ D7566 requirement	Results	
			Jet A-1	SIP
Density at 15 °C	kg/m^3	775.0–840.0	795.0	773.1
Heat of combustion	MJ/kg	>42.80	43.39	44.0
Aromatic compounds	%, vol.	<25.0	13.7	0
Smoke point	mm	<25.0	26	n/a

SIP fuel is characterized by energy values similar to conventional jet fuels. The heat of combustion is higher than the minimal requirement of specification; due to lower density the energy value is also slightly lower.

Due to the fact that aromatic compounds are absent in SIP fuel, its blending with fossil aviation fuel (until the minimal allowable limit of aromatics' content is met) may reduce the content of soot in exhaust gases. Thus, the content of aromatics in blended jet fuel is a limiting parameter for SIP component use. Such parameter as a smoke point of pure SIP fuel cannot be appropriately measured. However, blending SIP fuel with conventional one showed a significant improvement in this parameter. SIP fuel or pure farnesane has a comparatively high boiling temperature (247 °C). It is near to the final boiling point of fossil aviation fuel. Thus, the higher volume of SIP component is used in blends, and the heavier distillation range will be received. Farnesane fuel is also characterized by high flashpoint (about 100 °C) that is more than 2.5 times higher compared to conventional fuel.

Thus, blending SIP component with jet fuel leads to rising of the flashpoint of blended aviation fuel and the improvement of its fire safety (Hui et al. 2012; Allen et al. 2013; Pavlenko and Kharina 2018).

SIP fuel reveals a significantly low freezing point (< −100 °C) compared to the limit set by ASTM specification (<−47 °C). It means that using farnesane will provide a strong positive effect on the low-temperature properties of aviation fuels. At the same time, SIP fuel is characterized by high viscosity, and its value of viscosity at temperature − 20 °C does not meet standard specification requirements. Taking into account that both freezing point and viscosity influence fuel pumping through the fuel system, the use of SIP components in fuel blends is limited. Low values of acidity of farnesane fuel positively characterize its chemical stability. It is one more positive feature of farnesane fuel. On the other side, SIP fuel is characterized by a higher content of existent gums compared to conventional aviation fuels. However, tests for the thermal oxidation stability of SIP fuel were entirely successful. Thus, SIP fuel has positive effect on some jet fuel properties, but at the same time, we can observe the negative impact on other properties. It means that its content in jet fuel blends must be strictly checked and regulated.

1.6 ATJ-SPK Kerosene Pathway

ATJ-SPK is an abbreviation for alcohol to jet-synthesized paraffinic kerosene. This abbreviation is precisely related to synthetic fuel that does not contain aromatic compounds. The first step in ATJ fuel manufacturing is the production of alcohols with 2–5 carbon chain length. Today there are several technologies of alcohol processing: the single alcohol may be processed and can be processed as a mixture. These alcohols may be of straight or iso-structure. Generally, the ATJ fuel route includes two separate stages. The first is the manufacturing of alcohols and the second is the processing of alcohols to obtain fuel hydrocarbons. These stages are separate, do not depend on each other, and can be organized in separate locations. Conversion of alcohol feedstock to jet fuel goes through the stage of dehydration to olefin hydrocarbons. Next, olefins are purified from water and admixtures and come to the oligomerization stage to be transformed into larger unsaturated hydrocarbons. Then, they separate the received hydrocarbons of kerosene range and process it through the hydrogenation stage at specific catalysts to obtain n-paraffin, isoparaffins, and cycloparaffins of kerosene range.

Several kinds of alcohols serve as feedstock for ATJ-SPK production. Among them are ethanol, butanol, and iso-butanol. The feedstock for alcohols varies greatly; it can be different kinds of biomass that contain starch, sugars, cellulose, etc. (Maniatis 2018; Moses 2015; Starck et al. 2016).

ASTM has approved the application of ATJ-SPK to blending with conventional aviation fuel in 2016, but the approval was related only for ATJ-SPK produced from iso-butanol. Such a situation was explained by the lack of documents necessary for approval of other ATJ-SPK production sub-pathways. For today, other routes of this fuel production are also approved (Ratner et al. 2019).

The most developed route for ATJ fuel production is biotechnological fermentation of biomass to produce C2 to C6 alcohols. At the first stage, alcohols are dehydrated, and the simple alkanes are obtained, which then come through oligomerization. Later these hydrocarbons (olefins) are saturated with hydrogen (hydrogenated) and iso-alkanes are received.

In the case when biomass is used for alcohol production, they apply simple process of sugar fermentation by microorganisms. When they use edible biomass, such as corn or sugar cane, sugar monomers can be easily extracted – hot water treatment. When they use non-edible biomass, the process is more complicated, and special pre-treatment of feedstock may be applied. Alcohols are produced in special fermentators at small concentrations and mild temperatures. It is necessary to keep microorganisms alive. For alcohol dehydration different catalysts are applied. Ethanol processing takes place on zeolites, silica-alumina, heteropolyacids, and silicoaluminophosphates. Butanol and iso-butanol are processed on alumina and acidic catalysts. After the stage of dehydration, olefins pass through the oligomerization stage using typical homogeneous and heterogeneous catalysts. The technology also allows the production of aromatic compounds if necessary (Allen et al. 2013; Wormslev et al. 2017; Stephen and Periyasamy 2018). The distribution of hydrocarbons in ATJ-SPK fuel is presented in Table 1.6.

The presented hydrocarbon composition determines lower values of density in comparison to traditional aviation fuel (Table. 1.7). Moreover, some samples of ATJ-SPK fuel are characterized by values of density, which are lower than the

Table 1.6 Typical hydrocarbon composition of ATJ-SPK fuel

Fuel sample	n-Paraffins	Isoparaffins	Olefin	Cycloparaffins	Aromatic compounds	Total
ATJ-SPK	–	99.8	–	0.2	–	100

Table 1.7 Properties of ATJ-SPK fuel and fossil jet fuel

Property	Units of measurement	ASTM D1655/ D7566 requirement	Results		
			Jet A-1	ATJ-SPK 1	ATJ-SPK 2
Density at 15 °C	kg/m^3	775.0–840.0	795.0	757.1	760
Heat of combustion	MJ/kg	<42.80	43.39	44.1	44.0
Aromatic compounds	%, vol.	<25.0	13.7	0	0
Smoke point	mm	<25.0	21	27	n/a

requirements of the specification. Due to the chemical composition of ATJ-SPK, its energy characteristics differ from conventional fuels. It possesses higher heat of combustion and lower energy value. Taking into account low density values of ATJ-SPK, its maximal content may be limited to 70% vol.

Blending ATJ-SPK fuel with crude oil-derived jet fuel leads to the rising of energy properties of fuel blends, mainly heat of combustion. Introduction ATJ-SPK that does not contain aromatic compounds allows reduction of content in fuel blends, in case it is necessary. At the same time, adding of ATJ-SPK component is limited until the content of aromatics satisfies standard requirements. The distillation profile of ATJ-SPK fuel is very similar to conventional jet fuel but characterized with a slightly higher initial boiling point.

ATJ-SPK is characterized by a higher flashpoint that will have a positive effect on fire safety characteristics. ATJ- SPK fuel does not contain unstable hydrocarbons and heterogeneous and sulfur compounds. It determines the high chemical and thermal oxidation stability, which is even better compared to conventional jet fuels. It may be explained by the feedstock properties and production process that eliminates the formation of unstable compounds, except that ATJ-SPK fuel is characterized by the absence of sulfur compounds. However, it does not have adverse effect on the lubricating properties of jet fuel blends.

1.7 ATJ-SKA Kerosene Pathway

ATJ-SKA is understood as alcohol to jet-synthesized kerosene with aromatics. This abbreviation is used to describe alcohol to jet fuel that contains aromatic hydrocarbons. The overall technology for ATJ-SKA production is the same as for ATJ-SPK; the only difference is the additional stage of hydrocarbon aromatization. Depending on the peculiarities of technological process and technology, the aromatization stage may be organized separately or integrated into the overall process (IATA 2018; Yanovskii et al. 2020; Coppola 2014).

ATJ-SKA fuel has not passed approval by ASTM yet. Certification procedures are fulfilled by two fuel-producing companies – Swedish Biofuels and Biology. Till now, ATJ-SKA fuel produced by Swedish Biofuels technology has successfully passed lab tests, fit-for-purpose tests, estimation of toxicological properties, and tests for compatibility with operation materials. All these tests were successful. Next, fuel atomizer spray tests, APU combustor tests, and nozzle flow tests must be fulfilled (Moses 2015).

ATJ-SKA was developed so that it is an entirely synthetic fuel that meets the requirements of specification for traditional jet fuel. It was considered that all blends in any proportions with conventional jet fuel would also meet the standard requirements. The technology of ATJ-SKA fuel production results in fuel that contains iso-alkanes, n-alkanes, and some aromatic compounds. Due to the peculiarities of the production process, there is a possibility to vary the quantity of aromatics to satisfy fuel requirements.

ATJ-SKA fuel mainly contains saturated hydrocarbons of different structures and chain lengths: n-alkanes and iso-alkanes; smaller amounts of cycloalkanes also can be found. Such fuel contains about 15–17% vol. of aromatic hydrocarbons; however, its variety is limited (mainly alkylbenzenes, indanes, and tetrahydronaphthalenes). The content of aromatic compounds in ATJ-SKA fuel determines density values, which are very similar to the density values of traditional jet fuels. ATJ-SKA fuel's density typically meets the standard (Table. 1.8). Energy properties of ATJ-SKA fuel (heat of combustion and energy value) are very close to energy properties of conventional aviation fuels. As it may be seen from the experimental data, the content of aromatic compounds is very close to conventional fuel and meets the requirements of the specification.

The initial boiling point and final boiling point of ATJ-SKA fuel are close to those of conventional aviation fuels. ATJ-SKA fuel is characterized by higher flashpoint that positively influences the fire safety of aviation fuel. ATJ-SKA fuel does not contain unstable hydrocarbons and heterogeneous and sulfur compounds. All this determines the high chemical and thermal oxidation stability. The stability of ATJ-SKA fuel is even better compared to conventional jet fuels. It may be explained by the feedstock properties and production process that eliminates the formation of unstable compounds. It also explains the improved lubricating properties of synthetic fuel. For example, wear scar diameter of pure AJT-SKA is 0.606 mm, compared to 0.728 mm for conventional fuel. At the same time, it noticed the tendency – the higher ATJ-SKA content in fuel blend, the better lubricating properties.

Table 1.8 Properties of ATJ-SKA fuel and fossil jet fuel

Property	Units of measurement	ASTM D1655/ D7566 requirement	Results	
			Jet A-1	ATJ-SKA
Density at 15 °C	kg/m^3	775.0–840.0	795.0	785.9
Heat of combustion	MJ/kg	>42.80	43.39	43.4
Aromatic compounds	%, vol.	<25.0	13.7	15.8
Smoke point	mm	<25.0	26	23

1.8 CH Kerosene Pathway

CH is understood as the process of catalytic hydrothermolysis. During the procedure of certification by ASTM, this route was renamed into CHJ (for catalytic hydrothermolysis jet). Sometimes the route is also called catalytic liquefaction. The technological process of CH fuel production is realized in three stages. At the first stage, various oily feedstocks (triglycerides, fatty acids, and esters of fatty acids) are converted into fuel hydrocarbons: n-paraffins, isoparaffins, cycloparaffins, and aromatics. At the following stage, the product passes through hydrotreatment process for saturation of remaining alkenes and removal of oxygenates. At the last stage, a mixture of hydrocarbons is distilled for purification and separation of kerosene fraction. As it was mentioned, various oils and fats are used as a feedstock for the CH process. Moreover, both edible and non-edible materials may be successfully used. Thus, CH fuel is an entirely synthetic jet fuel, similar to HEFA fuel. Control of production process allows varying the content of aromatic compounds in fuel from 10% to 20% vol. According to the results of the studies, pure CH fuel is very similar in properties and chemical composition to fossil aviation fuel. Due to the peculiarities of the production process, the CH fuel contains a wide variety of kerosene fraction hydrocarbons, which provide similar density and distillation profiles (Ekici et al. 2020). Reports state that this kind of fuel may blend with conventional aviation fuel in high ratios, even as a complete 100% substitute. However, this can be possible after more testing procedures (Yanovskii et al. 2020; Yakovlieva et al. 2017).

The technological process of catalytic hydrothermolysis goes under the temperature range 450–475 °C and pressure about 210 bar with the water application; the presence or absence of catalysts may vary. Intermediate products (hydrocarbon, unsaturated compounds, oxygenates, and organic acids) pass through the processes of decarboxylation and hydrotreatment. The obtained product is a mixture of different C6–C28 alkanes and aromatics. Then it goes through the stage of distillation for separating gasoline, kerosene, and diesel fractions. The obtained CH jet fuel is of high quality. It meets the requirements of specification by parameters of energy properties, low-temperature properties, chemical stability, as well as environmental properties (Table 1.9).

Numerous long-term tests have been done since the CH fuel was developed. In 2012 the first test flight was successfully realized. In 2020 the new synthetic jet fuel route (CHJ) was included in ASTM D7566 specification as *Annex A6*. Although developers state that CH fuel may be used without blending, ASTM allowed its maximal content up to 50% vol. (Maniatis 2018; Hui et al. 2012).

Table 1.9 Typical characteristics of fossil fuel, CH fuel, and 50/50 blend

Property	Units of measurement	Results		
		Jet A-1	100% CH	50/50 Jet A-1/CH
Acid number	mg KOH/g	0.01	0.01	0.01
Aromatic compounds	% vol.	18	17	18
Olefins content	% vol.	0.8	0.9	0.8
Sulfur content	% mass	0.05	0.03	0.03
Flashpoint	C	37	45	42
Density at 15 °C	kg/m^3	801	804	802
Freezing point	C	−51	−44	−47
Viscosity at −20 °C	mm^2/s	4	4	4
Smoke point	mm	23	24	23
Heat of combustion	MJ/kg	43.2	43.3	43.3
Copper strip test	–	1a	1b	1b
Existent gums	mg/100 ml	<1	1	<1

1.9 HDCJ Kerosene Pathway

HDCJ is an abbreviation for hydroprocessed depolymerized cellulosic jet. This production process is also known as "pyrolysis"; however, it also includes other technological processes used for depolymerization. The HDCJ fuel production process is done in three stages. The first stage is the processing of feedstock – various kinds of lignocellulosic biomass (lignin, cellulose, and hemicellulose, which are made of polymers built of carbon, hydrogen, and oxygen). At this stage, biomass passes through depolymerization, where long-chain polymers are divided into smaller molecules. If the process is done via pyrolysis, it takes place in an oxygen-free medium. Other options are hydrothermal or catalytic approaches with possible combination with pyrolysis. At the next stage, intermediate products are hydroprocessed and transformed into hydrocarbons. Finally, the mixture of hydrocarbons is distilled, and a kerosene-like product is obtained (Hui et al. 2012; Yakovlieva et al. 2017; Wormslev et al. 2017).

Pure HDCJ fuel high content of aromatic compounds (about 50%) is much higher than is allowed by the specification for jet fuel (max. 25%). It means that it may be used only as a component for blending with conventional jet fuel until the specification requirements of aromatics content are met. Today they fulfill the test of fuel blends, which contain 70% (vol.) of conventional fuel and 30% (vol.) of HDCJ. A complex of lab tests, fit-for-purpose tests, and tests for compatibility with operation materials have been successfully done, and the next set of tests is planned for the nearest future (Prussi et al., 2019; Ratner, 2019).

1.10 HDO-SK Kerosene Pathway

HDO-SK is an abbreviation for hydrodeoxygenated synthesized kerosene. The technological process of HDO-SK fuel production includes four main steps. At the first stage, oxygenated compounds of plant origin come through hydrodeoxygenation; intermediate products such as alcohols, ketones, and other oxygenates are obtained. During this thermochemical process, a metal catalyst is applied. At the next stage, the above-mentioned products pass through dehydration, oligomerization, and hydrogenation. The mixture of n-paraffins, isoparaffins, cycloparaffins, and aromatic compounds is obtained. Compounds obtained after the process are various hydrocarbons typical for conventional jet fuels. The last two stages of fuel production are generally similar to the conventional fuel manufacturing: hydrotreatment and distillation. The final product results in hydrocarbons of various structures, similar to oil-derived jet fuel, which is called HDO-SK. As a feedstock for HDO-SK fuel production, various sugar products can be used: cellulosic material (e.g., wood residues) and commercial sugars (e.g., corn syrup).

The HDO-SK fuel is mainly composed of cycloparaffins (about 80% vol.) and a small content of aromatic compounds. It means that HDO-SK may be used in conventional jet fuels only as a component to provide a specific limit of aromatic compounds and desirable levels of cycloparaffins. There are no officially proposed concentrations of HDO-SK in fuel blends, but they have fulfilled test with 50% (vol.) content of HDO-SK. Lab tests of physical-chemical and operation properties as well as fit-for-purpose tests have been passed successfully. The test for fuel compatibility with operation materials will be done in the nearest future. The next set of tests is only at the stage of planning. Thus, we may predict that if all the tests are successful, the HDO-SK fuel will be considered by ASTM of certification (Maniatis 2018; Van der Westhuizen et al. 2011).

1.11 HDO-SKA Kerosene Pathway

HDO-SKA is an abbreviation for hydrodeoxygenated synthesized aromatic kerosene. Compared to HDO-SK, HDO-SKA fuel consists mainly of aromatic compounds, with no other hydrocarbons. The process of HDO-SKA fuel production consists of three main stages. Similar to HDO-SK, at the first stage, oxygenated compounds of plant origin pass through catalytic hydrodeoxygenation with the formation of intermediate alcohols, ketones, and other oxygenates. At the next stage, these products are condensed to obtain long-chain hydrocarbons. During this stage, the above-mentioned compounds are transformed into olefin intermediates, which then are transformed into aromatic compounds similar to those found in jet fuel. Also, the obtained mixture contains small amounts of n-paraffins, isoparaffins, and cycloparaffins. At the final stage of hydrocarbon distillation, the HDO-SKA fuel is received. The feedstock for the production process is similar to HDO-SK, comprising various kinds of sugars (Maniatis 2018; Yanovskii et al. 2020; Wormslev et al. 2017).

1.12 Conclusion

The ambitious aim to reach decarbonization of air transport sector and reduce its negative impact on environment requires implementation of new policies, new technologies, new products, legislations, and specific technical regulation. Taking the rate of industry and transport development into account, the timeframe for new SAF introduction must be shorter than it was several years ago. On the one hand, there must be a technological advance in design and construction of aircraft systems, jet engines in particular. On the another hand, strong attempts must be done in the field of development and implementation of high-quality SAF into commercial market.

Now we can observe a big difference in the cost of fossil aviation fuels and biofuels. This difference is mainly explained by the high price of the production process. Moreover, the tendency will be the same in the near years. Despite this, alternative aviation fuels still have to be used as they have a high potential for minimizing the negative impact on the environment from air transport. Moreover, governmental policies and industrial initiatives tend toward reducing the price of SAFs.

For today seven SAFs have already passed the procedure of certification and approval by the ASTM. Other fuels and their production processes are at the stage of testing and qualification. During the last years more and more attention has been paid to electrofuels and hydrogen fuels for aviation. Hydrogen fuels seem to be more realistic from a near-term perspective in contrast to electrofuels. Both of these energy sources are considered environmentally friendly as they produce zero emissions. However, much work has to be done toward the development of these technologies.

Changing the air transport sector to climate neutrality needs the combined efforts of scientists and researchers, producers, policymakers, public and governmental authorities, and other involved parties. It will engage new developments in aircraft design and SAF chemical technology, renovation of airports infrastructure, development of new legislative base for reaching decarbonization of civil aviation in the near-term future.

References

Allen C, Valco D, Toulson E et al (2013) Ignition behavior and surrogate modeling of JP-8 and of camelina and tallow hydrotreated renewable jet fuels at low temperatures. Combust Flame 160: 232–239

Coppola EN (2014) Evaluation of hydroprocessed esters and fatty acids (HEFA) synthetic kerosene containing aromatics (SKA) Readijet Renewable Jet Fuel. Chevron Report, USA

Ekici S, Altuntas O, Açıkkalp E et al (2016) Assessment of thermodynamic performance and exergetic sustainability of turboprop engine using mixture of kerosene and methanol. Int J Exergy 19(3:295–314

Ekici S, Orhan I, Karakoç TH et al (2020) Milestone of greening the flight path: alternative fuels. In: Walker T, Bergantino A, Sprung-Much N et al (eds) Sustainable aviation. Palgrave Macmillan, Cham, pp 243–253. https://doi.org/10.1007/978-3-030-28661-3_12
Hui X, Kumar K, Sung C-J et al (2012) Experimental studies on the combustion characteristics of alternative jet fuels. Fuel 98:176–182
IATA (2018) Aircraft technology roadmap to 2050. IATA
Maniatis K (2018) EU transport & renewable energy policies: the role of advanced biofuels in decarbonising transport. Paper presented at the EU-India conference on advanced biofuels, New Delhi, 6–8 March 2018
Moses CA (2015) Evaluation of synthesized aromatics co produced with iso-paraffinic kerosene for the production of semi-synthetic jet fuel. Technical Report CAM-TR-2015-1
Pavlenko N, Kharina A (2018) Policy and environmental applications of using HEFA+ for aviation. Paper presented at international council on clean transportation. https://theicct.org/sites/default/files/publications/Green-Diesel-Aviation_ICCT-Working-Paper_20180321_vF.pdf
Prussi M, O'connell A, Lonza L (2019) Analysis of current aviation biofuel technical production potential in EU28. Biomass Bioenergy 130:105371
Ratner SV, Yuri C, Hien NH (2019) Prospects of transition of air transportation to clean fuels: economic and environmental management aspects. Int Energy J 19(1)
Starck L, Pidol L, Jeuland N et al (2016) Production of Hydroprocessed esters and fatty acids (HEFA) – optimisation of process yield. Oil Gas Sci Technol 71(1):1–13
Stephen JL, Periyasamy B (2018) Innovative developments in biofuels production from organic waste materials: a review. Fuel 214:623–633
Van der Westhuizen R, Ajam M, De Coning P et al (2011) Comprehensive two-dimensional gas chromatography for the analysis of synthetic and crude-derived jet fuels. J Chromatogr A 1218(28):4478–4486
Wormslev E, Tang C, Eriksen C (2017) Sustainable fuels for aviation: an analysis of Danish achievements and opportunities. Danish Transport Authority, Denmark
Yakovlieva AV, Boichenko SV, Leida K et al (2017) Influence of rapeseed oil Ester additives on fuel quality index for air jet engines. Chem Technol Fuels Oils 53(3):308–317
Yanovskii L, Varlamova N, Popov I et al (2020) Manufacturing of coal-based synthetic jet fuels interchangeable with JET A-1 and T-8B petroleum fuels. Pet Chem 60(1):92–103

Chapter 2
Effect of Polymer Additives on the Rheological Properties of Heavy High-Viscosity Oil

Tetiana Yarmola, Viktoria Romanchuk, Volodymyr Skorokhoda, and Petro Topilnytskyy

Nomenclature

PPG	polypropylene glycol
PEG	polyethylene glycol
GC	gas condensate
GPD	Gas Processing Department

2.1 Introduction

Nowadays the reserves of light oil with low viscosity are exhausted. Therefore, there is an urgent necessity to introduce into operation the fields of heavy reserves, such as high-viscosity oils and natural bitumen. Oil fields of this type are usually characterized by a high concentration of metals and sulfur compounds, high values of density, viscosity, and high content of asphaltic bitumen. The distinctive features of this type of oils are the high enough content of metals and sulfuric compounds, high density, viscosity, and coking ability, as well as a significant amount of asphaltene and resinous compounds. Due to the mentioned features, these oils are tough to be dehydrated. Moreover, these oils can cause corrosion of equipment, which would lead to an emergency shutdown of the refinery (Polishchuk and Yashchenko 2005; Farmanzade et al. 2016; Zinov'yev et al. 2013a, b; Roschin et al. 2015; Bratychak and Gunka 2017; Topilnytskyy et al. 2014; Romanchuk and Topilnytskyy 2010; Gajek et al. 2012; Topilnytskyy et al. 2019).

T. Yarmola (✉) · V. Romanchuk · V. Skorokhoda · P. Topilnytskyy
Lviv Polytechnic National University, Lviv, Ukraine
e-mail: ixxt.dept@lpnu.ua; topoil@polynet.lviv.ua

S. Boichenko et al. (eds.), *Chemmotological Aspects of Sustainable Development of Transport*, Sustainable Aviation, https://doi.org/10.1007/978-3-031-06577-4_2

Rheological properties are essential parameters of oil field products. The study of rheological properties allows to justify and integrate the new technology, which could increase the efficiency of oil recovery during the operation of certain facilities. The peculiarity of the rheological properties of heavy oils in contrast to most oils is the dependence of dynamic oil viscosity on applied shear stress, as well as flow rate. Considering that oil is a non-Newtonian fluid, its colloidal state (disperse phase +disperse medium), intermolecular interactions occurred in it, and formation of the oil structure is significant (Farmanzade et al. 2016; Zinov'yev et al. 2013a, b). There are some peculiarities of non-Newtonian fluids. Their viscosity decreases under the influence of the velocity gradient V $dV/dx = G$, where G is the shear rate. In the range of high values of G (as a rule, at room temperature $G > 10^2\ s^{-1}$ for oil), when the viscosity very weakly depends on the shear rate, it is assumed that the colloidal solution becomes a Newtonian fluid.

Experts argue that such countries as Canada, Russia, and Venezuela have the largest reserves of heavy oils. The USA, China, Mexico, Brazil, and Iran have sizable reserves, and Ukraine has approximately 2% of the world's reserves. However, the extraction of such oils in Ukraine is insufficient and complicated. That is why there are almost no publications on their exploration (Polishchuk and Yashchenko 2005). One of the largest oil and gas fields in Ukraine (Yablunivske field) is situated in the Dnieper-Donets basin (Poltava region). Considering that this field is successfully being developed, it may serve as an additional source for the production of oil, gas, and gas condensate.

The capacity and economic efficiency of the pipeline depend on the oil properties. Viscosity is a main restriction of the required pumping speed. The decrease in oil viscosity reduces the hydraulic resistance of the pipeline and energy consumption for pumping. In the regions with low ambient temperatures, the oil viscosity reaches such values that the energy consumption for pumping substantially increases the cost of produced oil. In some cases, the pumping becomes impossible. To increase the efficiency of viscous and high-viscosity oil transportation, they are subjected to pretreatment. There are many ways to reduce oil viscosity, including polymer additives (Topilnytskyy et al. 2014; Romanchuk and Topilnytskyy 2010; Grinishin et al. 2013; Syunyayev et al. 1990; Bashkirtseva and Sladovskaya 2014; Pyshyev et al. 2016, 2017; Demchuk et al. 2018; Al-Ameri et al. 2013; Kasatkin 1961).

Chemical reagents used to reduce the oil viscosity provide the conditions that prevent the formation of collective colloidal structures, reduce gravity between colloidal particles, and preserve the colloidal component in the form of single particles.

The reagents are selected depending on the oil composition in such a way. They will not have an adverse effect on oil processing. Typical concentrations of reagents are usually from tens to hundreds of grams per 1 ton of oil.

The introduced synthetic polymer products can change oil viscosity and shear stress. Usually, esters, alcohols, and various polymers are used.

The Aim of the Research: to determine the rheological properties of resin-rich heavy oil with the introduction of polymeric additives, to reduce the oil viscosity, and to find ways of heavy oil processing.

2.2 Research Methods

To study the rheological properties, we used the high-viscosity oil of Yablunivske field from two wells:

Sample 1 – oil with a density of 977 kg/m^3 at 20 °C from well 88.
Sample 2 – oil with a density of 972 kg/m^3 at 20 °C from well 337.

To increase the oil recovery of wells and improve the transportation of oil due to the reduction of oil viscosity, we studied the effect of temperature, share rate, and addition of polymer additives on the rheological properties of the oil. For this purpose, we used Rheomat-30 viscometer produced by Contraves AG (Switzerland). The viscometer is equipped with a rotary-type adapter and coaxial cylinders. The rate gradient varied from 0 to 452 s^{-1} (a measuring system CM409.484 consisting of a cylinder d = 25 mm and chambers d = 23.8 mm with a total volume of 40 cm^3); the temperature varied from 20 °C to 70 °C. The Rheomat 30 Contraves is a rotary viscometer used to determine the viscosity of materials within a wide range. Its open concentric system allows measurements by immersion. The measuring cone and the tube are rigidly connected; the measuring unit is driven by a DC motor.

The UH-8 thermostat with a unique flow cell produced by MLW company (Germany) was used to maintain the experimental temperature. Heat carrier was demineralized water.

The method is based on determining the dynamic viscosity within the range of 0.1–4·10^5 Pa·s. The essence is to record the moment of resistance to rotation of the inner cone filled with the test material at different gradients of the strain rate, followed by the calculating shear stress and dynamic (effective) viscosity.

Dynamic (effective) viscosity was calculated according to the formula viscosity (η):

$$\eta = \eta_{\text{rep}} \cdot \alpha, \qquad (\text{Pa} \cdot \text{s})$$

where η_{rep} is a viscosity, with a value of viscosity corresponding to the switch position at the definite shear stress, Pa·s, and α is the indication on the device scale, %.

Shear stress (τ) was calculated from the ratio:

$$\tau = \eta \cdot D_{\text{rep}}, \qquad (\text{Pa})$$

where D_{rep} is a value of shear rate corresponding to the switch position at the definite shear stress, s^{-1}.

Polyethylene glycol (PEG, molecular weight 400, density $d_{20} = 1130$ kg/m^3), polypropylene glycol (PPG, molecular weight 400, density $d_{20} = 1010$ kg/m^3), and demulsifier of PM brand (the content of ethylene and propylene oxides copolymer is 70%, density 1000 kg/m^3) were added separately in the amount of 3% per sample to improve the rheological properties. Before adding, the oil was heated to 50 °C.

The dynamic viscosity and shear stress of samples from both wells were investigated at 20 °C, 30 °C, and 40 °C. After the addition of polymer additives the measurements were carried out again.

The possibility of obtaining residual bitumen from a mixture of oils from well 88 and well 337 of the Yablunivske field was examined.

2.3 Results and Discussion

The most critical physical and chemical properties of oils are presented in Table 2.1. Degassing oil of the Yablunivske field belongs to heavy, high-viscosity, high-sulfur oils. It is abnormally viscous. The values of coking ability and pour point are high; therefore, the content of asphaltenes and resins in the oil is also high. It means that difficulties will occur when transporting oil, especially during the cold season. Additional efforts regarding the reduction of pour points will be necessary. The quantity of chloride salts and water in the samples is also great. The reason is the absence of desalination and dehydration stages before oil processing. The kinematic viscosity at 50 °C is relatively high. The following technological parameters are determined and calculated according to the viscosity: oil fluidity in the reservoir during oil production, the type of invading fluent, the filtration rate in the reservoir, the conditions of transportation through the pipeline, the pump capacity, etc.

The rheological curves in the τ-η-D coordinates (Figs. 2.1, 2.2, 2.3, 2.4, 2.5 and 2.6) were plotted using the obtained results.

The addition of polymer additives significantly affects the rheological properties of oils. The figures show the experimental results obtained at four values of shear rate: 4.52, 21.0, 97.3, and 452 s^{-1}. The most significant effect is observed at a shear rate of 452 s^{-1}. The shear stress of sample 1 at 20 °C, 30 °C, and 40 °C are 1338.40, 564.03, and 261.94 Pa, respectively. When polypropylene glycol is added,

Table 2.1 Physicochemical properties of oil

Property	Sample 1	Sample 2
Density at 20 °C, kg/m^3	977	972
Kinematic viscosity at 50 °C, mm^2/s	326	386
Coking ability, %	9.8	7.3
Pour point, °C	14	8
Content of water, %	4.9	5.6
Content of chlorides, mg/dm^3	2900	6280
Content of sulfurous resins, vol. %	16.9	14.8

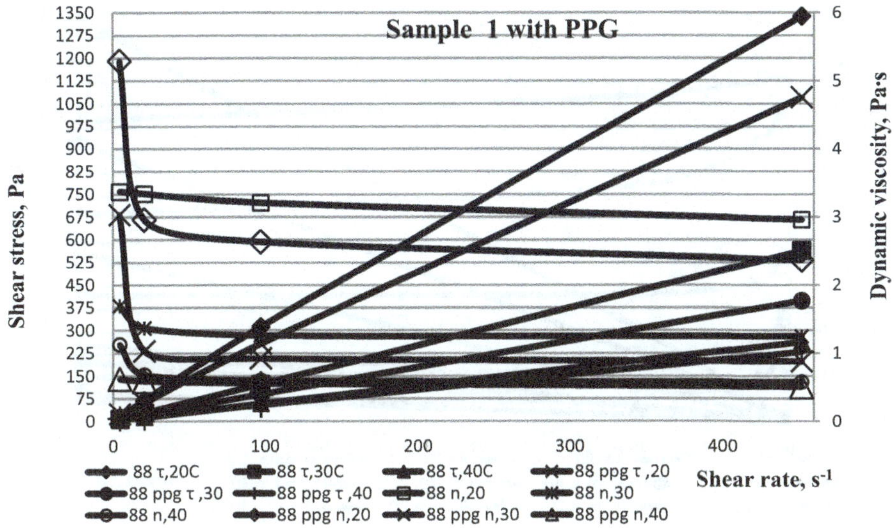

Fig. 2.1 Rheological properties of sample 1 with PPG at temperatures of 20 °C, 30 °C, and 40 °C

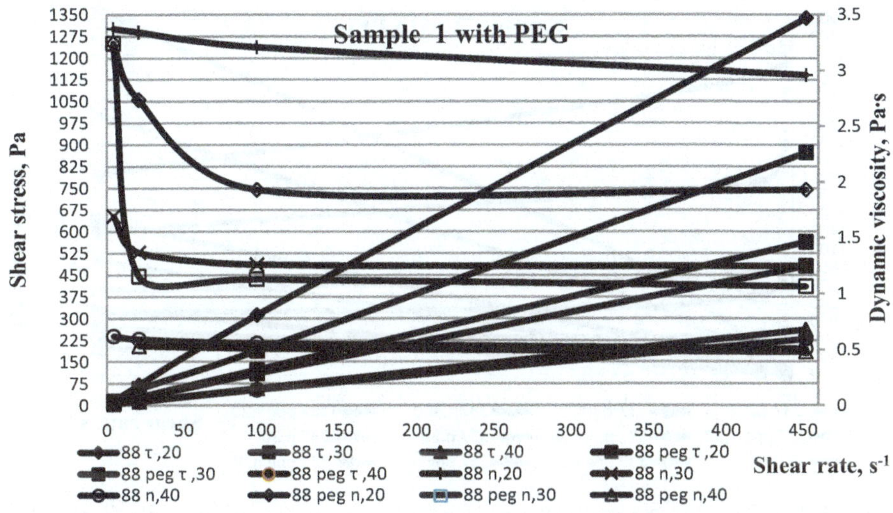

Fig. 2.2 Rheological properties of sample 1 with PEG at temperatures of 20 °C, 30 °C, and 40 °C

the shear stress is 1070.7, 398.33, and 229.44 PA, respectively. The addition of polyethylene glycol reduces the shear stress to 873.13, 481.18, and 2176.69 Pa, respectively. Moreover, the addition of the PM demulsifier to sample 1 reduces the shear stress to 924.11 Pa at 20 °C and 200.76 and 97.19 Pa at higher temperatures, respectively.

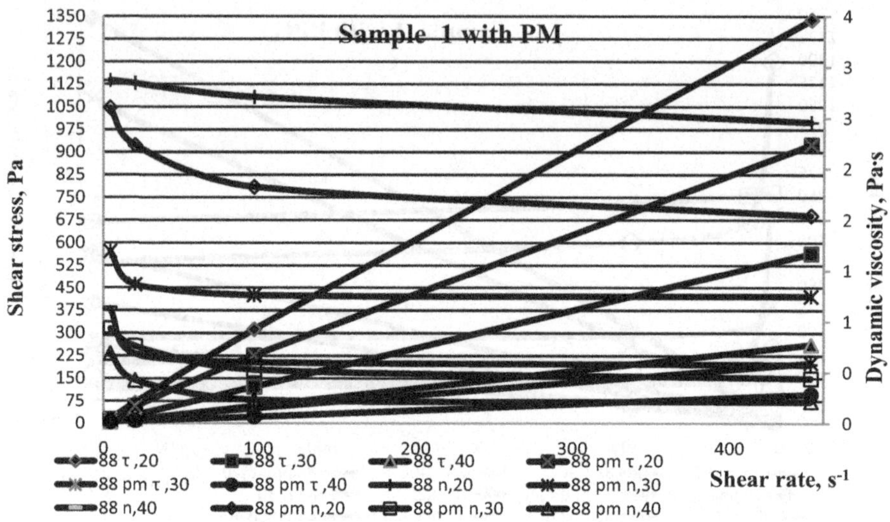

Fig. 2.3 Rheological properties of sample 1 with PM at temperatures of 20 °C, 30 °C, and 40 °C

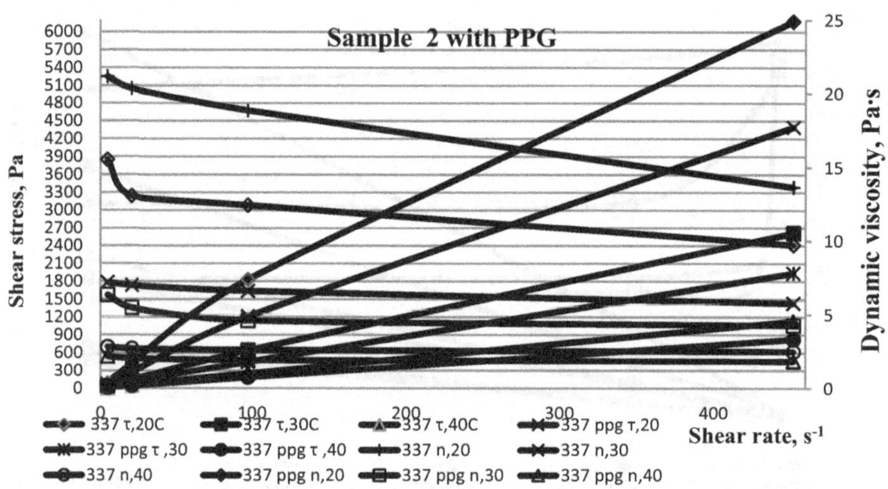

Fig. 2.4 Rheological properties of sample 2 with PPG at temperatures of 20 °C, 30 °C, and 40 °C

The addition of polymer additives to sample 2 also significantly affects the shear stress. Thus, at the shear rate of 452 s^{-1} the values of shear stress are 6177, 2613, and 1134 Pa at temperatures of 20 °C, 30 °C, and 40 °C, respectively. For this sample, the addition of polypropylene glycol shows slightly better results. The shear stress at the above-mentioned temperatures decreases to 4397.5, 1937.45, and 828.52 Pa, respectively. The addition of polyethylene glycol and PM demulsifiers also reduces the shear stress.

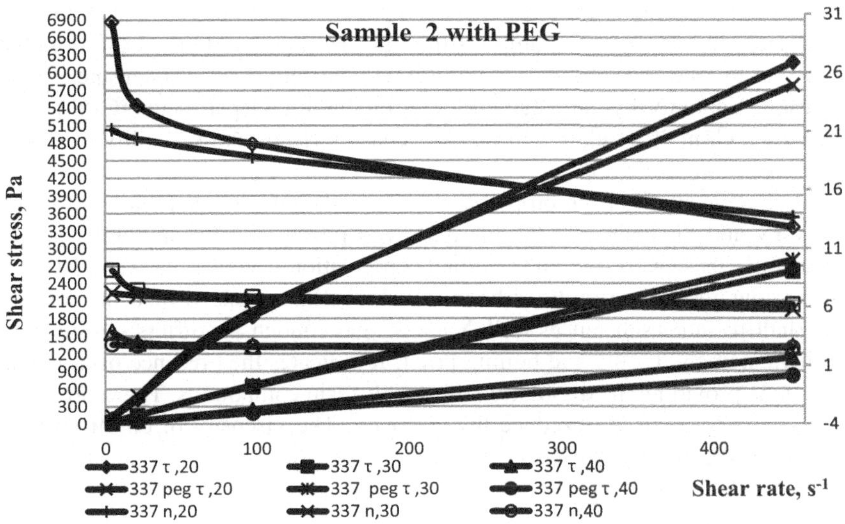

Fig. 2.5 Rheological properties of sample 2 with PEG at temperatures of 20 °C, 30 °C, and 40 °C

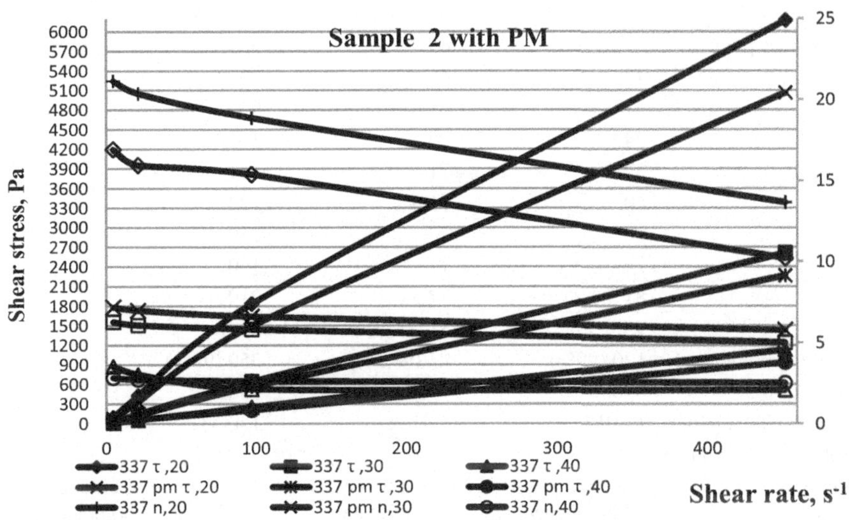

Fig. 2.6 Rheological properties of sample 2 with PM at temperatures of 20 °C, 30 °C, and 40 °C

The addition of polymer additives to high-viscosity oils also has a significant effect on reducing the dynamic viscosity. Thus, the dynamic viscosity of sample 1 at the temperatures of 20 °C, 30 °C, and 40 °C is 2.961, 1.248, and 0.579 Pa·s, respectively. When adding PPG, PEG, and PM demulsifier, the value decreases by 20–35% at a shear rate of 452 s^{-1}. For sample 2, the dynamic viscosity is 13.666, 5.781, and 2.51 Pa·s, respectively. When PPG and PEG are added, the dynamic viscosity decreases insignificantly, but a PM demulsifier reduces this value by 25%.

Even though polymer compounds reduce the viscosity of heavy oil, such viscosity is not enough to smoothly pump oil through the pipeline. One of the possible solutions is to add gas condensate (GC) to the oil. Gas condensates from Shebelinsky, Yablunivsky, and Yarivsky Gas Processing Departments (GPD) were used for research in 30% relative to oil. The characteristics and solubility of oils in them are shown in Table 2.2.

Based on an analysis of the properties of oils in the GC, it was found that in the Yablunivsky and Shebelinsky GC oils dissolved poorly and the mixture was not homogeneous. This GC differs in fractional composition from the Yarivsky GC and obviously contains alkane hydrocarbons in their compositions, which do not dissolve asphalt-resinous substances well. Yarivsky gas condensate differs significantly from others, namely, by the end boiling point. It indicates the presence of aromatic hydrocarbons providing the dissolution of asphalt-resin compounds. Therefore, 30% of Yarivsky GC was used to reduce the viscosity. The viscosity of oils from well 88 and 337 is shown in Table 2.3.

Table 2.2 Fractional distillation of GC from different GPD and oil solubility in them

	Distillation temperature of GC from different GPD, °C		
% of distillation	Yablunivsky	Shebelinsky	Yarivsky
Initial boiling point	45	65	56
10%	49	69	80
20%	51	61	92
30%	54	70	102
40%	57	71	111
50%	60	71,5	122
60%	66	73	139
70%	74	74	160
80%	81	77	210
90%	93	81	320
End boiling point	120 (99%)	93 (98.5%)	350 (92%)
Density, kg/m3	696	732	758
Refraction index, n20D	1.4015	1.4150	1.4353
Oil solubility in condensates			
Sample 1	The mixture of oil and condensate is inhomogeneous. During storage the bottom sediment is formed. The density of the mixture along the height of the cylinder is inhomogeneous		The mixture of oil and condensate is homogeneous
Sample 2			

Table 2.3 Change in oil viscosity with the addition of Yarivsky gas condensate in the amount of 30%

	Kinematic viscosity at 50 °C, mm^2/s	
Sample	Without GC	With GC
Sample 1	326	56.5
Sample 2	386	69.6

Thus, the addition of gas condensate from Yarivsky GPD reduced the viscosity of oils by more than five times.

The next step of investigations was to establish the technological possibility of obtaining residual bitumen using a mixture of oils from wells 88 and 337 of the Yablunivske field. To do this, a mixture of these oils (1:1) was prepared. The mixture was dehydrated at temperature of 50 °C with the addition of a demulsifier PM-1441 type A manufactured by JSC Barva. The mixture was distilled using an Engler apparatus to obtain fractions of sufficient quantity for analysis.

Distillation of the oil mixture	
Initial boiling point	51 °C
Up to 120 °C	30% (solvent) was distilled
From 120 °C to 200 °C	10% was distilled
From 200 °C to 340 °C	30% was distilled
Residue	0%

Analysis of the distillation data showed that the content of heavy gasoline fractions relative to the oils mixture is 14%, diesel fractions 43%, and the residue 43%. The obtained residue (after 330–350°C) was examined for compliance with the requirements for standard paving bitumen BD 60/90 (Table 2.4).

Table 2.4 Analysis of the residue after 330–350 °C for compliance with the paving bitumen BD 60/90 according to SOU 45.2–00018112-069: 2011

Index	Experiment 1	Experiment 2	Experiment 3	The standard for BD 60/90
Initial boiling point, °C	51	51	51	–
Distilled up to 120 °C, %	29	30	30	–
Distilled from 120 °C to 200 °C, %	11	10	9,5	–
Distilled from 200 °C to 330 °C, %	26	–	–	–
Distilled from 200 °C to 340 °C, %	–	29	–	–
Distilled from 120 °C to 350 °C, %	–	–	31	–
Residue, %	34	31	29,5	–
The softening temperature of the residue, °C	41	49	68	44–52
Penetration at 25 °C of the residue, 0.1 mm	100	65	20	61–90
Ductility of the residue, cm	110	105	60	100

2.4 Conclusions

We studied two oil samples from different wells of Yablunivske field and investigated their main properties.

Based on obtained results, the following procedure for the preparation and processing of oil mixture from Yablunivske field is recommended:

1. The addition of polymer additives PEG, PPG, copolymers of ethylene oxide, and propylene to high-viscosity oils significantly reduces the viscosity and shear stress and will help to reduce energy consumption in oil production and transportation.
2. To reduce the oil viscosity and production of oil, it is necessary to introduce a diluent into the reservoir; the diluent is gas condensate from Yarivsky GPD in 20–30%.
3. The residue after distillation will be a high-quality residual bitumen. The grade of bitumen will be determined by the temperature regime of the atmospheric column. The production of paving bitumen BD 130/200 is possible after fraction distillation up to 330 °C; deeper distillation will result in the production of bitumen BD 90/130 or BD 60/90. The determination of all indices will be carried out in the course of further researches.

References

Al-Ameri M, Grynyshyn O, Khlibyshyn Y (2013) Modification of residual bitumen from orhovytska oil by butonal polymeric latexes. Chem Chem Technol 7(3):323–326

Bashkirtseva NYU, Sladovskaya AYU (2014) Osobennosti primeneniya PAV v protsessakh transportirovki vysokovyazkikh neftey. Vestnik tekhnologicheskogo universiteta 17(14): 449–451

Bratychak MM, Gunka VM (2017) Khimiya nafty i gazu. Lvivska polytekhnika, Lviv

Demchuk Y, Sidun I, Gunka V, Pyshyev S, Solodkyy S (2018) Effect of phenol-cresol-formaldehyde resin on adhesive and physico-mechanical properties of road bitumen. Chem Chem Technol 12(4):456–461. https://doi.org/10.23939/chcht12.04.456

Farmanzade AR, Karpunin NA, Khromykh LN, Yevsenkova AA, Al′-Gobi G (2016) Issledovaniye reologicheskikh svoystv vysokovyazkoy nefti Pecherskogo mestorozhdeniya. Int Res J 3(45): 116–119. https://doi.org/10.18454/IRJ.2227-6017

Gajek A, Zakroczymski T, Topilnytsky P, Romanchuk V (2012) Protective properties and spectral analysis of nitrogen- and oxygen-containing corrosion inhibitors for oil equipment. Chem Chem Technol 6(2):209–219. https://doi.org/10.23939/chcht06.02.209

Grinishin OB, Al′-Ameri MSA, Khlibishin YY (2013) Kharakteristika i napravleniya pererabotki tyazholykh vysokosernistykh neftey. Vostochno-yevropeyskiy zhurnal peredovykh tekhnologiy 5(6):27–31

Kasatkin AG (1961) Osnovnyye protsessy i apparaty khimicheskoy tekhnologii. Goskhimizdat, Moskva

Polishchuk YM, Yashchenko IG (2005) Vysokovyazkiye nefti: analiz prostranstvennykh i vremennykh izmeneniy fiziko-khimicheskikh svoystv. Electron Sci J Oil Gas Business 1:1–31

Pyshyev S, Gunka V, Grytsenko Y, Bratychak M (2016) Polymer modified bitumen: Review. Chem Chem Technol 10(4s):631–636. https://doi.org/10.23939/chcht10.04si.631

Pyshyev S, Gunka V, Grytsenko Y, Shved M, Kochubei V (2017) Oil and gas processing products to obtain polymers modified bitumen. Int J Pavement Res Technol Open 10(4):289–296. https://doi.org/10.1016/j.ijprt.2017.05.001
Romanchuk V, Topilnytskyy P (2010) Investigation of reagents with different chemical compositions for protection of oil primary refining equipment. Chem Chem Technol 4(3):231–236. https://doi.org/10.23939/chcht04.03.231
Roschin PV, Zinoviev AM, Struchkov IA, Kalinin ES, Dziwornu CK (2015) Solvent selection based on the study of the rheological properties of oil. Int Res J 6–1(37):120–122
Syunyayev ZI, Safiyeva RZ, Syunyayev RZ (1990) Neftyanyye dispersnyye sistemy. Chimiya, Moskva
Topilnytskyy P, Romanchuk V, Boichenko S, Golych Y (2014) Physico-chemical properties and efficiency of demulsifiers based on block copolymers of ethylene and propylene oxides. Chem Chem Technol 8(2):211–218. https://doi.org/10.23939/chcht08.02.211
Topilnytskyy P, Paiuk S, Stebelska H, Romanchuk V, Yarmola T (2019) Technological features of high-sulfur heavy crude oils processing. Chem Chem Technol 13(4):503–509. https://doi.org/10.23939/chcht13.04.503
Zinov'yev AM, Ol'khovskaya VA, Kovalev AA (2013a) Obosnovaniye analiticheskoy modeli psevdoustanovivshegosya pritok nelineyno vyazkoplastichnoy nefti k vertikal'noy skvazhine. J Central Committee Rosnedra 2013(2):40–45
Zinov'yev AM, Ol'khovskaya VA, Maksimkin NM (2013b) Proyektirovaniye sistem razrabotki mestorozhdeniy vysokovyazkoy nefti s vnedreniyem modeli nen'yutonovskikh techeniya i rezul'tatov issledovaniya skvazhin na pritok. Petroleum Business J 2013(1):5–14

Chapter 3
Phenomenological Probabilistic Model of Friction Pair Wear Taking into Account Thermal Mechanical Stability of Boundary Layers

Oksana Mikosianchyk, Rudolf Mnatsakanov, Vitalii Tokaruk, and Olena Kharchenko

Nomenclature

TPM	Transition probabilities matrix
BLL	Boundary lubrication layer

3.1 Introduction

The most common types of the destruction of gear teeth are fatigue, abrasive, adhesive, corrosion-mechanical wear of contact surfaces, and destruction due to adhesion. Among the main factors that most significantly affect the mechanism, nature, and rate of wear of toothed wheel gears, it is fair to mention contact stresses, temperature, and properties of lubricants (Babak et al. 2019, 2020).

The complexity of evaluation of gear wear requires generalization of processes of different surface destruction nature in a single model.

When modeling wear processes, models of mechanics of destruction based on the kinetics of damage accumulation in surface layers under wear are often used as basic models. The wear rate is identified with the damage accumulation rate.

Analysis of the kinetics of surface layer destruction during wear allowed formulating principles of the delamination theory of wear, which describes the mechanism of gradual layer-by-layer destruction of the surface under friction due to damage accumulation in the surface layer through deformation processes. Along with a good enough qualitative description of wear processes, the theory produces essential

O. Mikosianchyk (✉) · R. Mnatsakanov · V. Tokaruk · O. Kharchenko
National Aviation University, Kyiv, Ukraine

S. Boichenko et al. (eds.), *Chemmotological Aspects of Sustainable Development of Transport*, Sustainable Aviation, https://doi.org/10.1007/978-3-031-06577-4_3

errors in quantitative estimates of wear. In particular, it has been established that real microcracks in the surface layer can be formed anywhere in a strongly deformed zone and can lie much deeper than the delamination theory forecasts. Microcracks exist at different depths, and their coalescence occurs in a way determined by many factors that affect the tribosystem. The action of these factors may not manifest itself in one plane. This leads to certain contradictions regarding the shape of particles and the mechanism of particle formation during the wear of the contact surfaces.

Despite the shortcomings, the theory explains two critical phenomena of wear, namely, the facts that microcracks, especially near inclusions, as well as their further development and coalescence, play an essential role in the generation of wear particles and that the mechanism of occurrence and development of adhesive wear, fretting wear, and fatigue wear can be the same.

In this regard, since the appearance of the first publications describing the delamination theory of wear in 1973, researches on its improvement and development have been going on (Kjer 1987).

In the process of development of the delamination theory of wear, the authors (Kapoor and Franklin 2000) have proposed a model that allows one to calculate the increment in shear deformation for each sublayer of the tribological layer under plastic deformation due to the action of periodic contact pressure.

The increment in deformation accumulated in the sublayers under friction is determined by many external and internal factors such as elastic and plastic properties of the material, contact pressure, coefficient of friction, and so on.

With regard to the above, two possible approaches to the description of wear taking into account the formation of a tribological layer on the friction surface in the process of self-arrangement have been suggested.

The first approach is based on using complicated defining equations, which include the above effects observed in the tribological layer. However, using models of this level to calculate the real tribocontacts of complex configuration is extremely difficult. In most cases, because of the complexity of the processes occurring in the friction zone and insufficient degree of their study at the physical level, they practically have not been implemented.

Another approach to the problem solution may be based on a phenomenological approach which uses simplified models and methods of analysis that give results that are not contradicting the experimental data. To analyze the kinetics of the wear process at the macrolevel, it is efficient to use phenomenological models which consider the relationship between surface wear characteristics and a number of parameters that characterize the properties of tribocontact and interaction conditions as to be known.

To describe the processes of accumulation of cumulative damages, Markov random processes with discrete time and states (Markov chains) have been used efficiently enough in Sorokatyi (2009).

The presented research results are interpreted in terms of the phenomenological probabilistic model of successive wear of tribological layer sublayers in tooth gearing on the basis of the methodology proposed in Sorokatyi (2002). The tribological layer wear is considered as a random Markov type process with discrete time and states. An essential feature of the model is that in order to determine the components of the TPM, it uses experimental dependences of wear intensity on a set of primary factors such as thermal-mechanical stability of lubrication layers on contact surfaces, type of lubricant material, hardening of friction pairs under thermocyclic ionic nitriding, local increase in contact temperature, and so on.

3.2 Procedure for Determining Successive Wear of Sublayers of the Tribological Layer

To use the terminology of mathematical apparatus of the theory of Markov random processes, we first define some concepts, in particular the system and the system state. Under the system we understand the tribological layer. For the system state, we take the layer wear that is a multiple of the sublayer thickness δz. Thus, the expression "the system is in the *i-th* state" means that the *i-1* sublayer is worn, and the wear of the tribological layer is $(i\text{-}1)\cdot\delta z$ (Fig. 3.1). Accordingly, the expression "the probability of transition of the system from state i to state $i + 1$" means the probability of wear of the i-th sublayer.

According to the mathematical apparatus of Markov chains, at time $t = 1$ the probabilities of finding the system in a particular state are defined as the product of the vector of initial states $[\boldsymbol{\pi}_\mathbf{i}]$ by TPM$[\mathbf{W}_\mathbf{ij}]$:

$$[\boldsymbol{\pi}_\mathbf{i}(t = 1)] = [\boldsymbol{\pi}_\mathbf{i}(t = 0)]\left[\mathbf{W}_\mathbf{ij}\right], \quad i,j = 1, 2, \ldots, K \tag{3.1}$$

where $[\boldsymbol{\pi}_\mathbf{i}(t = 0)]$ is the vector of initial states, $[\boldsymbol{\pi}_\mathbf{i}(t = 1)]$ is the vector of unconditional probabilities of finding the system in the i-th states ($i = 1, \ldots, K$) at the

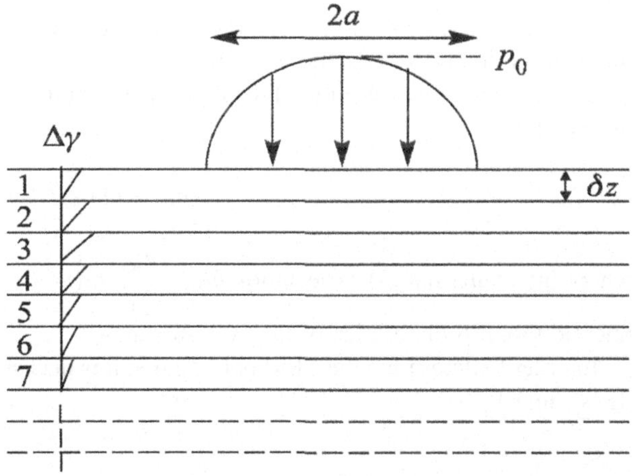

Fig. 3.1 Tribological layer of the contact surface under wear, composed of sublayers (according to Kapoor and Franklin 2000)

moment $t = 1$, $[\mathbf{W_{ij}}]$ is TPM, and K is the number of sublayers in the tribological layer.

The probabilities of system states at time $t > 1$ are defined as the product of the vector of unconditional probabilities at the moment $t - 1$, $[\boldsymbol{\pi}_i(t - 1)]$, by TPM, which determines the system behavior at time t:

$$[\boldsymbol{\pi}_i(t)] = [\boldsymbol{\pi}_i(t-1)]\,[\mathbf{W_{ij}}], \quad i,j = 1,2,\ldots,K. \tag{3.2}$$

The components of the vector of initial states are determined from the assumption that the system has no wear at the initial time, that is, it is in the first state:

$$[\boldsymbol{\pi}_i(t=0)] = [1,0,0,\ldots0]. \tag{3.3}$$

To describe the processes of accumulation of tribodamages, which refer to the class of cumulative damages, we efficiently used TPM with single-unit jumps and the presence of a complete operation state, which fully reflects the essence of the wear process, namely, the successive destruction of tribological layer sublayers (Sorokatyi 2009):

$$[\mathbf{W_{ij}}] = \begin{bmatrix} w_{11}(t) & w_{12}(t) & 0 & 0 & \ldots & 0 \\ 0 & w_{22}(t) & w_{23}(t) & 0 & \ldots & 0 \\ \ldots & \ldots & \ldots & \ldots & \ldots & \ldots \\ 0 & 0 & 0 & 0 & \ldots & 1 \end{bmatrix} \tag{3.4}$$

A complete operation state of the tribosystem means the state of complete wear of the tribological layer.

According to Sorokatyi (2009), the components $w_{ij}(t)$ are determined as follows:

$$w_{ij}(t) \cong \lambda_I(t)\,\Delta t, \tag{3.5}$$

where the wear intensity $\lambda_I(t) = V_I(t)/\delta z$, Δt is the time of one contact interaction of the gear teeth, and $V_I(t)$ is the wear rate at time t.

The layer wear is determined through the mathematical expectation $\overline{m}_t$ of finding the system:

$$z_t = (\overline{m}_t - 1)\,\delta z \tag{3.6}$$

where the mathematical expectation $\overline{m}_t = \sum_{i=1}^{K} i\,\boldsymbol{\pi}_i(t), \quad i = 1,2,\ldots,K$ and $\boldsymbol{\pi}_i(t)$ are unconditional probabilities of the system states.

In order to determine the TPM parameters, it is necessary to know the wear rate of the system $V_\mathrm{I}\,(t)$.

The linear wear rate was analyzed in this work as:

$$V_I(t)_h = \frac{\Delta h}{\Delta t}, \tag{3.7}$$

where Δh is the linear wear and Δt is the time of testing.

Measurement of sample wear was performed by the method of artificial bases using a PMT-3 device.

The linear wear of the contact surface was determined as the difference between the imprint depths before and after the wear process:

$$\Delta h = h_1 - h_2 = \frac{(d_1 - d_2)}{7} \tag{3.8}$$

where h_1 is the imprint depth for the initial surface, h_2 is the imprint depth after the experiment, d_1 is the imprint diagonal length from the square pyramid for the initial surface, and d_2 is the same after the experiment.

Load-kinematic, temperature, and tribological characteristics of contact surfaces (moment of friction, frequency (angular velocity) of sample rotation, initial (20 °C) and operation lubricant temperatures) were registered on the automated tribocomplex described in Mikocyanchyk et al. (2016). The maximum contact Hertz load (σ_{max}) was 350 MPa.

3.3 Procedure for Evaluation of Thermal-Mechanical Stability of the Boundary Lubrication Films in Critical Modes of Friction

Having developed a temperature method for the evaluation of lubricity of oils in the boundary lubrication mode, R.M. Matvievskyi used it for systematic studies of boundary layers of lubricants for various operation purposes (Matveevskij et al. 1985).

The obtained temperature dependences of the coefficient of friction distinctly demonstrate all transition temperatures revealed under friction:

- The first critical temperature, which refers to the fact that once it is exceeded, the rate of the destruction of boundary layer begins to dominate over the rate of its formation
- The temperature of chemical modification of the surface, at which the friction surfaces are separated by a modified layer formed as a result of tribochemical interaction of chemically active components of the lubricating medium with the materials of surface layers of triboelements
- The second critical temperature, over which the wear rate of the modified layer begins to dominate over the rate of its formation

It was established that with temperature rising over a certain value (i.e., transition temperature, individual for each oil under study), the coefficient of friction increased sharply. This temperature practically did not differ from the critical temperature obtained with varying acting factors for the same oil. At the same time, Khrushchev's hypothesis was confirmed, according to which the Teibor transition temperature, which characterizes transition of a continuous ("smooth") friction to a stick-slip one, is nothing more than a critical temperature of transition to Block's seizure in real friction units, which does not depend on the unit's operation mode (Mordiuk and Mikosianchuk 2017).

Also, R.M. Matvievskyi showed that the critical temperatures of lubricants determined by the proposed method decrease sharply during plastic deformations of samples under friction (Matveevskij et al. 1985). That is why his evaluation of critical temperatures was performed at the contact pressures that practically eliminate plastic deformation of the triboelement surface layers.

In real friction units of most mechanisms, the interaction of friction pairs occurs at pressures that lead to elastic-plastic deformation of contact surfaces and at such sliding velocities that cause the appearance of high gradients of shear rate of lubrication layers. That is why to evaluate the thermal-mechanical stability of boundary layers, we have developed a special procedure with the following actions:

- To perform the run-in of contact surfaces in nonstationary conditions of friction (start-stop mode)
- To create conditions for a sufficient supply of lubricant flow into the friction zone at a contact Hertz pressure 100–500 MPa in conditions of rolling with different degrees of sliding (from 3% to 40%)
- To provide stabilization of the main tribological characteristics of contact (thickness of lubrication layer for the stop, start-up period, and the period corresponding to the selected maximum rolling speeds of the leading and lagging surfaces; coefficient of friction; specific work of friction), which characterize the formation of a stable boundary layer from the lubricant material
- To cease the supply of lubricant to the friction zone and remove its remains from the contact surfaces
- To fix the change in the main tribological characteristics of contact before the appearance of the first signs of contact surfaces adhesion

Nonstationary friction conditions (start-stop mode) are based on the periodic repetition of cycles of the roller-roller friction pair in the start-stationary operation-braking-stop mode. This mode is programmed by the engine management system (Fig. 3.2).

In section I, there is a gradual increase in the rolling speed of the rollers, no sliding. In section II, the roller sliding occurs in the range from 3% to 40%. Section III corresponds to braking. Here, along the line AB, the predetermined degree of roller sliding gradually decreases to zero, and the rolling speed of the friction pairs decreases as well. Section IV corresponds to the stop. This research scheme was chosen as a prototype for the operation of gears with involute seizure,

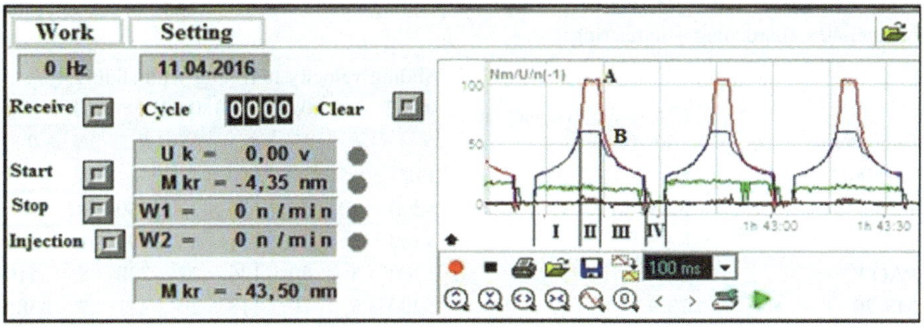

Fig. 3.2 Scheme of operation of friction pairs in nonstationary conditions: section I, start; section II, stationary work; section III, braking; section IV, stop

for which a pure rolling takes place in the pole zone, while maximal sliding does at the extreme points of seizure.

In the operation conditions of contact surfaces with no lubricant supply, the efficiency of contact lubrication properties is determined exclusively by the stability of boundary lubrication layers (BLLs) formed on friction-activated contact surfaces in hydro- and elastic-hydrodynamic lubrication conditions.

3.4 Evaluation of Thermal-Mechanical Stability of BLLs on Contact Surfaces Hardened by the Method of Thermocyclic ionic Nitriding

This work presents an evaluation of thermal-mechanical stability of the boundary films of lubricants under conditions that lead to elastic-plastic deformation of contact surfaces and at sliding velocities that cause the appearance of high gradients of the shear rate of lubrication layers. Once BLLs destruct, there occurs a local temperature rise on discrete spots due to frictional interaction. The investigated contact model "roller – roller" is considered as a heat source in the form of an additive set of point heat sources connected along the contact line. A criterion for assessing the conditions for the occurrence of seizure is the critical temperature in the contact zone of triboelements calculated according to Block:

$$t_{\sum cr} = t_0 + \vartheta, \tag{3.9}$$

where t_0 is the temperature of the contact surfaces at the contact zone entrance and ϑ is the increment of temperature at the moment when the seizure occurs (Mikosyanchyk et al. 2019).

Base oils of different viscosities and lubricants with different multifunctional additives were used to evaluate the thermal-mechanical stability of boundary films. Table 3.1 presents the results for thermal-mechanical stability of boundary layers

Table 3.1 Kinetics of the destruction of BLLs depending on the sliding velocity of friction pairs for steel 40× (hardening + tempering))

Lubricant	Kinematic viscosity at 100 °C, mm²/s	Sliding velocity in rolling with sliding, m/s							
		0.062		0.607		0.93		1.151	
		N	*ϑ*	*N*	*ϑ*	*N*	*ϑ*	*N*	*ϑ*
T-1500U	8[a]	170	15	2	110	–	–	–	–
I-20	3.85	>200	10	20	115	1	160	–	–
I-40	8.9	>200	8	51	111	8	190	1	270
PAO-8	8.3	>200	5	80	126	22	230	8	410
MS-20	23.9	>200	5	70	118	20	211	7	385
I-40 +1% Anglamol-82		>200	5	67	110	15	203	5	313
I-40 +2% Anglamol-82		>200	5	76	119	15	214	8	386
I-40 +3% Anglamol-82		>200	5	81	121	19	218	10	389
TSp-14 hip (SAE 90 API GL-4)	14.9	>200	5	81	124	20	225	10	393
TAD-17i (SAE 80w90 API GL-5)	19.2	>200	5	86	128	29	220	11	427
M10H$_2$к (SAE 30 API CC)	11.2	>200	5	70	109	17	215	5	321
Arian-extra (SAE 15w40 API CF-4/SG)	14.7	>200	5	79	116	19	219	9	389

N, number of runs before the first signs of adhesion
ϑ, local temperature increment, °C
[a]kinematic viscosity at 50 °C, mm²/s

formed on the contact surfaces of steel 40× (hardening + tempering (400 °C)). Analogous data for steel 40× (thermocyclic ionic nitriding) are listed in Table 3.2.

The analysis of the thermal-mechanical stability of the boundary layers has shown that the critical temperature of the lubrication layer destruction depends on the base viscosity, the presence of additives in the lubricant, and the type of material of contact surfaces. Comparison of the results presented in Tables 3.1 and 3.22 shows a more noticeable increase in the local temperature in tribocontacts and decrease in the period of tribocontact operation before the first signs of adhesion of contact surfaces for steel 40X as compared to the samples whose wear resistance was increased due to thermocyclic ionic nitriding.

Let us consider the kinetics of the formation and destruction of the BLLs formed by lubricants on nitrided surfaces for various operational purposes (Table 3.2).

The studied transformer oil T-1500U, characterized by the lowest viscosity, forms boundary layers, which efficiently protect the contact surfaces from adhesion at a shear rate gradient of $6.2 \cdot 10^4$ s^{-1}. In the nonstationary mode of operation, the first signs of adhesion of contact surfaces were detected at the 90th min of tribocontact operation (Fig. 3.3). The maximum temperature increment was 10 °C (Fig. 3.4).

Table 3.2 Kinetics of the destruction of BLLs depending on the sliding velocity of friction pairs for steel 40× (thermocyclic ionic nitriding)

Lubricant	Kinematic viscosity at 100 °C, mm²/s	Sliding velocity in rolling with sliding, m/s							
		0.062		0.607		0.93		1.151	
		N	*ϑ*	*N*	*ϑ*	*N*	*ϑ*	*N*	*ϑ*
T-1500U	8[a]	180	10	4	75	1	120	–	–
I-20	3.85	>200	8	30	90	3	140	1	185
I-40	8.9	>200	5	55	100	10	160	3	250
PAO-8	8.3	>200	4	85	120	28	210	13	400
MS-20	23.9	>200	4	75	115	22	190	10	370
I-40 +1% Anglamol-82		>200	4	70	105	18	180	8	300
I-40 +2% Anglamol-82		>200	4	80	115	23	190	12	370
I-40 +3% Anglamol-82		>200	4	84	117	25	195	13	375
TSp-14hip (SAE 90 API GL-4)	14.9	>200	4	84	118	25	200	13	380
TAD-17i (SAE 80w90 API GL-5)	19.2	>200	4	90	122	30	213	15	410
M10H₂к (SAE 30 API CC)	11.2	>200	4	75	105	18	187	8	310
Arian-extra (SAE 15w40 API CF-4/SG)	14.7	>200	4	83	110	22	193	13	375

N, number of runs before the first signs of adhesion
ϑ, local temperature increment, °C
[a]kinematic viscosity at 50 °C, mm²/s

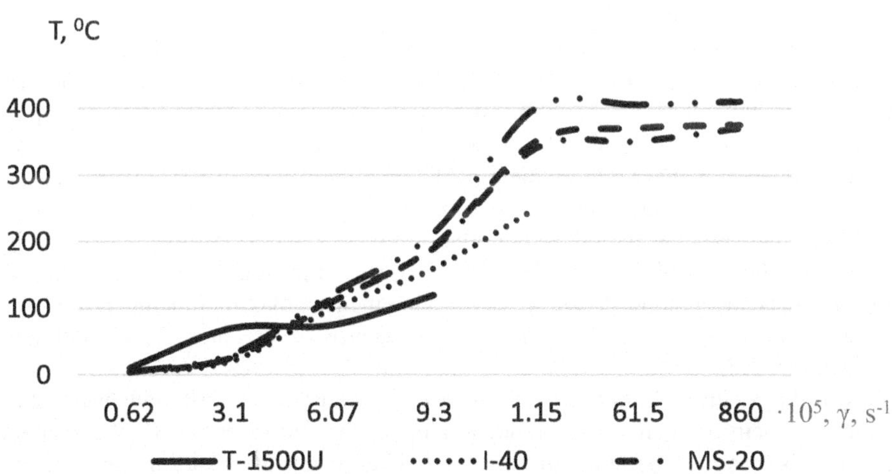

Fig. 3.3 Local temperature (T) in contact for the destruction of BLLs against the shear rate gradient (γ)

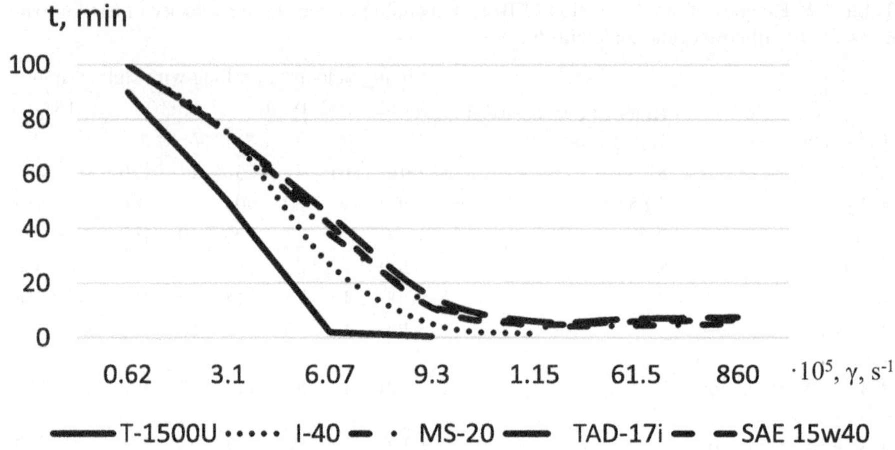

Fig. 3.4 Operation time of friction pairs (t) before the first signs of adhesion against the shear rate gradient (γ)

The increase in the shear rate gradient to $3.14{\cdot}10^5$ s^{-1} halves the run-in time before the first signs of adhesion of triboelements. The increase in local temperature at this moment is 70 °C. At $\gamma > 6.07{\cdot}10^5$ s^{-1}, these boundary layers of lubricant completely lose their protective properties. A sharp increase in temperature and adhesion of the contact surfaces takes place.

The most viscous oil MS-20, which does not contain additives, forms BLLs characterized by mechanical resistance to shear rate gradients to $6.07{\cdot}10^5$ s^{-1}. The first signs of adhesion appear only at the 38th min of operation, when the local temperature increment is 115 °C. The protective properties of the boundary films of this lubricant with an increase in γ to $8.6 \cdot 10^7$ s^{-1} deteriorate sharply: the first signs of adhesion appear at the 5th minute of operation.

The introduction of the antiwear additive Anglamol-82 into the base (active elements: 10.5% S, 2.8% Cl, 0.3% P) increases the thermal-mechanical stability of the formed boundary layers in nonstationary friction conditions. For example, whereas in the case of using oil I-40 without additives, the first signs of adhesion were detected at a shear rate gradient of the boundary films of $6.07{\cdot}10^5$ s^{-1} at the 27th min of operation after cessation of lubricant supply, the addition of 1% or 3% of additives increases the period before adhesion by 1.3 and 1.55 times, respectively (Fig. 3.5). The presence of an additive in the base also increases the heat resistance of the boundary layers to destruction: a local increment in maximum temperatures for adhesion is over 5 °C and 17 °C in the presence of 1% and 3% of additive, respectively (Fig. 3.6).

In rolling with sliding, the local contact temperatures increase with increasing sliding velocity, which causes reduction in the operation period of the contact surfaces before the first signs of adhesion. When using the base oil I-40 without additives at a 40% sliding, adhesion of friction pairs in the conditions of cessation of lubricant supply occurs after 1.5 min. Here γ is $6.0{\cdot}10^6$ s^{-1}, and the local temperature

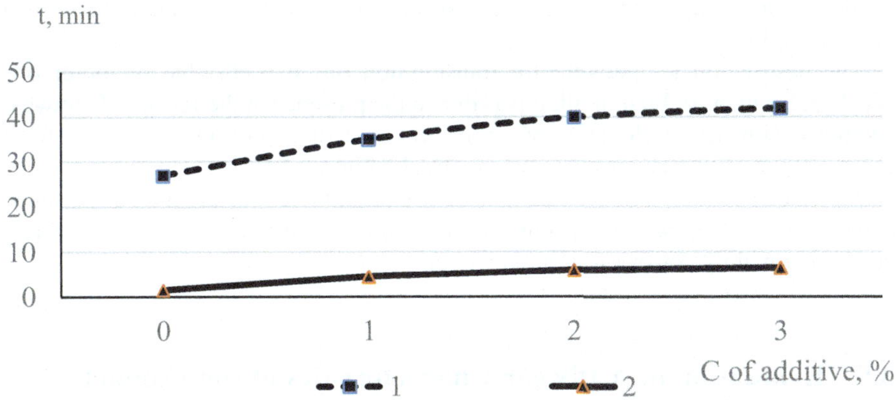

Fig. 3.5 Operation time of the friction pairs (T) before the first signs of adhesion of contact surfaces against the additive concentration (C): (1) $\gamma - 6.0 \cdot 10^5\ s^{-1}$; (2) $\gamma - 6.0 \cdot 10^6\ s^{-1}$

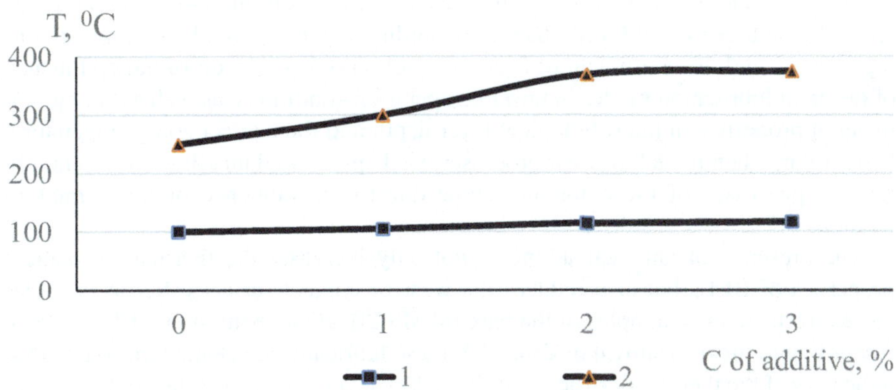

Fig. 3.6 Local temperature (T) in contact at the first signs of adhesion of contact surfaces against the additive concentration (C): (1) $\gamma - 6.0 \cdot 10^5\ s^{-1}$; (2) $\gamma - 6.0 \cdot 10^6\ s^{-1}$

increase is 250 °C. The addition of 3% Anglamol-82 increases the run-in time of triboelements by 4.3 times, and the temperature of the beginning of boundary layer destruction increases by 125 °C (Figs. 3.5 and 3.6).

Thus, the introduction of an antiwear additive provides the formation of more resistant to thermal-mechanical destruction of boundary layers characterized by efficient properties in terms of preventing adhesion of contact surfaces in critical conditions of friction.

The efficiency of antiwear additives is confirmed by the results obtained in the study of thermal-mechanical stability of the boundary layers formed by active components of transmission and motor oils for various operation purposes. The local temperature rise in contact at the first signs of adhesion, which corresponds to the destruction temperature for the BLLs formed by these oils, is high in the studied range of sliding.

The destruction temperature for transmission oils exceeds that of motor oils (Table 3.2, Fig. 3.3). First of all, this is due to the presence in the former of antiwear additives with active elements of sulfur, chlorine, and phosphorus, which form chemically modified layers characterized by the highest heat resistance.

Motor oils contain another functional class of additives (detergents, dispersants, antiwear ones, etc.), which are aimed at preventing varnish and soot formation and decreasing mainly corrosion wear and so on.

3.5 Evaluation of Antiwear Characteristics of the Contact Using the Phenomenological Model

In evaluating the contact antiwear characteristics, we took into account that the properties of the tribological layer significantly differ from the bulk material properties. In the tribological layer, changes in hardness, deformation hardening (softening occurs), and the coefficient of friction are observed as a result of the dominance of different lubrication modes in unstable operation conditions, as well as changes in material properties in the tribological layer depth and the contact zone temperature. In addition, when modeling wear processes, we kept in mind that they are essentially random processes of the cumulative type due to the influence of many random factors.

The presence of antiwear additives not only increases the thermal-mechanical properties of BLLs but also reduces the wear of contact surfaces. Let us consider these effects on the examples of the base oil MS-20 and transmission oil TSp-14hip. According to the data shown in Table 3.2, these lubricants are characterized by close indicators of the thermal destruction of boundary layers in a wide range of their shear rate gradients.

However, the wear of nitrided contact surfaces of steel 40× for the 5 km friction path significantly depends on the type of lubricant. At 20% sliding, the wear of the leading and lagging surfaces is on average 1.5 times lower when the transmission oil TSp-14hip is used than that for mineral oil MS-20 (Fig. 3.7). An increase in the sliding degree to 40% causes an increase in wear, but with TSp-14hip oil, wear of contact surfaces is lower by 1.7 times compared to that of MS-20 oil.

It was also established that the wear resistance of contact surfaces primarily depends on the nature of the formed boundary layers on friction-activated contact surfaces. Mineral oil MS-20 of selective purification forms self-generating organic films on contact surfaces, which are products of polymerization of hydrocarbons of different classes (paraffinic, naphthenic, aromatic) as a result of activation of mechanochemical processes in the friction zone. Their formation mechanism can be represented as follows: the increasing shear rate gradient of the lubricating layers to $6.0 \cdot 10^5\ s^{-1}$ (20% sliding) and $6.0 \cdot 10^6\ s^{-1}$ (40% sliding) creates the preconditions

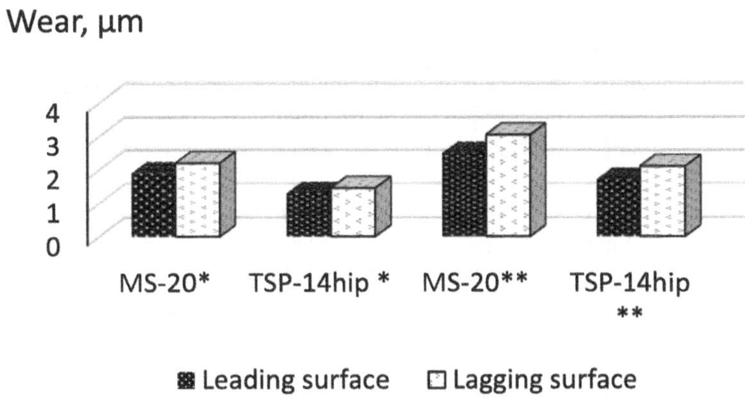

Fig. 3.7 Wear of contact surfaces of steel 40× (thermocyclic ionic nitriding) in rolling conditions with sliding degree 20% (*) and 40% (**)

for the formation of active radicals due to the rupture of carbon bonds in lubricant molecules during mechanical destruction.

At the same time, friction-activated contact surfaces intensify the interaction with the formed active oil components, resulting in the formation of a lubricating layer through the polymerization of active radicals.

According to the data of (Akhmetov 2002), the energy of the carbon-carbon bond rupture for hydrocarbons is within 250–500 kJ/mol.

To analyze the energy characteristics of the friction contact, a method for determining the specific work of friction in the contact under nonstationary load conditions was used.

The specific work of friction was calculated by integrating the area bounded by the friction moment curve for one test cycle and choosing an arbitrary range of integration along the abscissa with a coordinate of the tribocouple operation time. This resulted in determining the rotation angle of the contact surfaces by their fixed speed at a certain selected operation time and taking into account the kinetic energy of rotating parts according to the formula:

$$\mathrm{A} = \left[\left| \int_0^{t_i} M_i(t) \cdot 2\pi n_i(t) d(t) - \frac{1}{2} \sum_0^{i=n_{ti}} J\mathrm{p}_i \cdot \omega_i^2 \right| \right] / F \tag{3.10}$$

where M is the friction moment; n, ω are the rotation frequency and angular velocity, respectively – for one contact surface during sliding and for rolling with sliding, the average value of these parameters for two triboelements is taken; t is the cycle duration; Jp is the polar moment of inertia for tribological unit rotating parts which affect the accuracy of measuring the friction moment in the contact; and F is the nominal contact area by Hertz.

Analysis of the kinetics of changes in the specific work of friction in contact lubricated with mineral oil MS-20 showed that for 20% and 40% sliding, it is 2 kJ/mm^2 and 4.3 kJ/mm^2, respectively. Such values of the specific work of friction indicate a high energy consumption by the triboelements in the conditions of rolling with sliding, which is a leading factor for the activation of polymerization processes in the lubricant.

Transmission oil TSp-14hip, in contrast to MS-20, forms a different type of boundary layers. First of all, they are chemisorbed or chemically modified layers formed on friction-activated contact surfaces through the interaction of active additive components with the metal surface. Such boundary layers are characterized by high antiwear and anti-scoring properties. It should be noted that these boundary layers are characterized not only by low shear resistance with increasing shear rate gradient of lubricating layers but also by reduction in the propagation of elastic-plastic deformation deep into the metal. Analysis of the microstructure of the near-surface nitrided steel 40× layers after friction in conditions of rolling with 40% sliding revealed that the amorphous metal layer thickness is 80 μm, which is 1.4 times smaller than that for lubricant from mineral oil MS 20 (Fig. 3.8), which is due to deformation processes.

In the deformed volume of metal, a nonequilibrium structure-phase state of the deformed layer, containing local zones difficult to etch, is observed. The area of such zones for transmission oil TSp-14hip is 40% smaller than for MS-20. First of all, this is due to the high thermal-mechanical stability of the boundary layers of chemisorption nature, which reduce the degree of localization of tangential stresses deep into the metal and bring them to the surface (Liashenko et al. 2019).

Let us consider the wear rate of contact surfaces in extreme conditions leading to adhesion of friction pairs for the samples of steel 40× (hardening + tempering) and steel 40× (thermocyclic ionic nitriding) lubricated with transmission oil TSp-14hip for hypoid gears. The phenomenological probabilistic model of successive wear of tribological layer sublayers was used for this analysis.

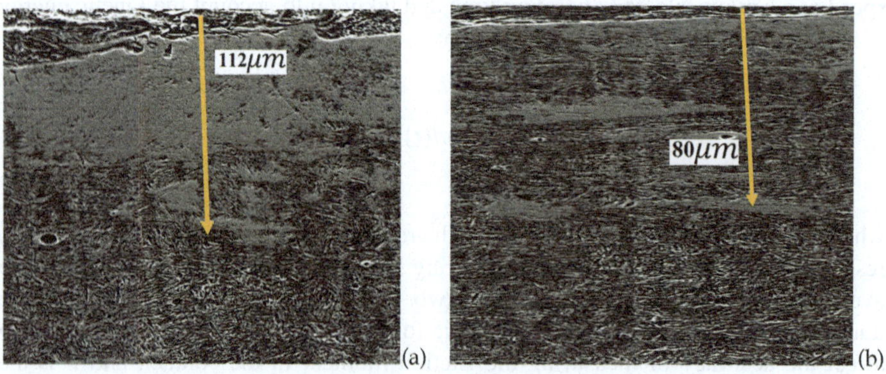

Fig. 3.8 Microstructure of nitrided near-surface layers of the lagging surface of steel 40× (×400): (**a**) lubrication with MS-20 and (**b**) lubrication with TSp-14hip

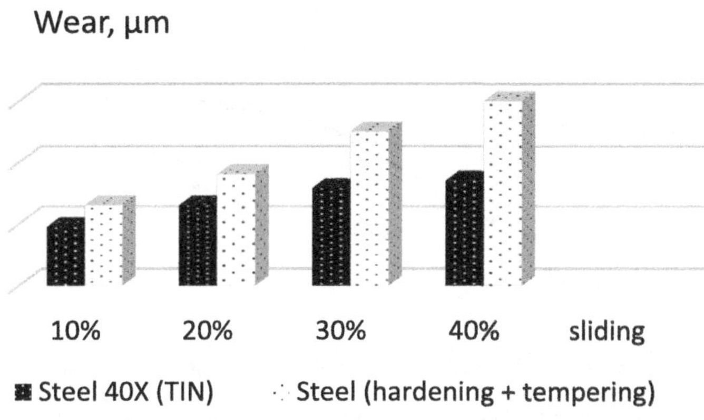

(a)

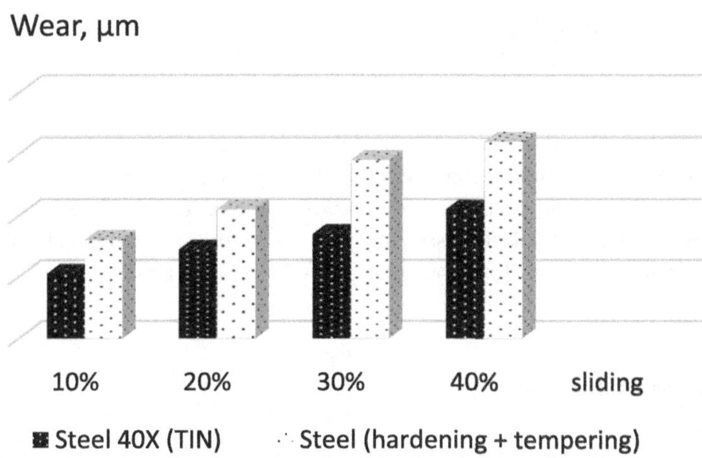

(b)

Fig. 3.9 Linear wear of friction pairs in the conditions of rolling with different degrees of sliding: (**a**) leading surface and (**b**) lagging surface

Based on the obtained results for wear of the leading and lagging surfaces (Fig. 3.9) and the time of triboelement operation before the first signs of adhesion, the wear rate of the contact surfaces is presented in Fig. 3.10.

The analysis of the results allows us to conclude that the use of thermocyclic ionic nitriding reduces the wear rate of contact surfaces in critical modes of friction, which leads to adhesion of the triboelements by 1.58 times on average. In hydrodynamic and elastic-hydrodynamic modes of lubrication, using this technique for hardening the surface layers reduces the wear of friction pairs by 2.3 times.

Thus, the used phenomenological probabilistic model of successive wear of the sublayers of the tribological layer allows one to forecast the wear of the contact surfaces depending on the operation time of friction pairs. The obtained

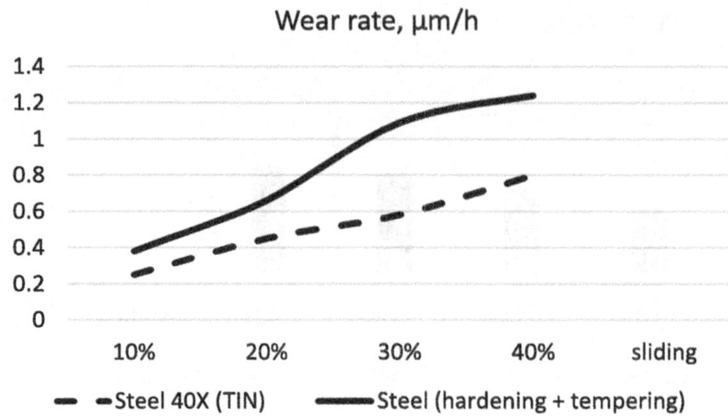

Fig. 3.10 Kinetics of dependence of the wear rate of lagging surface on the sliding degree in the friction contact under rolling with sliding

experimental results along with the introduction of automated methods and means for control of tribological parameters and testing of contact surfaces in modes as close as possible to the performance conditions make it possible to extrapolate the regularities of tribological processes to real friction units, which is part of the task to minimize wear and to prolong the service life of friction pairs.

3.6 Conclusions

1. An increase in the thermal-mechanical stability of the boundary layers formed on the contact surfaces of steel 40× (thermocyclic ion nitriding) was shown to be 1.7 times (on average) than for steel 40× (hardening + tempering).
2. The dependence of the wear of nitrided steel 40× contact surfaces on the type of lubricant has been established. In the case of using the transmission oil TSp-14hip, at 20% and 40% slipping, the wear of contact surfaces is lower than that of the mineral oil MS-20 by 1.5 and 1.7 times, respectively.
3. Lubricant-formed chemisorbed or chemically modified layers provide reduction in the propagation of elastic-plastic deformation deep into the hardened nitrided layers of metal during friction in the conditions of rolling with 40% sliding: the amorphous metal layer thickness was 80 µm, which is 1.4 times smaller than that in the case of using mineral oil MS-20 as a lubricant.
4. The phenomenological probabilistic model of successive wear of tribological layer sublayers was used to describe the processes of tribodamage accumulation referring to the class of cumulative damage, which allows forecasting the degree of wear of contact surfaces depending on the operation time of friction pairs. The use of the thermocyclic ionic nitriding technique to harden friction pairs decreases

the wear rate of contact surfaces in the boundary mode of lubrication, on average, by 1.58 times and in hydro- and elastic hydrodynamic modes of lubrication by 2.3 times.

References

Akhmetov SA (2002) Physicochemical techniques for deep processing of oil and gas: textbook for universities. Gilem, Ufa, p 672

Babak VP, Shchepetov VV, Harchenko SD (2019) Antifriction nanocomposite coatings that contain magnesium carbide. J Friction Wear 40(6):593–598

Babak VP, Shchepetov VV, Harchenko SD (2020) Wear resistance of self-lubricating coatings due to the formation of carbide graphite. Mech Adv Technol 89(2):99–105

Kapoor A, Franklin FJ (2000) Tribological layers and the wear of ductile materials. Wear 245:294–215

Kjer A (1987) A lamination wear mechanism based on plastic waves. Paper presented in proceedings international conference on wear of materials, ASME New York, pp 191–198

Liashenko BA, Stotsko ZA, Kuzin OA, Kuzin MO, Mikosianchyk OA (2019) Determination of the optimal parameters of the structure of functional gradient materials using mathematical modeling approaches. J Achiev Mat Manuf Eng 92(1-2):13–18

Matveevskij RM, Bujanovskij IA, Baginskij VV (1985) Verbindungen von Elementen der 4 Hauptgruppe des Periodensystems als er Zusatze für Ole. Schmierungstechnik 5:132–135

Mikosyanchyk O, Mnatsakanov R, Zaporozhets A, Kostynik R (2016) Influence of the nature of BLLs on the adhesion component of friction coefficient under rolling conditions. East Eur J Enterp Technol 4-1(82):24–31

Mikosyanchyk OO, Mnatsakanov RH, Lopata LA, Marchuk VE, Yakobchuk OE (2019) Wear resistance of 30KhGSA steel under the conditions of rolling with sliding. Mater Sci 55(3): 402–408

Mordiuk BM, Mikosyanchyk OO (2017) Influence of shear component of load under the friction on a structure–phase state and wear of surface layer of steel 1045. Metallofizika i Noveishie Tekhnologii 39(6):795–813

Sorokatyi RV (2002) Modeling of behavior of tribosystems by method of triboelements. Trenie i iznos 23(1):16–22

Sorokatyi RV (2009) Method of triboelements: monograph, p 242 [in Russian]

Chapter 4
Revisiting the Synthesis of Fatty Acid Alkyl Esters of Lower Monohydric Alcohols by Homogeneous Base-Catalyzed Transesterification of Vegetable Oils

Serhii Konovalov, Stepan Zubenko, Lyubov Patrylak, Anjela Yakovenko, Volodymyr Povazhnyi, and Kateryna Burlachenko

Nomenclature

BD	biodiesel
DG	diacylglycerides
EL	ester layer
FABE	fatty acid butyl esters
FAEE	fatty acid ethyl esters
FAME	fatty acid methyl esters
FFA	free fatty acids
GL	glycerol layer
KOBu	potassium n-butoxide
KOEt	potassium ethoxide
MG	monoacylglycerides
R_{AO}	molar ratio of alcohol to oil
RO	rapeseed oil
RSO	refined sunflower oil
TE	transesterification
TG	triacylglycerides
WFO	wasted frying oil
Y_{ef}	effective yield of esters
Y_t	total yield of esters

S. Konovalov (✉) · S. Zubenko · L. Patrylak · A. Yakovenko · V. Povazhnyi · K. Burlachenko
V.P. Kukhar Institute of bioorganic chemistry and petrochemistry of National Academy of Sciences of Ukraine, Kyiv, Ukraine

S. Boichenko et al. (eds.), *Chemmotological Aspects of Sustainable Development of Transport*, Sustainable Aviation, https://doi.org/10.1007/978-3-031-06577-4_4

4.1 Introduction

Biomass-derived transport fuels, such as biogasoline and biodiesel (BD), became the widespread alternative for traditional fossil fuels over the past two decades. In 2019 the world production of these two types of biofuels increased by 3% up to 1,842,000 barrels of oil equivalent per day (boe/d), according to the recent British Petroleum Statistical Review of World Energy (BP 2020). The annual growth of biogasoline production amounted to 1.8% (up to 1,143,000 boe/d), while the biodiesel production increased by 5.1% (up to 699,000 boe/d). Such growth of BD production was driven mainly by Indonesia. Biofuel consumption increased even more rapidly. The growth amounted to 6% in 2019. At the world scale, bioethanol consumption made up 63% of transport biofuels, but the share of BD has risen continually. Namely, biodiesel's share was 23% in 2009 but rose to 37% last year. BD is also the dominant transport biofuel in Europe and Asia Pacific regions (BP 2020).

When saying biogasoline, bioethanol is implied in most cases. At the same time, biodiesel is almost exclusively fatty acid methyl esters. The latter's production is grounded on the catalytic transesterification (TE) of vegetable oils or animal fats by methyl alcohol. As methanol, in most cases, is still produced from fossil raw materials, FAME cannot be considered a totally biorenewable fuel. Using bioalcohols for fatty acid alkyl ester production improves the biorenewability and sustainability of the BD obtained. In addition to the apparent ethanol (Sanli et al. 2019; Joshi et al. 2012), biobutanol (*n*-butyl alcohol) is the object of great interest (Joshi et al. 2012; Patrylak et al. 2019). Fatty acid ethyl and butyl esters (FAEE and FABE), as well as esters of other lower monohydric alcohols, possessed some different physicochemical properties compared to FAME. In general, when increasing the length of alcohol moiety in a molecule of ester, cold flow properties are considerably improved, which makes biodiesel a more viable fuel in cold weather conditions (Joshi et al. 2012; Hájek et al. 2017). Besides that, some enhancement of lubricity and heating value (Joshi et al. 2012) and decreasing the volatility of BD (Homan et al. 2017) are observed as alcohol moiety grew longer. Iodine number is also decreased slightly (Joshi et al. 2012) due to the lower content of unsaturated olefinic bonds per mass unit of alkyl esters. The latter may somewhat contribute to the long-term stability of BD. At the same time, increasing the ester head group length causes the higher kinematic viscosity of BD (Joshi et al. 2012), which is not favorable for its utilization. However, the relatively high viscosity of such products may be advantageous in another application, i.e., as biocomponents of sustainable lubricating compositions.

The alkaline-catalyzed route is still one commonly used in the industrial production of biodiesel (FAME). It relies on the utilization of soluble alkaline catalysts, namely, alkali metals, hydroxides, or alkoxides (only K and Na are of practical significance). These reagents have high activity, affordability, and availability (Fonseca et al. 2019). Alkaline-catalyzed transesterification takes place in a short (~60 min) time due to the fast reaction rate. Although it is the most widely used method, alkaline catalysis has disadvantages, including the generation of the extensive amount of wastewater (in the case of conventional BD purification by water washing), the impossibility of catalyst reuse, and the neutralization of free fatty acids (FFA) (Fonseca et al. 2019). The latter factor limits the applicability of conventional alkaline-catalyzed transesterification for oils, having a relatively significant acid value. It is often considered that FFA content in oil should not be higher than 1% (Sanli et al. 2019). In another case, oil pretreatment by acid-catalyzed esterification of FFA is needed before alkaline transesterification. It is usually carried out using strong mineral acids as catalysts.

Alkaline transesterification of oils by monohydric alcohols other than methanol is quite challenging. The reason is mainly the high sensibility of the process to the presence of water in reaction media. As generally known, the actual active catalytic species in alkaline TE are alkoxides. There always exists the equilibrium (Eq. 4.1) between hydroxide and alkoxide of corresponding alcohol, where A is the alkaline metal atom and R is the alkyl moiety of an alcohol molecule.

$$\begin{gathered}\mathrm{ROH} + \mathrm{AOH} \rightleftharpoons \mathrm{ROA} + \mathrm{H_2O} \\ \mathrm{ROH} + \mathrm{OH^-} \rightleftharpoons \mathrm{RO^\cdot} + \mathrm{H_2O}\end{gathered} \tag{4.1}$$

Monohydric alcohols possess weaker acidity compared to water. As a result, the latter's presence shifts the equilibrium (Eq. 4.1) toward the hydroxide formation. The higher the water concentration, the lower the alkoxide content. It is also generally known that acidity of alcohols decreases while growing the length of the alkyl chain. So, in the case of alcohols, other than methanol, the alkoxide concentration should be lower, which decreases the efficiency of alkaline TE. On the other hand, an increase in carbon chain length results in an alcohol polarity decrease. So, the miscibility of alcohol and oils increases significantly. Better reagents' miscibility reduces the mass transfer limitation in the course of transesterification. Such limitation is one of the biggest hurdles that need to be overcome in the methanolysis process technology (Homan et al. 2017). But better reagents' miscibility also intensifies the alkaline hydrolysis (saponification) side reactions. Such reactions lead to alkaline catalyst losses and lower alkyl esters' yield. Formed soaps also strongly contribute to stable products' emulsions formation. The quick and effective self-separation by gravity, which is inherent to alkaline methanolysis, could not be observed in the case of utilization of the longer-chain alcohols. The homogenous phase forming may be due to the much higher miscibility of longer-chain alcohols with corresponding esters, causing glycerol to become miscible in the ester phase technology (Homan et al. 2017). The bulk of the byproduct glycerol separation into the so-called glycerol layer (GL) makes the following BD downstream processing significantly simpler. Besides glycerol, the vast majority of an alkaline catalyst, soaps, and some mono- and diacylglycerides (MG and DG) also transfer into GL composition. So, the presence of self-separation in alkaline TE processes is very desirable.

Related papers on FAEE production, regardless of feedstock used, clearly indicate that alkaline-catalyzed ethanolysis is more complex compared to methanolysis due to forming of stable emulsion, preventing separation of reaction product technology (Sanli et al. 2019). Sanli et al. (2019) revealed that even low FFA content in used frying sunflower oil as 0.8% makes ethanolysis products inseparable by their own when using NaOH and KOH as catalysts. However, syntheses over sodium and potassium methoxides resulted in the self-separation of TE products. As for the alkaline butanolysis process, literature information on the issue of self-separation is very limited. Hájek et al. (2017) carried out butanolysis of rapeseed oil (about 1% FFA, 0.065% of H_2O) using 1.2–1.5% of KOH relative to the oil mass. Almost complete conversion values were achieved under high excess of alcohol (molar ratio of butanol to oil 12–15) after 2–4 h of reaction both at 30 and 80 °C. When using a lower ratio of butanol to oil, equal to 6, products contained a significant amount of unconverted glycerides. In order to enforce the phase separation, reaction products were neutralized by strong mineral acids or by bubbling CO_2 gas followed by butanol removal by vacuum distillation. Treatment by strong acid resulted in high acid values in the range 1.6–4.8 mg KOH/g of ester layer (EL) obtained. In the case of CO_2, a yield of the upper phase (EL) after separation was about 80–100% (relative to oil mass) for different sets of reaction conditions. However, the obtained products were still out of range of biodiesel requirements for free and total glycerol content, as well as for the concentration of soaps. Alkaline transesterification by longer-chain alcohols can proceed quickly when the corresponding alkoxides are used. In such a case, the reagent mixture contains a very high excess of actual alkoxide catalyst from the very beginning of the reaction.

Homan et al. (2017) carried out TE of edible grade soybean oil (<0.5% FFA) by *n*-propyl and *n*-butyl alcohols over sodium *n*-propopoxide and sodium *n*-butoxide, respectively. Both alkoxides were obtained by reaction of Na with corresponding alcohols. As high yields as 95% of propyl esters and 97% of butyl esters were obtained within 5 min reaction time at 22 °C, the molar ratio of alcohol to oil is 6 and 0.66% catalyst load. However, the reaction product formed a homogeneous phase and did not tend to any self-separation. Purification through alcohol removal by rotary evaporation followed by extraction by green, deep eutectic solvents based on choline chloride was applied to achieve the esters' concentration required by BD standards.

Similar observation relating to the very high reaction rates of alkaline butanolysis was made in the previous study of our group (Patrylak et al. 2019). Alkaline TE of refined sunflower oil was carried out over potassium butoxide, obtained from the KOH solution in *n*-butanol through water removal by Dean-Stark distillation as described in the patent of Ukraine (Zubenko et al. 2016). At the butanol-to-oil molar ratio 4.5, synthesis temperature 15 °C, and 0.95% butoxide load in terms of potassium (relative to oil), FABE yield was about 68% after 2 min and about 78% after 60 min of syntheses. However, after the settling of products overnight, FABE yield increased up to 92–98%, almost independently of the reaction time. Moreover, the observed products' self-separation had a very non-common characteristic for the processes of alkaline transesterification of oils. The bulk of both glycerol byproduct and alkaline catalyst transferred to the composition of glycerol layer. The latter's formation started directly in the course of syntheses on the bottom of the reactor and continued during the settling of reaction products. The formed GL was almost transparent and contained only trace amounts of FABE, butanol, and soaps, which is uncommon for the GLs formed in traditional methanolysis or ethanolysis. The conclusion about the preferential recovery of the alkaline catalyst as potassium glyceroxide, which supports the sedimentation of reaction glycerol, was made. In the case of KOH utilization as the catalyst, the reaction was slower and the products' mixture remained homogeneous without any signs of self-separation.

As for the feedstock used for biodiesel production, high-quality edible grade oils, having low FFA content, are preferable from the technological point of view. However, the price of such feedstock is high, which results in the high net cost of BD obtained. It is generally known that biodiesel fuel production is noncompetitive with petroleum-derived diesel in economic terms. In most countries, having the developed biodiesel industry, its production is driven by only non-economic mechanisms. Such mechanisms may include legislative established mandatory fuel blending rates and tax preferences. The utilization of non-edible low-grade oils, especially wasted frying oils (WFO), can partially improve the economic aspects of biodiesel production. However, together with decreasing the feedstock price, more complex downstream processing may be needed to obtain biodiesel, fulfilling the requirements of the standards.

The study aims to make insight into the peculiarities of alkaline synthesis and self-separation of fatty acid esters of lower monohydric alcohol over corresponding alkoxides as catalysts, obtained from only hydroxide and corresponding alcohol by way of the methods, pending by Patents of Ukraine (Zubenko et al. 2016, 2021). A particular aspect of the work was to gain the efficient alkaline syntheses of fatty acid esters without acid-catalyzed pretreatment stage from oil samples, containing as high as 2.5–2.8% FFA.

4.2 Main Part

Materials Some characteristics of vegetable oils used in the current study are given in Table 4.1. The oil feedstock includes the samples of used frying sunflower oil (marked as WFO), containing 0.40–2.80% FFA, the sample of low-grade rapeseed oil with high FFA content (RO), as well as the sample of refined sunflower oil (RSO). The latter was purchased from a local market and used as received. Rapeseed oil (RO sample) was produced by JSC "Nizhynsky zhyrcombinat" (Nizhyn, Ukraine) and used after storage for more than 10 years in a sealed barrel. Samples of oil WFO-1, WFO-2, and WFO-4, used in deep frying, were obtained from local catering establishments. The sample of wasted frying sunflower oil WFO-5 was kindly provided by local reseller. Sample WFO-3 is a blend of the samples WFO-2 and WFO-4. The acid number of wasted frying oils could increase slightly during prolonged storage. So, the acidity of each sample (FFA content in terms of oleic acid), measured directly before the corresponding synthesis, will be emphasized separately.

All samples of wasted frying oils contained a relatively insignificant amount of water (no higher than 0.065%), so no special drying was carried out. It should be emphasized that the water content values, measured by Karl-Fischer coulometric titration and by

Table 4.1 Some characteristics of oil feedstock

Sample of oil	Acidity: Acid value, mg KOH/g	Acidity: FFA content, $\%_{eq}$ of oleic acid	Saponification value, mg KOH/g	Ester value, mg KOH/g	Iodine value, g I_2 /100 g	Viscosity at 40 °C, mm^2 /s	Water content[a], %
WFO-1	0.80	0.40	188	187.2	107	37.43	–/0.05
WFO-2	5.60	2.80	–[b]	–	106	43.33	–/0.04
WFO-3	4.99	2.50	–	–	107	40.25	–/0.05
WFO-4	2.30	1.15	–	–	107	36.99	0.065/0.06
WFO-5	1.42	0.71	–	–	90	58.56	0.03/0.04
RSO	0.07	0.03	190	189.9	118	31.55	0.04/0.03
RO	5.71	2.85	186	180.4	95	37.74	–/0.03
Fatty acid composition							
WFO-1	6.9% C16:0; 4.9% C18:0; 30.7% C18:1; 54.7% C18:2; 0% C18:3; 2.8% – others						
WFO-2	6.9% C16:0; 2.8% C18:0; 33.2% C18:1; 53.7% C18:2; 0% C18:3; 3.4% – others						
WFO-3	6.8% C16:0; 3.2% C18:0; 31.2% C18:1; 54.9% C18:2; 0% C18:3; 3.9% – others						
WFO-4	7.1% C16:0; 3.5% C18:0; 27.7% C18:1; 59.3% C18:2; 0% C18:3; 2.4% – others						
WFO-5	7.8% C16:0; 5.4% C18:0; 27.3% C18:1; 54.5% C18:2; 0% C18:3; 5.0% – others						
RSO	6.7% C16:0; 2.8% C18:0; 33.0% C18:1; 54.6% C18:2; 0% C18:3; 2.9% – others						
RO	5.3% C16:0; 1.9% C18:0; 62.8% C18:1; 18.8% C18:2; 7.6% C18:3; 1.5% C20:1 0.4% C22:1; 1.7% – others						

[a] The first value is measured by Karl-Fischer titration, the second value is measured by Dean-Stark method

[b] Characteristic wasn't measured for corresponding sample

Dean-Stark method, are pretty close. So, the Dean-Stark method, which does not require sophisticated analytic equipment and reagents, looks like enough suitable alternative to evaluate even such low water content in oils before alkaline TE.

Some values of kinematic viscosity of the wasted oils' samples are higher than the sample of refined oil, which may be due to polymerization of oil in frying process. The sample WFO-5, having the highest viscosity, is also characterized by the lowest iodine value. This may be due to the deeper polymerization degradation.

Fatty acid composition was typical in case of both refined and wasted sunflower oils' samples, and none of them does contain triunsaturated linoleic acids. Samples of rapeseed oil contain mainly monounsaturated oleic acid and a trace amount of erucic acid.

Technical grade bioethanol (Ukraine, purchased as commercial alternative component of automobile gasoline) contained no higher than 0.2% of water and 0.75% of organic denaturating admixtures (GC area assay). Technical grade rectified ethyl alcohol (Ukraine) was used after selective dehydration over KA molecular sieves. Both reagents were stored over KA molecular sieves. Water content when used in syntheses was no higher than 0.1%. *n*-Butyl alcohol (Turkey) was technical grade (GC area assay – 99.37%, water content about 0.15%).

Synthetic zeolite KA-Y/3A (Russian Federation, dynamic water vapor capacitance – 150 mg/cm^3) was used for selective water removal.

Technical grade potassium hydroxide (China), containing 89.7% KOH (titration of freshly opened reagent batch), was used to synthesize alkoxides of corresponding alcohols as transesterification catalysts. The primary admixture of the reagent was the water. Also, reagent grade sodium hydroxide was used for the preparation of butoxide for analytical purposes.

Reagent grade *p*-toluenesulfonic acid and indicator bromothymol blue were used in the acid-base titration of TE products.

Reagent grade orthophosphoric acid (85-% water solution) was used to neutralize transesterification products in the purification stage. Dry washing of ethyl esters was carried out using meta-kaolin, prepared by calcination of kaolin powder (Prosyaniv deposit, Ukraine). Technical grade anhydrous sodium sulfate was used for drying esters after washing.

N,N-dimethylformamide (DMFA, 99.9%) and reagent grade *n*-hexane were used as solvents for analytic purposes.

Potassium bromide-bromate, sodium thiosulfate (fixanals, Ukraine), and analytical grade potassium iodide were used in iodine value measurement. Samples were dissolved in pharmaceutical-grade chloroform. The water solution of corn starch (0.5%) was used as indicator.

Methyl palmitate (reagent grade, GC area assay 98.65%), pharmaceutical-grade glycerol, analytic grade ethyl alcohol, N-methyl-N-(trimethylsilyl)-trifluoroacetamide (MSTFA), pyridine solution of tricaprin, 1,2,4-butanetriol, glycerol, 1-monoolein, 1,3-diolein, and triolein, as well as high-pure helium and hydrogen gases, were used in gas chromatographic analyses.

Synthetic Methods Alkaline syntheses of fatty acid alkyl esters were conducted in conic flasks (250 or 500 cm^3). Stirring (500 rpm) and heating (if necessary) were provided with a magnetic stirrer with a water bath. Maintaining the lower-ambient temperature of reaction was provided by circulation of chilled water through the water bath. When synthesis was carried out at a relatively low temperature (10–30 °C), the reaction flask was plugged with a glass plug. Otherwise, it was equipped with a reverse water condenser and calcium chloride drying tube to prevent the evaporation of alcohol from reaction media and atmosphere moisture access. At first, the oil (usually 100.00 g) was put into a reaction flask, then the predetermined masses of catalyst solution and alcohol (if necessary) were quickly added and stirring was started. Products after reaction were immediately transferred into the separation funnel, where settling and self-separation by gravity occurred. The main technological variables in syntheses were the molar ratio of alcohol to oil (R_{AO}, mol/mol), temperature (t, °C), amount of alkaline catalyst (n_{cat}, % relative to the mass of oil), and reaction time (τ, min). The temperature of settling, when self-separation occurred, may also impact on the composition of products obtained. Both transesterification and saponification reactions can also continue during this period. Settling temperature was not the variable in the current study. It took place in a relatively wide range of season ambient temperatures (~15–30 °C), specified for each synthesis.

Methods of purification of synthesized crude ethyl esters (upper ester-enriched layer) included removing ethanol by distillation under low pressure, washing with water, dry washing with adsorbent, and vacuum distillation. Removing of alcohol by distillation was carried under mild vacuum (waterjet pump, 3–4 kPa) in the round-bottom flask (250 cm^3) under continuous stirring and heating of the mixture by a water bath. Distillation was started at ambient temperature, followed by gradually raising of bath temperature up to 90 °C. Afterward, the products spontaneously divided into two layers. The bulk of soaps and residual glycerol transferred to the bottom layer. Water washing was performed by crude esters mixed with a fixed volume of hot (60–80 °C) tap water, followed by non-intense shaking and phase separation. The procedure was repeated until a neutral transparent wash water layer was formed. In another case, the first washing step was carried out by a 2-% water solution of H_3PO_4. Esters after washing were dried over anhydrous Na_2SO_4. So-called dry washing was carried out as a batch process in a conic flask, loaded with a sample and 5% adsorbent, for 30 min at 50 °C under continuous stirring (500 rpm). Afterward, the adsorbent was removed by filtration through a paper filter. Batch distillation of esters was carried out under vacuum (about 0.3–0.4 kPa) and continuous stirring (500 rpm, oval magnetic anchor) in a round-bottom flask (250 cm^3). Heating was provided by an oil bath. Its temperature was gradually increased from 190 °C to 240 °C. The temperature range of ethyl esters' fraction condensation was 146–168 °C.

Potassium hydroxide, potassium ethoxide (KOEt), and potassium *n*-butoxide (KOBu) were used as transesterification catalysts in the current study. The alkoxides were synthesized as alcohol solutions from only KOH and corresponding alcohol by methods, pending by Patents of Ukraine (Zubenko et al. 2016, 2021).

To shift the reaction equilibrium (1) in the alkoxide formation direction, water should be constantly removed from the reaction media. In the case of *n*-butoxide preparation, the property of *n*-butyl alcohol to form heterogeneous azeotrope was used for this purpose. As generally known, the *n*-butanol azeotrope mixture with water (boiling point – 92.4 °C) contains 55.5% of alcohol. It divides into two phases after condensation due to partial miscibility of *n*-butyl alcohol and water. The alcohol-enriched upper phase contains 20.1% of n-C_4H_9OH and 79.9% of H_2O, while the concentration of alcohol in the water-enriched lower phase is only 7.7%. The synthetic procedure for KOBu included water-butanol azeotrope removal by distillation from butyl alcohol hydroxide solution. Heterogeneous azeotrope condensed and gathered in Dean-Stark trap. The upper alcohol-enriched layer was continuously returned to the reactor from the top of the trap, while the water-enriched phase was collected in its bottom. Distillation was continued until the volume of the latter stopped increasing. Knowing this volume enables us to estimate the yield of butoxide.

Contrary to n-butanol, ethyl alcohol does not form the heterogeneous azeotrope with water. The synthetic procedure for KOEt included the selective dehydration of homogeneous water-ethanol azeotrope over the molecular sieve KA-Y/3A. The solution of KOH in ethyl alcohol, placed in a round-bottom flask (500 cm^3), was boiled on an oil bath, which provides reflux and condensation of water-ethanol azeotrope through zeolite in Soxhlet extractor. Selectively dehydrated ethanol was periodically returned into the reaction flask. Such an approach prevents the contact of zeolite with a highly alkaline solution. Therefore, the resulting ethoxide solution is not contaminated by the products of zeolite dissolving with alkali. Zeolite can be utilized many times after proper dehydration. The number of repeated distillation-condensation cycles needed may depend on the water content in initial reagents (KOH and ethyl alcohol). KOEt solution in ethanol, prepared in the described way, can be also used for the synthesis of alkoxides of higher boiling alcohols, including *n*-butoxide. For this purpose, the corresponding alcohol is added, and ethanol is removed by distillation.

In the case of the not total conversion of the hydroxide, the obtained catalytic solutions contained both alkoxide and KOH. There were no reliable methods to distinguish between hydroxide and alkoxide in their composition. Only in KOBu preparation the yield estimation is possible from the volume of the lower water-butanol phase. The amount of water in this phase is substantially higher than it can be formed by reaction (1). The one possible source of this water is reagent KOH. If treating all admixtures of KOH reagent as water, the yield of KOBu after reaching the equilibrium is always about 70–80%. So, the alkaline catalyst load, regardless of the alcohol used, is expressed in terms of initial KOH (as %eq KOH). For example, 1.00 $\%_{eq\ KOH}$ load of butoxide in the case of 80% butoxide yield means the load of 1.60 g of neat KOBu and 0.2 g of KOH in the composition of catalytic solution for each 100 g of oil.

Analytic Methods The composition of transesterification products was analyzed by gas chromatograph Agilent 7890A series. It was equipped with split/splitless and cool-on-column inlets, corresponding capillary columns, and a flame-ionization detector (FID). High-pure helium was the carrier gas. Three different analytic techniques were used.

Technique 1. The concentration of alkyl esters was analyzed by a technique based on the standard European method of determination of methyl esters in biodiesel EN14103. When analyzing esters, other than FAME, not only methyl heptadecanoate or methyl nonadecanoate can be used as internal standards. Far more affordable methyl palmitate, being one of such alternatives, was used in the current study. Samples for the analyses were prepared as follows. About 10 drops (0.15–0.25 g) of transesterification products were put into a 10 cm^3 vial. Afterward, about 0.02 g of methyl palmitate was added. Masses of the sample and standard were measured with analytic accuracy (four-digit weighing). The sample and standard were dissolved in 5 cm^3 of *n*-hexane and subjected to gas chromatographic analysis. It was conducted using J&W HP-5 capillary column ((5% phenyl)-methyl polysiloxane, 30 m length, 0.32 mm internal diameter, 0.25 μm film thickness). Conditions of analyses were the following: injection volume 1 μl, inlet temperature (split/splitless) 250 °C, inlet excess pressure 83 kPa (constant pressure mode), split ratio 38, FID temperature 250 °C.Chromatographic column oven temperature programs were the following: 210 °C/15 min - heating 5 °C/min up to 280 °C -280 °C/ 10 min (in the case of FAEE) or 210 °C/27 min - heating 5 °C/min up to 320 °C

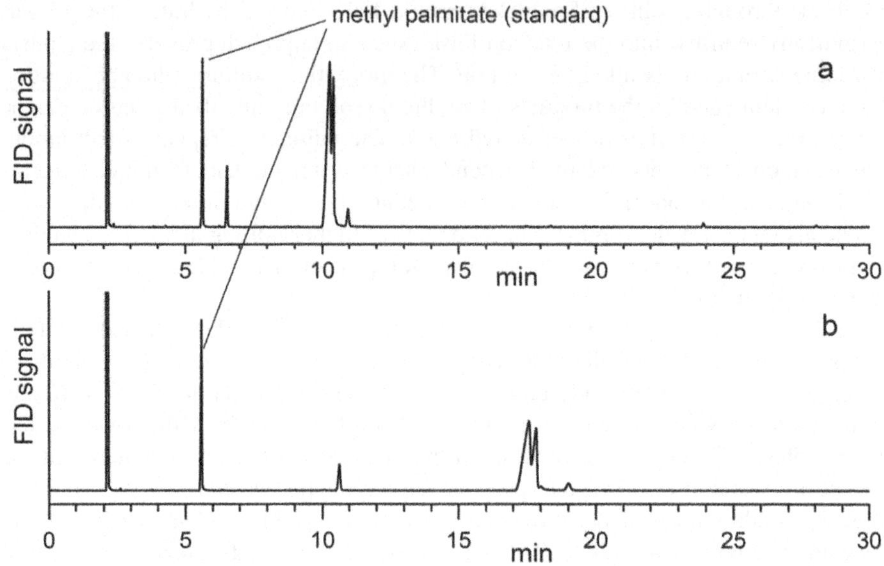

Fig. 4.1 Chromatograms of FAEE (**a**) and FABE (**b**), obtained on the base of wasted frying sunflower oil

-320 °C/10 min (in the case of FABE). Such conditions provide enough full separation of chromatographic peaks of standard and analyzed esters (Fig. 4.1). The latter's concentration was calculated from the known masses of the sample and standard and the area ratio of corresponding peaks, calculated from the chromatogram.

Technique 2. Concentration of glycerol, mono-, di-, and triacylglycerides (TG) in purified products was analyzed by standard method ASTMD 6584. It relies on derivatization (trimethylsilylation) of OH-groups by MSTFA. 0.1 cm^3 of the sample was put into 10 cm^3 vial and weighted (four digits). Afterward, 100 μl of standard № 1 (1,2,4-butanetriol, 1000 μg/cm^3 in pyridine), 100 μl of standard № 2 (tricaprin, 8000 μg/cm^3 in pyridine), and 100 μl of MSTFA were added. The mixture was manually shaken, and after 20 min at ambient temperature, 8 cm^3 of *n*-hexane was added. The solution was subjected to gas chromatographic analysis using DB-5HT capillary column ((5% phenyl)-methyl polysiloxane, 15 m length, 0.32 mm internal diameter, 0.1 μm film thickness) equipped with 1 m length and 0.53 mm diameter fused silica retention gap. Conditions of analyses were the following: injection volume 1 μl, inlet (cool-on-column) heating – track by oven, inlet excess pressure 53 kPa (constant pressure mode), FID temperature 380 °C, chromatographic column oven temperature program 50 °C/1 min (15 °C/min) up to 180 °C (30 °C/min) up to 380 °C (380 °C/10 min). The concentration of the glycerol, MG, DG, and TG in the sample was calculated from calibration plot, built using the range of standard solution of glycerol, mono-, di-, and triolein in pyridine as described in ASTMD 6584.

Technique 3. In order to evaluate the fullness of self-separation, glycerol content in ester layers was measured. Analysis was carried out without derivatization. The same methyl palmitate was used as the internal standard for the semiquantitative determination of glycerol content. Also, the concentration of alcohol can be measured in the same procedure. 0.6 cm^3 of the sample analyzed was put into 10 cm^3 vial, then 5 cm^3 of the 0.210% solution of methyl palmitate in DMFA was added. Masses of sample and standard solution were measured by four-digit weighing. The solution was subjected to gas chromatographic analysis using J&W HP-5 capillary column. Conditions of analyses were the following: injection volume 1 μl, inlet (split/splitless) temperature 250 °C, split ratio 20, FID temperature 300 °C, carrier gas flow (constant flow mode) 3.0 cm^3/min during 9 min – increase 3 cm^3/min up to 9 cm^3/min . Chromatographic column oven temperature program was the following: 35 °C/9 min – heating 10 °C/min up to 80 °C – 80 °C/1 min – heating 5 °C/min up to 380 °C – 380 °C/7.5 min. The concentrations of ethanol and glycerol were calculated from the areas of corresponding peaks and the area of the peak of methyl palmitate, taking into account the early built calibrations of sensitivity of FID to standard and components analyzed. The concentration limit of glycerol semiquantitative determination in the used analysis condition is about 0.5%, which is enough to evaluate the efficiency of self-separation of TE products. Besides the measurement of the mentioned component concentration, the used technique provides the as-complete-as-possible chromatogram of TE products. The latter contains the peaks of all compounds that can be visible in used gas chromatographic conditions.

The acid number of oils esters was measured through titration by sodium butoxide solution in *n*-butanol. About 10–15 g of the sample was weighed in titration flask and dissolved in the same amount of *n*-butanol. Afterward, dissolved sample was titrated by sodium butoxide *n*-butanol solution (~1 $\%_{eq}$ NaOH) with indicator bromothymol blue.

The concentration of the bulk potassium in the composition of both alkaline catalyst and soaps was measured by acid-base titration. TE products were titrated by organosoluble *p*-toluenesulfonic acid. About 3–8 g of the sample was weighted in a titration flask and dissolved in 20–30 g of DMFA. The flask content was titrated by 1% or 2% solution of *p*-toluenesulfonic acid in DMFA with indicator bromothymol blue until the color changed from blue to orange.

The iodine value of oils was measured by the following procedure. The oil sample (0.1–0.2, four-digit weighting) was dissolved in 5 cm^3 of $CHCl_3$ and 20 cm^3 of 0.1 N water solution of potassium bromide-bromate, and 5 cm^3 of 45-% sulfuric acid was added. After intense shaking mixture was kept for 5 min in darkness, then 25 cm^3 of 10-% water solution of KI was added. After another 5 min in darkness, intense shaking, and dissolution in 25 ml of distilled water, the excess of formed iodine was titrated by 0.1 N $Na_2S_2O_3$ with corn starch as an indicator.

The water content in oil was determined by two methods. One method was the heterogenic azeotrope distillation using benzene to form heterogeneous azeotrope (Dean-Stark method). Another method was the coulometric Karl-Fischer titration.

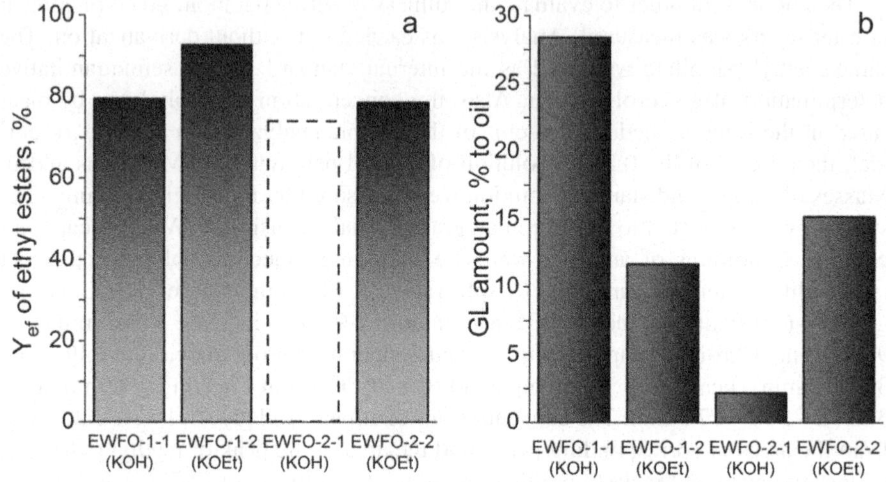

Fig. 4.2 The effective yield of FAEE (**a**) and amount of separated GL (**b**) after ethanolysis of samples of wasted frying oils

FTIR spectra of transesterification products were recorded by Shimadzu IRAffinity-1S infrared spectrometer, equipped with ATR accessory Specac GS 10801-B.

^{1}H NMR spectra were recorded by Bruker AVANCE DRX-500 NMR spectrometer at 500 MHz relatively internal standard TMS (tetramethylsilane), and samples were dissolved in $CDCl_3$.

The density of esters was measured using areometers according to EN ISO 3675. Kinematic viscosity of FAEE was determined as described in EN ISO 3104 using a glass capillary viscometer.

Transesterification of Oils by Ethanol In order to reveal the difference between using potassium hydroxide and ethoxide as alkaline catalysts of ethanolysis, the series of syntheses on the base of wasted oils' samples WFO-1 and WFO- 2 (0.40% and 2.80% of free fatty acids) was carried out. Catalyst loading includes $1.5\%_{eq\ KOH}$ plus excess for acid neutralization ($0.2\%_{eq\ KOH}$ for each 1% of FFA), which corresponds to 1.58 and $2.06\%_{eq\ KOH}$ for WFO-1 and WFO-2. Other conditions were the following: reagent ratio $R_{AO} = 6.0$, temperature $= 30$ °C, and reaction time $= 30$ min. The effective yields of FAEE and the amounts of separated GL in corresponding syntheses are presented in Fig. 4.2. From now on, the code name of each synthesis includes the name of the oil sample (according to Table 4.1), E or B as identification of implied alcohol (ethanol or *n*-butanol correspondingly), and the number of synthesis on the base of corresponding oil. Ethyl alcohol used in the current series of syntheses was dehydrated technical grade ethanol, while bioethanol was used in all further experiments.

In our studies, we use the term of the effective yield of esters (Y_{ef}) as the primary indicator of the efficiency of the syntheses. Y_{ef} takes into account formed esters only in the composition of the upper ester-enriched layer after phase separation. If there is no self-separation, the value of the Y_{ef} is regarded as 0. On the contrary, the value of total yield (Y_t) considers esters in the whole reaction mixture (without separation) or in the composition of all layers. Both Y_{ef} and Y_t are calculated as the ratio of the esters' amount in reaction products to theoretically possible amount, regarding oil (less FFA) as the pure TG of known fatty acid composition. It is generally known that low-grade wasted oils contain the so-called polar admixture, which cannot be fully converted to monoalkyl esters (Vieitez et al. 2014). Thus, the yield calculated in the described way may be slightly underestimated. It is also worth emphasizing that alkaline TE is always accomplished with saponification, and the yield cannot be quantitative even at 100% conversion of oil.

The significant importance of self-separation for further downstream processing of alkaline TE products was already underlined. For a more detailed study of the peculiarities of the self-separation during ethanolysis of relatively high-acidic wasted oils (sample WFO-3, Table 4.1) over KOEt, another series of syntheses at $R_{AO} = 6.0$ and $R_{AO} = 4.5$ and prolonged reaction time (100 min) was carried out (Table 4.2). It is worth mentioning that 2.50% concentration of FFA in the sample is usually considered as too high to carry out alkaline TE. There is the well-established convention in research papers that the esterification pretreatment is needed for oils, containing more than 1.0% FFA (Sanli et al. 2019). Both syntheses at higher excess of ethanol (EWFO-3-1 and EWFO-3-2) resulted in partial self-separation, but from one-third to a half of the theoretically possible mass or free glycerol (about 10% of the mass of initial oil) remained in the composition of the EL. A higher amount of separated GL resulted in a lower concentration of glycerol in EL composition. On the other hand, it also causes the lower value of effective yield due to the transfer of FAEE into the GL composition and, probably, due to a higher saponification rate at the higher reaction temperature. As for the syntheses at $R_{AO} = 4.5$, self-separation was observed only at higher catalyst load and lower temperature (synthesis EWFO-3-4). The concentration of glycerol, ethanol, and residual potassium in EL was substantially lower than in the case carrying out the reaction at a higher R_{AO} ratio. But effective yield also appeared to be sufficiently lower. Yield lowering is partially due to losses of ethyl esters into the GL. The latter's amount was even higher than in synthesis EWFO-3-1.

It looked interesting to find out whether it is possible to carry out alkaline ethanolysis with both effective phase separation and high esters yield at an ethanol-to-oil ratio 4.5. Figure 4.3 illustrates the results of the syntheses, carried out at such ratio, and a variable other technological parameter. The data for the reference synthesis EWFO-3-10 carried out at harsh conditions (very significant excess of ethanol and ethoxide, reflux) is also presented. The total yield of ethyl esters (Fig. 4.3a) appeared to be almost independent of the conditions applied. Only in the case of synthesis EWFO-3-7, carried out at the lowest load of ethoxide catalyst (1.0 $\%_{eq\ KOH}$), that it was about 79%. In all other syntheses, it was in the range 85–88%. It looks like this value is close to the maximum achievable yield for oil

Table 4.2 Comparison of ethanolysis of WFO-3 at different ethanol excess

	Conditions				Results						
Sample	R_{AO}	t, °C	τ, min	n_{cat}, $\%_{eq}$ KOH	Y_{ef}, %	EL, % to oil	GL, % to oil	FAEE in EL, %	Glycerol in EL, %	Ethanol in EL, %	K in EL, %
EWFO-3-1	6.0	60	100	1.4	78.7	102.6	29.2	78.8	3.4	8.3	0.34
EWFO-3-2	6.0	35	100	2.0	84.6	120.0	12.2	72.3	5.4	11.2	0.32
EWFO-3-3	4.5	60	100	1.4	(87.9)[a]	-	1.2	–		–	–
EWFO-3-4	4.5	35	100	2.0	71.7	83.9	38.4	87.5	1.2	4.2	0.20

[a] The value of Y_t in non-separated products of synthesis is given

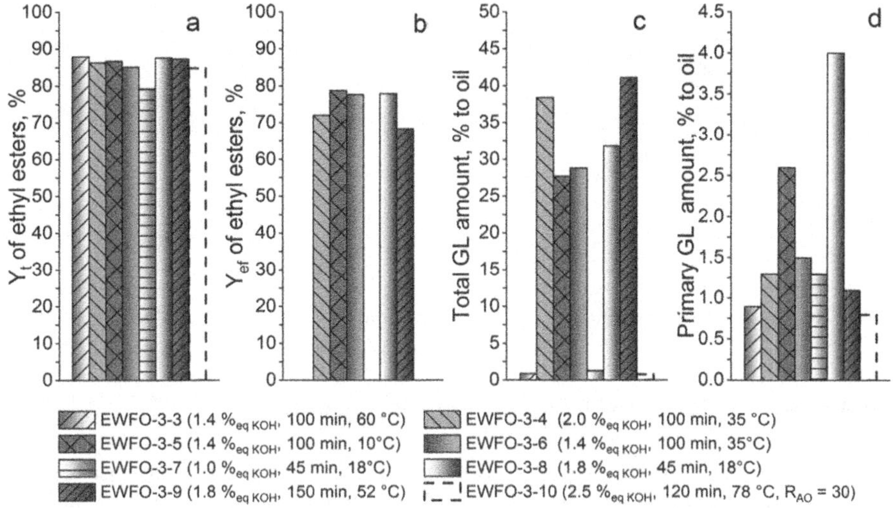

Fig. 4.3 The results of the FAEE syntheses on the base of WFO-3 oil sample at variable conditions

sample WFO-3. Even in synthesis, which was carried out at tenfold excess of ethyl alcohol over stoichiometry (EWFO-3-10), the Y_t value was no higher than 85%. Conducting the reaction at a high temperature with a high excess of ethanol, a large amount of catalyst for a prolonged reaction time must provide close to 100% oil conversion. However, such conditions are also strongly favorable for the saponification reaction, which may decrease ethyl esters' yield.

More complicated patterns were found out for the phase separation of the TE products. In most syntheses, self-separation was observed. Besides the discussed EWFO-3-10, it was absent when the reaction was carried out at the lowest catalyst load (1.0 $\%_{eq\ KOH}$, EWFO-3-7). The latter synthesis was also characterized by the lowest total yield of ethyl esters. However, in the case of the synthesis at the highest temperature (60 °C, EWFO-3-3), resulting in higher FAEE yield, self-separation also was not observed.

When reaction (EWFO-3-8) was carried out during a short time at low temperature and high load of catalyst, a substantial amount of glycerol separated directly in the course of the syntheses. It deposited in the form of dense non-fluent light yellow residue on the bottom of the reactor and remained there after pouring the reaction mixture into the separating funnel. Then, such a primary glycerol layer will be referred to as GL1, while the bottom glycerol-enriched layer formed during settling will be marked as GL2. The IR spectra of the sample of GL1 (Fig. 4.4) do not contain the absorption bands of stretching vibrations of carbonyl group around 1700 cm^{-1}, which indicates the absence of significant concentration of ethyl esters, acylglycerides, and soaps in its composition. The weaker intensity of the band around 3300 cm^{-1}, corresponding to the stretching vibration of hydroxyl groups, is due to the presence of an alkaline catalyst in their composition. The latter is,

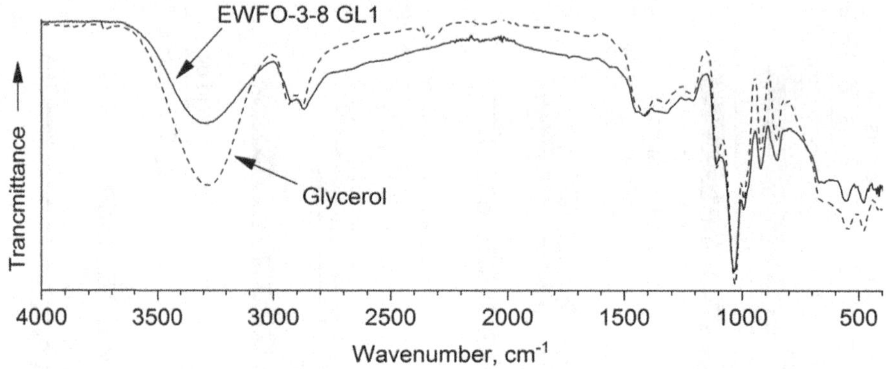

Fig. 4.4 IR spectra of GL1 (synthesis EWFO-3-8) and pure glycerol

probably, presented by potassium glyceroxide. It can be formed in the absence of a significant amount of water in reaction media due to the stronger acidity of glycerol compared to monohydric alcohols. Similar observation was made early by our group for alkaline butanolysis (Patrylak et al. 2019).

Separation of the reaction glycerol in the form of such a pure glycerol layer is uncommon for the alkaline transesterification process, especially in the case of TE of oil, containing as high as 2.50% of FFA. The authors of the current study have not met any literature information regarding such peculiarities of the alkaline ethanolyis. However, it is similar to that observed in alkaline butanolysis (Patrylak et al. 2019) and was discussed earlier.

It should be emphasized that the formation of about 4% of GL1 (relative to oil mass) means the removal of about 40% from theoretically possible reaction glycerol amount. A meaningful amount of GL1 was also formed in synthesis EWFO-3-5, which was carried out at low temperature (10 °C). In all other cases, the residue on the bottom of the reaction flask was deposited in minor amounts. However, it is also marked as the glycerol layer, including syntheses without self-separation.

When the reaction was carried out at conditions, which favor the GL1 formation, the effective yield of esters was the highest (about 77–78%). The same value of the Y_{ef} was achieved at intermediate values of technological parameters (EWFO-3-6). In other cases, the lower value of effective yield is mainly due to the transfer of FAEE into GL2 composition, forming in more significant amounts (Fig. 4.3c). The latter may be caused by a higher saponification rate at high reaction temperature and a load of the alkali. The ELs after all syntheses contained 0.13–0.21% of K, which is about 15–20% from its loading in the catalyst composition.

The experiments revealed that low reaction temperature and short reaction time result in higher effective yield and effective self-separation, including the formation of a significant mass of primary glycerol layer. However, the main driver of the effective alkaline ethanolysis is the high load of KOEt. It looks like ethoxide acts not only as catalyst, but also as a precipitation reagent, enforcing the glycerol separation. As the concentration of the actual active catalytic species, in this case, is very high, the short reaction time and a low temperature don't cause the lower yield of esters.

Table 4.3 Brief description of purification methods applied

Sample	Purification method
EWFO-3-P1	Six-time washing with hot water (60–70 °C, $V_{FAEE}/V_{H2O} = 2$); drying over Na_2SO_4
EWFO-3-P2	One-time washing with H3PO4 diluted water solution ($V_{FAEE}/V_{H3PO4} = 2$, acid concentration was taken to completely bind all K in the form of dihydrophosphate) followed by one-time washing with hot water (60–70 °C, $V_{FAEE}/V_{H2O} = 2$); drying over Na_2SO_4
EWFO-3-P3	Removal of ethanol by vacuum distillation (~0.3–0.4 kPa; 20–90 °C); washing of the upper layer (after self-separation); washing with hot water (as EWFO-3-P1); drying over Na_2SO_4
EWFO-3-P4	Removal of ethanol by distillation under atmospheric pressure (80–190 °C); washing of the whole mixture (no self-separation) with hot water (as EWFO-3-P1); drying over Na_2SO_4
EWFO-3-P5	Neutralization with 85-% H_3PO_4 (to dihydrophosphate) followed by vacuum distillation (~0.1 kPa; 190–240 °C)

Purification of Ethyl Esters In order to evaluate the possibility of isolation of the fuel-grade ethyl esters from the obtained ester-enriched layers, a number of conventional purification methods were applied. All ester layers after syntheses were mixed. The resulting mixture contained 84.8% of FAEE, 4.1% of ethanol, and 1.7% of free glycerol. The concentration of potassium was 0.21%. If assuming that all K is in the form of soaps, the latter's concentration was about 1.7%.

Table 4.3 gives a brief description of the purification methods applied. Direct washing with water occurred without extensive emulsification. The water after the last washing step was clear and had neutral pH. On the contrary, the bottom layer after the washing with H_3PO_4 was transparent and slightly acidic (pH $\approx$ 4); the following washing with water resulted in the separation of the white emulsion, having a pH of about 8–9. This fact indicates that neutralization was not complete. Ethanol removal from the ethanolysis products needed heating at least 70°C even under vacuum. The additional self-separation occurred after ethanol removal. In the case of atmospheric distillation, far higher temperatures appeared to be needed. The low intense boiling continued even at 190 °C, which is likely the result of ethanol forming by reverse transesterification of ethyl esters with glycerol, MG, and DG. Product after atmospheric distillation became dark yellow. No self-separation was observed. The following washing was complicated with intense emulsification. Direct vacuum distillation after the neutralization with H_3PO_4 proceeded efficiently without any foaming, and distillate was clear and colorless.

Table 4.4 compares the composition and some properties of products obtained with the requirements of the biodiesel standards EN 14214 and DSTU 7178. DSTU 7178 is the local Ukrainian standard for fatty acid ethyl esters as biodiesel fuel. Its requirement for the quality of BD is mainly identical to that of European EN14214 (FAME as biodiesel), besides some higher allowed MG concentration ($\leq$0.80% against $\leq$0.70%in EN 14214). The concentration of ethyl esters in all purified samples was lower than the required 96.5%. The highest ester content was found

Table 4.4 Composition and properties of ethyl esters after purification by different methods

Characteristic	Value for sample					EN 14214/ DSTU 7178 requirements
	EWFO-3-P1	EWFO-3-P2	EWFO-3-P3	EWFO-3-P4	EWFO-3-P5	
Ethyl esters content, %	90.3	89.6	86.4	90.5	94.1	≥96.5
Ethanol content, %	0.03	0.23	0.01	0.03	0.01	≤0.20
Kinematic viscosity at 40 °C, mm^2/s	4.63	4.92	5.32	4.73	4.31	3.5–5.0
Density at 15 °C, g/cm^3	0.882	0.882	0.884	0.881	0.878	0.860–0.900
Acidic number, mg KOH/g	0.32	2.76	0.57	0.22	4.40	≤0.5
Mass yield of product relative to initial oil, %	0.656	0.794	0.561	0.598	0.744	–

to be in the sample, purified by vacuum distillation. However, it was also characterized by the acid number, which many times exceeds the allowed value 0.5 mg KOH/g. This fact is due to the distillation of FFA, formed as products of soaps' neutralization by strong acid, in the same temperature range as ethyl esters. Vacuum distillate EWFO-3-P5 was also characterized by the lowest values of kinematic viscosity and density and the lowest ethanol content. The yield of the purified sample was the second highest in the series after EWFO-3-P2, being washed with neutralization. The relatively high yield, in this case, is due to not enough full washing after neutralization. Sample EWFO-3-P2 also had high acidity and contained some residual ethanol. The removing of ethanol by distillation under normal pressure appeared to be unsuitable technique. The sample after water washing was characterized by the lowest ester content, highest viscosity, and highest density among the samples of current series. Removing ethanol under vacuum followed by water washing (EWFO-3-P3) provided the composition and properties of esters close to that obtained after direct EL water washing. The lower yield is due to losses of the product as a result of the additional phase separation.

No gas chromatographic determination of unconverted glycerides was carried out for these samples. However, the estimation from the ^1H NMR spectra (not shown) shows that the concentration of MG in purified samples of ethyl esters is not higher than 1.5–2%. The main impurities are believed to be dimerized ethyl esters or glycerides. Such compounds are not visible in standard gas chromatographic methods applied.

The problem of the low ester content in BD, obtained from wasted frying oils, is well-known from literature data. Vieitez et al. (2014) studied the composition of 24 samples of wasted frying oils (origin – sunflower or rice bran) by adsorption column chromatography and high-performance size-exclusion liquid chromatography. It was shown that besides non-polar fraction (TG), samples contained 17–43% of the so-called polar fraction, including dimers, oligomers, or polymers of TG. Such compounds cannot be fully converted into monoalkyl esters. For example, the transesterification of one hypothetical molecule of dimerized TG gives four molecules of monoalkyl esters and one molecule of dimerized alkyl ester. Biodiesel standards do not set requirements for the content of such compounds. However, they may cause inappropriate low content of esters in biodiesel fuel. In the case of the work of Vieitez et al. (2014), TE of wasted frying oil samples by methanol (either alkaline or two-stage acid-alkaline process followed by water washing) resulted in ester content higher than 96.5% in the case of only 1 of the 24 samples studied. All other products were out of range of biodiesel standards in ester content.

The results obtained demonstrate that washing with water, which is a common industrial method in the production of methanol-based biodiesel, cannot provide enough high content of ethyl esters in products of alkaline ethanolysis of WFO. At the same time, this method is efficient in the case of ethanol transesterification of high-quality food-grade oil. In order to prove its efficiency, the sample of refined sunflower oil RSO was transesterified over KOEt at the following conditions: $R_{AO} = 6$, $n_{cat} = 1.5$ $\%_{eq\ KOH}$ KOH, 13 °C, 60 min. The upper layer after phase separation was washed with water and dried as described earlier. The obtained product contained 98.2% FAEE, 0.06% ethyl alcohol, about 0.1% FFA (acid number 0.2 mg KOH/g), 0.47% MG, 0.05% DG and no TG, and free glycerol. The content of all listed components is within the requirements of EN 14214 and DSTU 7178.

Vacuum distillation looks a promising method for the isolation of high-pure ethyl esters from wasted frying oil TE products. However, neutralization with strong acids, leading to breakage of soaps, results in a high acid number of distillate (Table 4.4). On the other hand, the high content of the soaps may cause the foaming and jellification of the mixture during batch distillation. If the residual active catalyst is present, the reverse TE reaction and saponification are possible in the course of distillation. The prior removal of ethanol by distillation, followed by additional phase separation, may simplify the further vacuum distillation of ethyl esters. Such a way of purification was investigated in three separate syntheses on the base of different oil samples (Table 4.5). The syntheses were carried out at conditions, providing the highest FAEE yield and effective phase separation in previous experiments ($R_{AO} = 4.5$, 18 °C, 45 min, and catalyst load 1.1–1.4 $\%_{eq\ KOH}$ plus excess for FFA neutralization). Y_{ef} in synthesis on the base of the WFO-3 sample (EWFO-3-11) was close to that obtained earlier (Fig. 4.3b). In the case of ethanolysis of the lower acidic oil samples (including the sample of refined oil RSO), Y_{ef} was 4–10% higher. All syntheses were accompanied by the formation of a significant amount of GL1. In the case of synthesis EWFO-4-1 on the base of the oil sample, having relatively low FFA content, more than half of the reaction glycerol transferred into GL1 composition. At the same time, the mass of formed GL2 was lower than in synthesis EWFO-3-11 (2.50% FFA in initial oil). This fact resulted in some higher content of ethanol and K in the obtained ester layer. Also, despite the higher effective yield of ethyl esters, its concentration in EL was slightly lower.

The oligomer compounds cannot be distilled and remained in the cube residue. Some additional polymerization was also possible at high temperatures during the distillation. As no neutralization of the reaction products was carried out, the reverse TE reaction was also possible. The obtained products also fit the biodiesel standards

Table 4.5 Conditions and results of syntheses of FAEE for further vacuum distillation with prior ethanol removal

Characteristic	Sample **EWFO-4-1**	**EWFO-3-11**	**ERSO-1**
Conditions of syntheses			
Oil sample (% FFA)	WFO-4 (1.15)	WFO-3 (2.50)	RSO (0.03)
R_{AO}	4.5	4.5	4.5
n_{cat}, $\%_{eq\ KOH}$ [a]	1.60 (0.23)	1.60 (0.50)	1.31 (0.01)
τ, min	45	45	45
t, °C	18	18	18
Results of syntheses			
Y_{ef} of FAEE, %	81.7	77.5	87.0
EL, % relative to oil	101.4	92.3	105.2
GL1, % relative to oil	5.7	2.7	5.0
GL2, % relative to oil	16.1	27.7	12.3
FAEE concentration in EL, %	81.7	85.5	87.8
Ethanol concentration in EL, %	7.0	5.1	5.9
K concentration in EL, % [b]	0.27 (24.7)	0.19 (16.0)	0.15 (17.3)

[a] The part of the catalyst for FFA neutralization is indicated in brackets
[b] The percent relative to the loaded K in composition of catalyst is indicated in brackets

Table 4.6 Results of FAEE purification by vacuum distillation with prior ethanol removal

Characteristic	Sample EWFO-4-1	EWFO-3-11	ERSO-1	EN14214/DSTU7178 requirements
Yield of upper layer after ethanol removal, % relative to oil	95.4	87.5	99.1	–
Yield of distillate, % relative to oil	76.6	64.5	83.9	–
Kinematic viscosity at 40 °C, mm^2/s	4.07	4.28	4.16	3.5–5.0
Density 15°, g/cm^3	0.878	0.878	0.878	0.860–0.900
Acid value, mg KOH/g	0.14	0.08	0.08	≤0.50
Distillate composition				
FAEE concentration, %	97.6	99.2	98.9	≥96.5
Ethanol concentration, %	0.04	0.05	0.04	≤0.20
Glycerol concentration, %	0.050	0.128	0.042	≤0.02
MG concentration, %	0.08	0.08	0.06	≤0.70/0.80
DG concentration, %	0.01	0.00	0.01	≤0.20
TG concentration, %	0.00	0.00	0.00	≤0.20

for the MG, DG, TG, and ethanol content. Their kinematic viscosity, density, and acid value are also within the range of requirements for BD. Moreover, the concentration of MG was very low (<0.10%), and concentrations of DG and TG were almost negligible. At the same time, the glycerol content was 2–6 times higher than

Table 4.7 Conditions and results of syntheses of FAEE for further vacuum distillation without prior ethanol removal

Characteristic	Sample EWFO-4-2	EWFO-4-3	EWFO-1-3	EWFO-5-1
Conditions of syntheses				
Sample of oil (% FFA)	WFO-4 (1.38)	WFO-4 (1.38)	WFO-1 (0.70)	WFO-5 (0.71)
R_{AO}	4.5	5.5	5.5	5.0
n_{cat}, $\%_{eq}$ KOH[a]	1.38 (0.28)	1.38 (0.28)	1.24 (0.14)	1.24 (0.14)
τ, min	45	45	45	45
t, °C	18	15	15	18
Results of syntheses				
Y_{ef} of FAEE, %	82.9	86.6	88.1	79.2
EL, % relative to oil	102.0	112.6	112.7	109.0
GL1, % relative to oil	2.2	3.1	5.1	1.6
K concentration in GL1, %[b]	3.81 (6.3)	5.21 (11.7)	6.60 (27.0)	6.21 (8.3)
GL2, % relative to oil	17.4	11.5	9.6	14.6
FAEE concentration in EL, %	84.3	79.8	81.6	75.8
Ethanol concentration in EL, %	5.8	10.5	9.1	9.0
Glycerol concentration in EL, %	0.7	2.2	1.8	2.3
K concentration in EL, %[b]	0.19 (20.2)	0.34 (40.8)	0.28 (37.3)	0.41 (40.0)

[a] The part of the catalyst for FFA neutralization is indicated in brackets
[b] The percent relative to the loaded K in composition of catalyst is indicated in brackets

0.02% allowed by BD normative. The yield of the final product in the case of the oil sample WFO-3 appeared to be lower than in the case of distillation with prior neutralization (see EWFO-3-P5 in Table 4.4).

It was essential to investigate the possibility of carrying out direct batch vacuum distillation of the ester layers without any intermediate treatment. For this purpose, other TE experiments were conducted (Table 4.7). In this experiment series the concentration of glycerol in ELs was measured. It appeared to be significantly higher in the case of higher excess of ethanol, even in lower FFA content in oil (about 0.70% for WFO-1 and WFO-5 samples). The concentration of residual K also increased when increasing the R_{AO} ratio. The effective yield was slightly higher in the case of more significant excess of alcohol and lower oil acidity. However, in synthesis on the base of the WFO-5 oil sample, Y_{ef} appeared to be the lowest in the series. This fact is probably due to deeper oil polymerization rate, which is indirectly indicated by lowest iodine value and highest viscosity of the WFO-5 sample (Table 4.1). The results of the titration of the GL1 are of particular interest. Even in the case of deposition of insignificant amount of GL1, the concentration of K in its composition was about 3–5 times higher than in the whole reaction mixture. 6% to 27% of the initially loaded potassium transferred to GL1 composition. Similar to the synthesis EWFO-3-8, the IR spectra of GL1 almost did not contain the absorption

Table 4.8 Results of FAEE purification by vacuum distillation without prior ethanol removal

Characteristic	Sample				EN14214/ DSTU7178 requirements
	EWFO-4-2	EWFO-4-3[a]	EWFO-1-3[b]	EWFO-5-1	
Yield of distillate, % relative to oil	77.6	–	63.5	59.7	–
Acid number of distillate, mg KOH/g	0.20	n/a	n/a	0.36	$\leq$0.50
FAEE concentration, %	98.7	n/a	96.2	98.2	$\geq$96.5
Ethanol concentration, %	0.05	n/a	n/a	0.04	$\leq$0.20
Glycerol concentration, %	0.008	n/a	n/a	0.034	$\leq$0.20
MG concentration, %	0.01	n/a	n/a	0.03	$\leq$0.70/<0.80
DG concentration, %	0.00	n/a	n/a	0.01	$\leq$0.20
TG concentration, %	0.00	n/a	n/a	0.00	$\leq$0.20

[a] Proper distillate was not obtained; no analyses were carried out
[b] Distillate was contaminated; only FAEE content was measured

band of a carbonyl group. This fact indicates that potassium in the GL1 is not in the composition of soaps but in the composition of alkaline catalyst. We can assume that K in GL1 is in the composition of potassium glyceroxide as observed earlier in alkaline butanolysis (Patrylak et al. 2019). Thus, the GL1 after alkaline ethanolysis may be considered for reuse in the catalysis of TE reactions.

Only in the case of synthesis EWFO-4-2, resulting in the most efficient phase separation (lowest glycerol and K content in the EL), that it was possible to carry out the vacuum distillation properly. Vacuum distillate samples were clear and colorless. The concentration of the ethyl esters was high enough, and MG, DG, and TG were almost absent (Table 4.8), and the acid value was also in range of BD requirements. The glycerol content appeared to be very low (in the range of normative). It can be assumed that glycerol and MG underwent the reverse TE reaction in the presence of the residual alkaline catalyst. The yield of purified esters was almost the same as in the case of the distillation of the products of the EWFO-4-2 synthesis with prior ethanol removal.

In the case of the EWFO-4-3 synthesis products, the process was complicated by boiling irregularity, foaming, and jellification of the distilling mixture. As a result, the big part of the latter was thrown into the receiver of the distillate. In the case of the product EWFO-1-3, similar problems were observed, but the distillation was stopped before the bulk of the mixture was thrown into the distillate receiver. Such problems are more likely due to potassium soaps and glycerol content in the ester layer. Also, acylglycerides, formed in reverse TE of glycerol with ethyl esters, cannot be distilled. However, the double bonds of their acyl chains may undergo polymerization at high temperatures. In the case of product EWFO-5-1, containing a higher amount of K and glycerol, the distillation was also quite complicated, but the process was stopped before contaminations got into the distillate. Distilled ethyl esters fit the BD requirements for ester, MG, DG, and TG concentration and acid value, while glycerol content was higher than the allowed 0.02%.

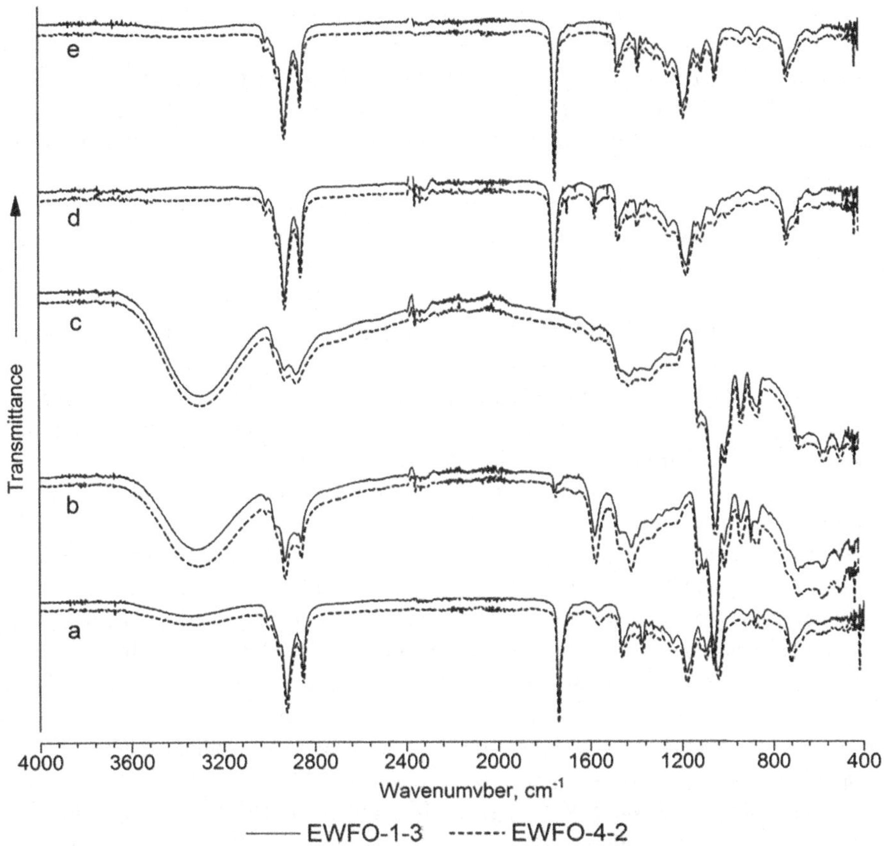

Fig. 4.5 IR spectra of EL (**a**), GL2 (**b**), GL1 (**c**), residue after distillation (**d**), and vacuum distillate (**e**) of corresponding syntheses

Figure 4.5 compares the IR spectra of the ester and glycerol layers and IR spectra of the vacuum distillate and cube residues, obtained in syntheses EWFO-4-3 and EWFO-1-3. The absorption bands of carbonyl groups of ethyl esters and acylglycerides (~1730 cm^{-1}) and soaps (~1560 cm^{-1}) are the main object of interest. The spectra of the ester layers (Fig. 4.5a) are characterized by intense absorption around 1730 cm^{-1}, while the intensity of the corresponding bands of soaps is relatively low. The IR spectra of GL1 are similar to the spectrum of GL1 after synthesis EWFO-3-8 (Fig. 4.4). Both GL1 spectra do not contain the adsorption band of esters' C=O stretching vibration, and the band of soaps' carbonyl group is of very low intensity (Fig. 4.5c). On the contrary, the latter band is intense in the IR spectra of GL2 (Fig. 4.5b). This observation indicates the transfer of the bulk of soaps into GL2 composition. It is slightly confusing that the intensity of the band around 1730 cm^{-3} is weak, while the chromatographic analyses of GL2 samples have shown the content of esters about 15–20%. This fact may be due to the inhomogeneity of the GL2 samples through the height of the layer. It should be mentioned that in the ATR technique, the drop of the sample is put on the surface of the horizontal diamond prism. The IR spectra are recorded from the bottom part of the drop, which directly contacts the surface of the latter. As a result, the top part of the sample drop, probably enriched with ethyl esters, may be in not direct contact with the prism.

The IR spectra of the distillate samples and cube residues after distillation are almost identical, besides the presence of soaps' characteristic band in the spectra of the latter (Figs. 4.5c, d). Also, the spectrum of the contaminated distillate EWFO-1-3 contains the broad weak of hydroxyls' stretching vibrations (around 3300 cm^{-1}). At the same time, the cube residue EWFO-4-1 contains only about 6% of ethyl esters. Other components are not visible in gas chromatographic analyses. They may be the polymerized ethyl esters and polymerized acylglycerides, formed in reverse TE reactions. The existence of the more complex polymer structures may be assumed, which contain potassium soap moieties in their composition.

Generally, the investigations indicate that the reaction conditions of ethanolysis, combining the low temperature, short synthesis time, low ethanol excess, and large catalyst (KOEt) load, are the most favorable for the effective yield of esters, product self-separation, and carrying out of the vacuum distillation of the ester layers formed.

In addition, the possibility of recovering the fuel-grade ethyl esters by dry washing procedure by meta-kaolin was estimated. The latter was obtained from the kaolin powder by calcination at 750 °C for 4 h and utilized in dehydrated form. New transesterification of the WFO-4 sample, containing 1.20% FFA, was carried out (code name EWFO-4-4). This time the higher excess of ethanol ($R_{AO} = 6.0$), temperature (30 °C), and longer time of synthesis (120 min) were applied. The load of the catalyst (KOEt) was 1.49 $\%_{eq\ KOH}$. This value included 0.25 $\%_{eq\ KOH}$ for FFA neutralization. Used conditions resulted in an even lower effective yield (80.6 %) than in the case of the previous synthesis on the base of the same oil sample (see Tables 4.5 and 4.7). Ester layer (102% relative to oil mass) contained 82.1% FAEE, 10.6% ethanol, 1.7% glycerol, and 0.26% K. The amount of GL1 was only 1.6%.

Removing ethanol under vacuum resulted in additional phase separation. The yield of the target upper layer appeared to be low (Table 4.9) due to the transfer of the large part of the ethyl esters together with impurities into the bottom layer. The content of the residual glycerides in upper layer was within the range of standards, but FAEE concentration was still lower than 96.5%. The concentration of glycerol and ethanol was also out of the range of BD requirements. A bulk of potassium transferred to the bottom layer in the form of soaps. Dry washing resulted in a decrease of ethanol and glycerol content to the allowed level. The MG concentration after dry washing appeared to be slightly higher and was out of range of EN 14214 but met the requirements of DSTU 7178. At the same time, there was no increase of the FAEE concentration, which indicates the inefficiency of the treatment with the adsorbent to remove oligomerized admixtures. However, it may be helpful for the post-treatment of the vacuum distillate to remove the residues of glycerol or other low-molecular polar compounds. The latter, probably, can be formed at high temperatures during distillation.

Table 4.9 Results of FAEE purification through dry washing with prior ethanol removal

Characteristic	Sample		EN14214/DSTU7178 requirements
	EWFO-4-4 upper layer after ethanol removal	EWFO-4-4 after dry washing	
Upper layer after ethanol removal by distillation			
Mass yield, % relative to oil	50.7	42.4	-
FAEE concentration, %	93.3	92.2	≥96.5
Ethanol concentration, %	0.26	0.03	≤0.20
Glycerol concentration, %	0.056	0.011	≤0.02
MG concentration, %	0.63	0.74	≤0.70/0.80
DG concentration, %	0.12	0.10	≤0.20
TG concentration, %	0.04	0.03	≤0.20

It was interesting to compare the chromatograms of ester layers and ethyl esters after downstream processing (removing of ethanol, vacuum distillation, and dry washing). The chromatograms obtained by gas chromatographic technique №. 3 (the temperature of the column up to 320 °C, dissolving the sample in DMFA) are given in Fig. 4.6. All given chromatograms of ester layers contain the set of low-resolved signals in the high-temperature region (after 50 min). These peaks are not related to FAEE and more likely belong to oligomeric ester compounds with various cross-linkage. However, these signals are not suitable for quantitative determination, as the used inlet temperature (250 °C) is not high enough to provide the proper volatilization of such components. The use of higher temperature is restricted with concerns of thermal stability of analyzed components. However, chromatograms enable us to qualitatively estimate the efficiency of the purification methods in removing oligomeric admixtures. Dry washing after prior ethanol removal (synthesis EWFO-4-4) had not any effect on the discussed peak appearance. On the contrary, the vacuum distillate chromatograms are almost free of the peaks of such high molecular weight components. Only the chromatogram of the contaminated distillate EWFO-1-3 contains the number of weak peaks in a broad range of retention times.

Transesterification of Oils by n-butanol The already discussed observations of our group (Patrylak et al. 2019) show that effective phase separation of alkaline butanolysis products is possible using KOBu as catalyst and conducting the reaction at low excess of butanol and decreased temperature. In the current study we compared the alkaline butanolysis of different low-grade oils (wasted frying oil samples WFO-1, WFO-4, WFO-5, and rapeseed oil sample RO) at fixed set of reaction conditions. The conditions have been chosen on the basis of the previous study (Patrylak et al. 2019). They were similar to the most favorable conditions of

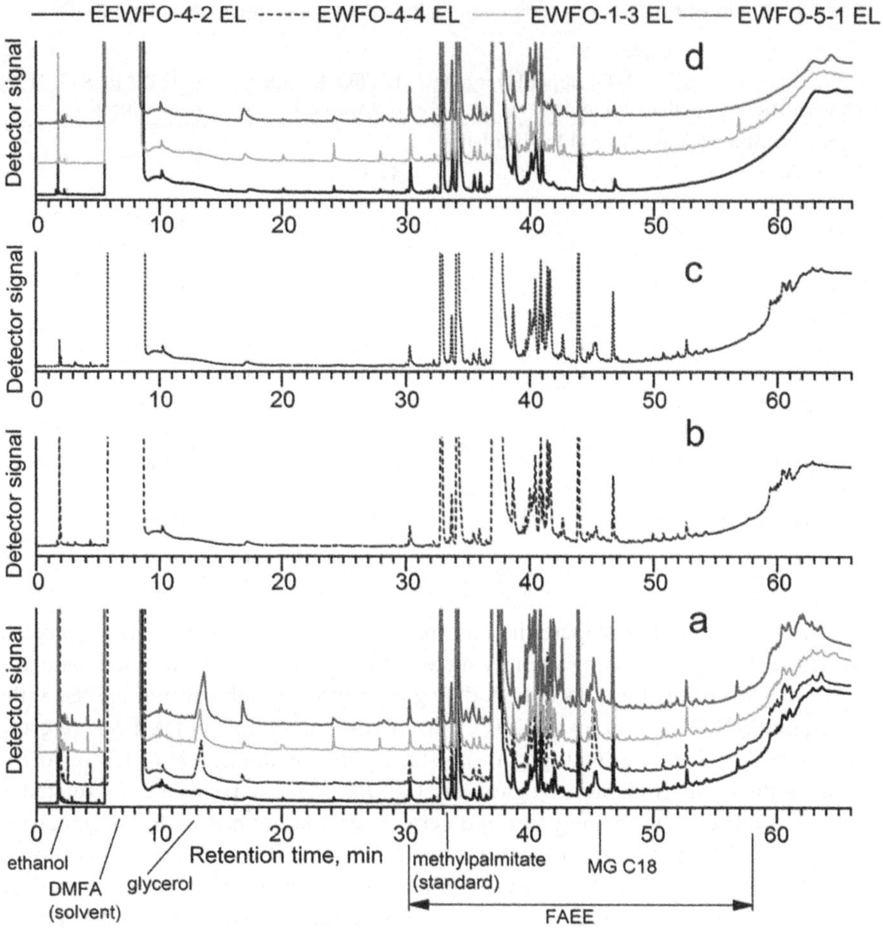

Fig. 4.6 Chromatograms of ester layers (**a**), products after ethanol removing by distillation (**b**), products after dry washing (**c**), and distilled ethyl esters (**d**)

alkaline ethanolysis, found in this work. Two catalytic solutions of butoxide were used in syntheses. One of them (marked as KOBu) was obtained directly from KOH solution in butanol by the Dean-Stark method. Another one, marked as KOBu-Et, was prepared from KOEt solution as described earlier.

Conditions and results of alkaline butanolysis of wasted frying oils' samples are given in Table 4.10. Let us, first of all, emphasize that the self-separation characteristic of butanolysis products is more complex than in the case of ethanolysis. The light yellow primary glycerol layer GL1, formed during syntheses, was characterized by high alkali concentration (about 10% in terms of K). This corresponds to the recovery of about 40% of the catalyst in GL1 composition. Also, about half of the

Table 4.10 Conditions and results of alkaline butanolysis of wasted frying oils' samples

Characteristic	Sample BWFO-1-1	BWFO-1-2	BWFO-4-1	BWFO-5-1
Conditions of syntheses				
Sample of oil (% FFA)	WFO-1 (0.57)	WFO-1 (0.57)	WFO-4 (1.15)	WFO-5 (0.71)
Catalytic solution	KOBu-1	KOBu-Et	KOBu-1	KOBu-Et
R_{AO}	4.5	4.5	4.5	4.5
n_{cat}, $\%_{eq\ KOH}$[a]	1.52 (0.12)	1.52 (0.12)	1.65 (0.25)	1.54 (1.40)
τ, min	20	20	20	20
t, °C	15	15	15	15
Results of syntheses				
Y_{ef} of FABE, %	87.7	89.2	84.2	80.9
EL, % relative to oil	130.0	127.5	124.6	130.0
GL1, % relative to oil	5.8	6.4	6.6	5.5
GL1', % relative to oil	1.6	1.4	0.5	3.3[b]
GL2, % relative to oil	1.4	3.5	3.4	-[b]
FABE concentration in EL, %	78.8	81.7	76.6	70.6
Butanol concentration in EL, %	10.7	10.9	9.1	10.1
Glycerol concentration in EL, %	2.1	1.9	2.3	2.6
K concentration in EL, %[c]	0.38 (45.4)	0.30 (35.2)	0.45 (48.8)	0.45 (53.9)
K concentration in GL1, %[c]	10.6 (40.4)	10.3 (43.1)	10.4 (43.4)	9.8 (36.8)

[a] The part of the catalyst for FFA neutralization is indicated in brackets
[b] No clear phase boundaries between GL1' and GL2 were observed
[c] The percent relative to the loaded K in the composition of catalyst is indicated in brackets

theoretically possible mass of reaction glycerol transferred to the latter. Contrary to ethanolysis, in most syntheses, the formation of two distinct glycerol layers was observed, having clear phase boundaries, after product settling. The upper of them was similar to the GL2 of ethanolysis and would be marked from now on by the same abbreviation. The lower of them had the same nature as GL1 and would be marked as GL1'.

The higher yield of butyl esters was obtained in syntheses on the base of the WFO-1 oil sample, having the lowest acidity. Catalytic solution KOBu-Et utilization provided slightly higher FABE yield and higher FABE concentration in EL. The latter also contained less glycerol and K. The lower Y_{ef} was obtained in synthesis on the base of WFO-4 sample. Concentration of glycerol and potassium was also slightly higher. Transesterification of the WFO-5 sample resulted in an even lower FABE yield, despite lower FFA content in oil. This fact is, most likely, due to the highest polymerization rate of the oil WFO-5 among the investigated samples. Generally, the ester layers after butanolysis appeared to contain a higher concentration of K compared to ethanolysis products. This fact can be explained by the separation of the lower amount of the GL2 after settling. In the case of all syntheses,

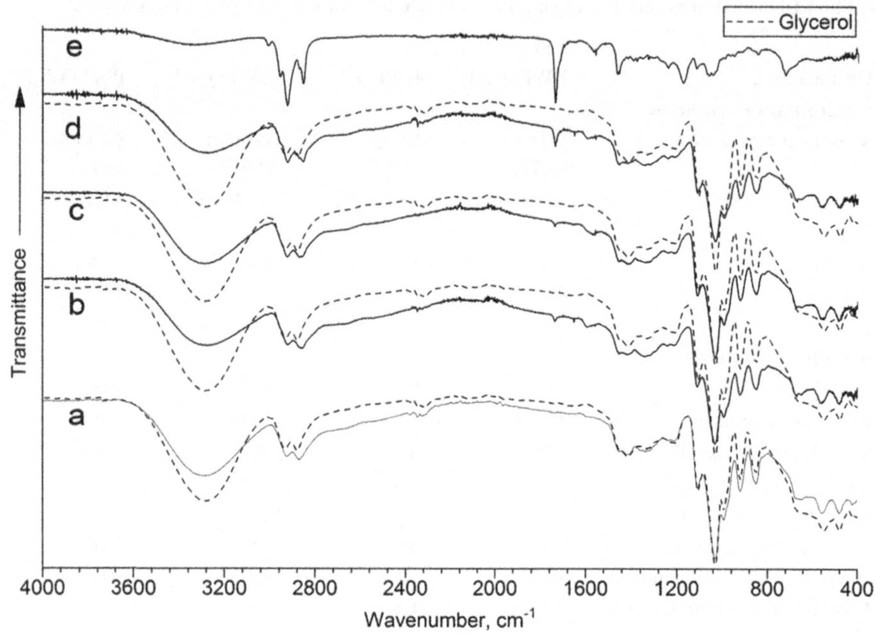

Fig. 4.7 IR spectra of the GL1' after butanolysis of refined sunflower oil (**a**), built according to Patrylak et al. (2019) in comparison with the IR spectra of GL1 (**b**), GL1' (**c**), GL2 (**d**), and EL (**e**) after BWFO-4-1 synthesis

about 25–30% from the theoretically possible reaction glycerol left in the composition of EL.

It was interesting to compare the IR spectra of each separate layer of the reaction products of alkaline butanolysis. The corresponding IR spectra of BWFO-4-1 synthesis products versus the spectrum of the glycerol layer after butanolysis of refined sunflower oil (according to data (Patrylak et al. 2019)) and the spectrum of pure glycerol are given in Fig. 4.7. The spectra of the GL1 and GL1' are identical (Figs. 4.7b, c). Both contain a weak band around 1740 cm^{-1}, which indicates the presence of minor admixtures of butyl esters. The higher intensity of the absorption at this frequency is observed in the spectrum of the GL2 (Fig. 4.7d). The spectra of all species of glycerol layers also contain the weak band around 1590 cm^{-1}, which more likely corresponds to the stretching vibration of soaps' carbonyl group. However, it is slightly shifted toward the higher frequencies compared with the corresponding band in the spectra of ethanolysis products (Fig. 4.5). It should be emphasized that the discussed bands are not presented in the spectrum of glycerol layers, obtained in butanolysis of refined oil, which indicates the higher purity rate of the latter. The recovery of the significant part of alkaline catalyst in composition of GL1 (more likely in the form the of potassium glyceroxide) enables its reuse as TE catalyst.

Table 4.11 Conditions and results of alkaline butanolysis of low-grade rapeseed oil

Characteristic	Sample BRO-1	Sample BRO-2
Conditions of syntheses		
Sample of oil (% FFA)	RO (2.85)	RO (2.85)
Catalytic solution	KOBu-1	KOBu-Et
R_{AO}	4.5	4.5
n_{cat}, $\%_{eq}$ KOH[a]	2.00 (0.57)	2.00 (0.57)
τ, min	20	20
t, °C	15	15
Results of syntheses		
Y_{ef} of FABE, %	80.5	88.8
EL, % relative to oil	125.2	129.0
GL1, % relative to oil	1.9	1.9
GL1', % relative to oil	2.5	4.9[b]
GL2, % relative to oil	6.2	-[b]
FABE concentration in EL, %	71.7	76.6
Butanol concentration in EL, %	11.4	11.9
Glycerol concentration in EL, %	3.7	2.9
K concentration in EL, %[c]	0.86 (77.3)	0.75 (69.2)
K concentration in GL1, %[c]	1.46 (2.0)	3.02 (4.1)

[a] The part of the catalyst for FFA neutralization is indicated in brackets

[b] No clear phase boundaries between GL1' and GL2 were observed

[c] The percent relative to the loaded K in the composition of catalyst is indicated in brackets

TE of the low-grade rapeseed oil containing 2.85% of FFA was carried out at the same reaction conditions using both KOBu and KOBu-Et catalytic solutions (Table 4.11). In these syntheses, phase separation took a longer time. GL1' separated during few hours after synthesis, while GL2 appeared only after about 3–4 days of settling. This fact may be related to the continuing of TE and saponification reactions in the course of the settling. It also should be mentioned that the yield of FABE the next day after synthesis (before GL2 formation) was almost the same as the values given in Table 4.11. The results obtained prove the higher efficiency of the KOBu-Et catalytic solution. Y_{ef} of FABE and FABE content in EL were noticeably higher, while the concentrations of glycerol and potassium in EL were slightly lower. However, the efficiency of phase separation in both cases was not high. About 35–45% from the theoretical amount of the reaction glycerol and the bulk of the potassium left in the EL. The mass of formed GL1 and alkali content in their composition was also low.

Figure 4.8 compares the chromatograms of ester layers after butanolysis of wasted frying oil and rapeseed oil. As in the case of ethanolysis (Fig. 4.6), high-temperature region of the chromatogram contains pure resolved signals, which may be attributed to the oligomeric compounds. In the case of rapeseed oil, which was

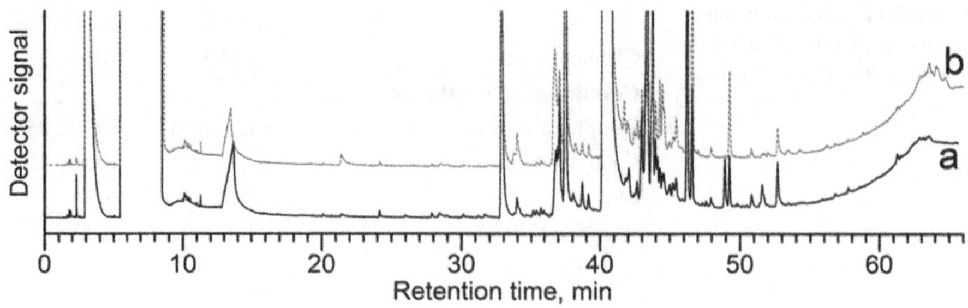

Fig. 4.8 Chromatograms of ester layers obtained in syntheses BRO-2 (**a**) and BWFO-1-1 (**b**)

stored for more than 10 years, weaker signals in the same region also can be observed. The presence of such oligomeric components should cause complications of the isolation of fuel-grade butyl esters from butanolysis products. The same approach, as for ethyl esters, including the vacuum distillation with prior butanol removing, may be recommended for their downstream processing. It also should be emphasized that conventional water washing may be unsuitable for the purification of butanolysis product due to *n*-butyl alcohol partially mixing with water.

4.3 Novelty

The main aspects of the novelty of this work consist in carrying out the effective alkaline ethanolysis and butanolysis of the low-grade oils, which is commonly regarded as unsuitable for one-step alkaline transesterification due to high acid number (up to 5.0–5.7 mg KOH/g). The high efficiency of alkaline transesterification was achieved by utilization of the appropriate alkoxides (ethoxide in ethanolysis and *n*-butoxide in butanolysis) and an especial approach to the combination of the reaction conditions. The latter included the slightly decreased temperature, short reaction time, limited excess of alcohol, but a high load of the alkoxide with proper excess for neutralization of free fatty acids. Among others, carrying out the transesterification in such a way results in the deposition of a high-pure primary glycerol layer, consisting predominantly of glycerol and alkali, which was not reported earlier. In the case of butanolysis, the formation of such glycerol layer is more extensive and its alkalinity is higher. This observation opens the prospects for the reuse of such glycerol layer in alkaline catalysis of transesterification by various alcohols, but additional studies are needed. The described way of alkoxide preparation for carrying out the transesterification reactions was not also reported in literature. The study results also demonstrate that conventional methods of biodiesel purification are ineffective for products of transesterification of wasted frying sunflower oil due to the presence of oligomeric components. So, in order to obtain the fuel-grade esters, vacuum distillation was utilized. The latter is overlooked as biodiesel downstream processing method in related literature sources.

4.4 Conclusions

The current investigation results demonstrate the efficiency of the early developed patent-pending methods of preparation of alkoxides as the catalyst of alkaline transesterification of oils. Catalytic solutions of alkoxides can be easily produced using only corresponding alcohol and hydroxide.

It has been shown that alkaline ethanolysis of used frying oils over potassium ethoxide results in high yields and effective phase separation even in the case of concentration of free fatty acids in oil up to 2.50%. The new approach for carrying out the alkaline ethanolysis of relatively high-acidic wasted oils was developed. It consists in carrying out the reaction for a relatively short time (45 min) at slightly decreased temperature (10–18 °C), limited excess of alcohol (molar ratio to oil 4.5–5), and high load of ethoxide catalyst (0.9–1.3% in terms of KOH plus excess for neutralization of acids). Such conditions provide not only high effective yield of esters (in ester layer) but also ensure the better phase separation. The recovery of the significant part of glycerol byproduct (~20–50% from theoretically possible) directly during the synthesis as the residue on the bottom of the reactor was observed. This primary glycerol layer consists mainly of neat glycerol and some alkali and almost does not contain the admixtures of ethyl esters and soaps. Such peculiarities of alkaline ethanolysis have not ever been reported in literature.

It was revealed that butanolysis of wasted frying oils' samples (0.57–1.15% of free fatty acids) at earlier established reaction conditions (butanol-to-oil molar ratio 4.5, 15 °C, 20 min, 1.4% of potassium butoxide in terms of KOH plus excess neutralization of acids) results in high effective yield of esters (81–89%) and formation of the distinct glycerol layers. Similar to ethanolysis, the primary glycerol layer consisted mainly of glycerol, but contained more alkali (about 10% in terms of KOH). Almost a half of the loaded catalyst was recovered in the composition of the primary glycerol layer (more likely in the form of potassium glyceroxide). This fact opens the prospects for at least one-time partial reuse of the alkaline catalyst, which is uncommon for the homogeneous catalytic transesterification of oils. Transesterification of the higher acidic rapeseed oil (2.85% of free fatty acids) resulted in 89% yield of butyl esters, but self-separation was less efficient.

It was demonstrated that fatty acid esters, produced from the wasted frying sunflower oil, cannot meet the requirements of biodiesel standards for ester content ($\geq$96.5%) due to the presence of oligomerized admixtures. The latter cannot be removed by conventional purification methods, including variable techniques of water washing and dry washing. It was grounded that the vacuum distillation of ethyl esters, early purified by conventional methods or directly of the ester layer after phase separation, is necessary to obtain the fuel-grade esters. Vacuum distillate samples contained more than 98% of esters, while the concentration of acylglycerides was very low. Only the content of glycerol in most cases was slightly out of the biodiesel specification range. Thus, the final treatment of the distillate by dry washing can be recommended.

References

BP (2020) Statistical Review of World Energy. British Petroleum https://www.bp.com/content/dam/bp/business-sites/en/global/corporate/pdfs/energy-economics/statistical-review/bp-stats-review-2020-full-report.pdf

Fonseca JM, Teleken JG, Almeida VC et al (2019) Biodiesel from waste frying oils: methods of production and purification. Energ Conv Manage 184:205–218

Hájek M, Skopal F, Vávra A et al (2017) Transesterification of rapeseed oil by butanol and separation of butyl ester. J Clean Prod 155:28–33

Homan T, Shahbaz K, Farid MM (2017) Improving the production of propyl and butyl ester-based biodiesel by purification using deep eutectic solvents. Sep Purif Technol 174:570–576

Joshi H, Moser BR, Walker T (2012) Mixed Alkyl Esters from cottonseed oil: improved biodiesel properties and blends with diesel fuel. J Am Oil Chem Soc 89:145–153

Patrylak LK, Zubenko SO, Konovalov SV et al (2019) Alkaline transesterification of sunflower oil triglycerides by butanol-1 over potassium hydroxide and alkoxides catalysts. Voprosy Khimii i Khimicheskoi Tekhnologii 5:93–103

Sanli H, Alptekin E, Canakci M (2019) Production of fuel quality ethyl ester biodiesel: 1. laboratory-scale optimization of waste frying oil ethanolysis, 2. pilot-scale production with the optimal reaction conditions. Waste Biomass Valori 10(7):1889–1898

Vieitez I, Callejas N, Irigaray B et al (2014) Acid value, polar compounds and polymers as determinants of the efficient conversion of waste frying oils to biodiesel. J Am Oil Chem Soc 91:655–664

Zubenko SO, Okhrimenko MV, Patrylak LK (2016) A method of preparing a catalyst for the production of biodiesel fuel [In Ukrainian]. UA Patent 108271, 11 July 2016

Zubenko SO, Konovalov SV, Patrylak LK (2021) A method of preparing a catalyst for the production of fatty acids' esters [In Ukrainian]. UA Patent 123473, 7 April 2021

Chapter 5
Cultivating Microalgae in Wastewaters for Biofuel and Fertilizer Production

S. Shamanskyi, S. Boichenko, I. Nezbrytska, and L. Pavliukh

Nomenclature

CO2	carbon dioxide
PBR	photobioreactor
CH4	methane
KOH	potassium hydroxide

5.1 Introduction

The determining factor that affects the productivity of microalgae biomass is the presence of optimal concentrations of macronutrients and micronutrients in their dew environment. The use of synthetic media for mass cultivation is often not economically justified due to their high cost compared to other available substrates. In this case, the most acceptable is using wastewater of various origins: agricultural, domestic, municipal, and industrial. The use of such an environment can significantly reduce the cost of cultivation due to the high availability and availability of nutrients necessary for the growth of algae, including nitrates, phosphates, ammonium, urea, some vitamins (biotin and thiamine), as well as metals (calcium, potassium, magnesium, iron, and others). The CO_2 saturation of wastewater is sufficient for the efficient cultivation of microalgae. It is shown (Solovchenko et al. 2013; Santos and Pires 2018) that microalgae grown on the effluents of livestock complexes are characterized by high biomass and lipid accumulation rates (Table 5.1).

S. Shamanskyi (✉) · S. Boichenko · I. Nezbrytska · L. Pavliukh
Kyiv National University of Construction and Architecture, Kyiv, Ukraine

National Academy of Sciences of Ukraine, Kyiv, Ukraine

S. Boichenko et al. (eds.), *Chemmotological Aspects of Sustainable Development of Transport*, Sustainable Aviation, https://doi.org/10.1007/978-3-031-06577-4_5

Table 5.1 Biomass and lipid productivity of microalgae grown in farm waste. (Solovchenko et al. 2013)

	Type of microalgae	Biomass Productivity, DW	Lipid productivity
Pig manure, high content NO_3^-	*Botryococcus braunii*	700 mg/L/day	69 mg/L/day
Cow manure on a styrofoam support	*Chlorella* sp.	2.6 g/m^2/day	230 mg/m^2/day
Pig wastewater, digested	*Scenedesmus* sp.	6 mg/L/day	0.54 mg/L/day
Pig wastewater, max. load	*Rhizoclonium hieroglyphicum*	10.7 g/m^2/day	72 mg/m^2/day
Dairy farm wastewater, aeration + CO_2	*Rhizoclonium hieroglyphicum*	17.9 g/m^2/day	210 mg/m^2/day
Cow manure, 25% dilution	Consortium: *Chlorella* sp., *Micractinium* sp., *Actinastrum* sp.	59 mg/L/day	17 mg/L/day

Prospective environments for the accumulation of microalgae biomass are domestic wastewater, ethanol production waste, and rice production waste. When *Chlorella pyrenoidosa* was grown on rice straw hydrolyzate, the increase in algal biomass in 48 h reached 2.83 g/l, and the lipid concentration was 56.3%.

Microalgae can function as autotrophs, using inorganic substances for the synthesis of organic. They can also function as heterotrophs, using organic matter available in the environment. Some species are capable of mixotrophy – the simultaneous use of inorganic substances that are assimilated during photosynthesis and organic compounds that come from outside. As a rule, microalgae strains capable of mixotrophic type of nutrition are cultivated in wastewater, as a significant amount of both inorganic and organic compounds is included in the composition of effluents. Sewage from different sources is characterized by different content of these substances, so it is necessary to conduct a preliminary screening of crops for cultivation in the available wastewater. *Chlorella*, *Botryococcus*, and *Scenedesmus* are the most adapted to different environmental conditions. *Chlorella* is characterized by the highest biomass productivity.

One of the possible ways to solve the problem of adaptation of microalgae cells to different wastewater compositions is the use of immobilized culture inoculum. Immobilization of microalgae cells and their use at the initial stage of cultivation can provide the ability to regulate their concentration, as well as increase resistance to high concentrations of organic or inorganic substances (Moreno-Garcia et al. 2017). Fixation of microalgae cells on a three-dimensional gel lattice, made of natural polymers (agar, cellulose, alginate, carrageenan) or synthetic polymers (polyacrylamide, polyurethane, polyvinyl, polypropylene), is a commonly used immobilization technique.

In addition to monocultures, consortia of microalgae are cultivated in wastewater. The advantage of using algae consortia is that they can withstand changes in environmental conditions and are resistant to the invasion of other species or organisms (fungi, bacteria, and others) that have entered the wastewater.

5.2 Systems for the Cultivation of Microalgae

Two types of cultivation systems are used to realize the biosynthetic capabilities of microalgae strains: closed and open (Sreekumar et al. 2016; Ficara et al. 2014). Open systems include ponds, usually of small depth, equipped with a paddle wheel for mixing water (Fig. 5.1). It is necessary to maintain the optimal level of light, temperature, pH, and CO_2 concentration, to ensure normal living conditions of microalgae in the cultivation process. Despite the relative ease of operation, the use of open cultivation systems does not consistently achieve high productivity of microalgae biomass, which is associated with significant water loss due to evaporation, contamination, and photoinhibition of crops and CO_2 diffusion.

The type of closed microalgae cultivation systems includes photobioreactors, which allow maintaining the stability of cultivation conditions, particularly temperature, pH, and CO_2 water saturation. In addition, photobioreactors provide a large surface area and better illumination. Consequently, they provide high yields and microbiological purity of the culture. Liquid evaporation is minimized through the use of a closed system. In contrast to ponds, closed systems allow achieving high rates of biomass accumulation.

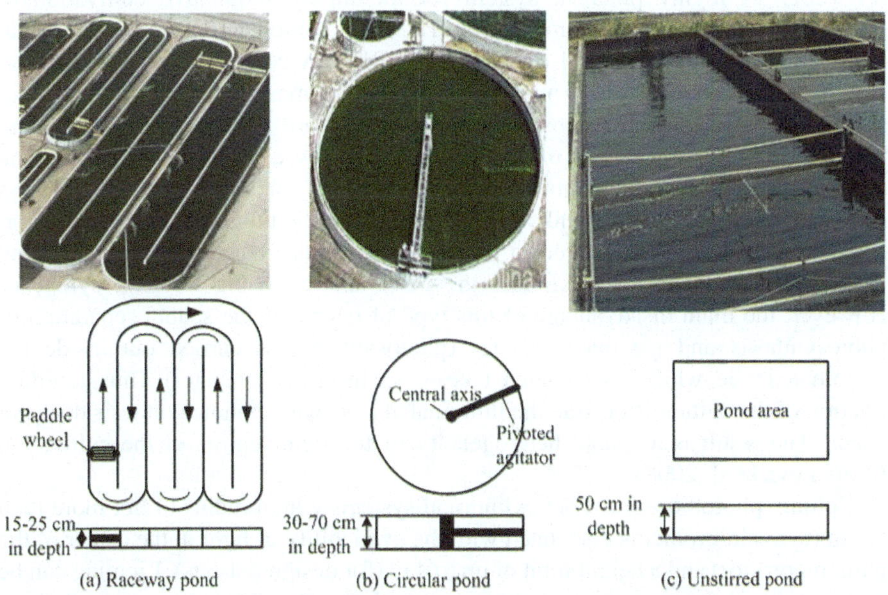

Fig. 5.1 Three different designs of open pond systems. (Sreekumar et al. 2016)

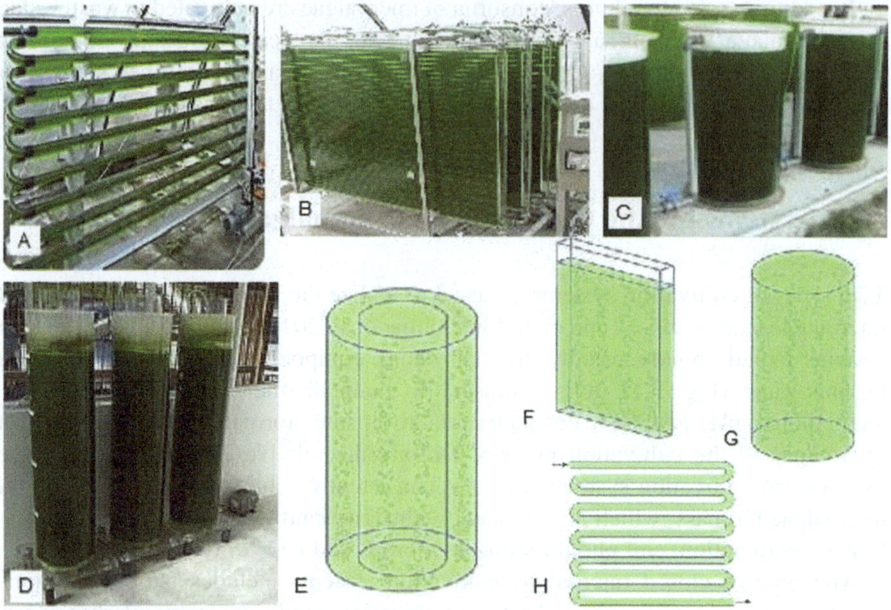

Fig. 5.2 Different types of photobioreactors for algal cultivation. (Ficara et al. 2014): (**a**) tubular, (**b**) plate, (**c**) column, and (**d**) annular. E–H – diagrammatic representation of photobioreactors: (**e**) annular, (**f**) plate, (**g**) column, (**h**) tubular

There are tubular, column (bubble and gas lift), and panel photobioreactors (Fig. 5.2). A highly practical system for the highly productive cultivation of microalgae is tubular photobioreactors. These photobioreactors are divided into horizontal, vertical, and spiral (Ficara et al. 2014). A common feature of tubular systems is the presence of two modules: light and gas-mass exchange. Photosynthesis mainly takes place in transparent tubes which are connected to a gas-mass transfer tank. The suspension of microalgae constantly circulates through the tank and transparent tubes. Fresh nutrient medium and gas mixture enters the gas-mass exchanger and from there, thanks to the pumps, into the tubular part of the reactor. Microalgae can be well lit even in the center in tubular reactors due to the small diameter of the tubes. It maximizes the outer area available for photosynthesis. However, the main disadvantage of this type of reactor is the spatial separation of photosynthesis and gas-mass transfer (photosynthesis is carried out inside the illuminated tube, while gas exchange takes place in a gas exchanger). Thus, a certain fraction of the culture is outside the illuminated part, where photosynthesis does not occur. The result is a spatial heterogeneity of the light regime of the cultivation (Zolotareva et al. 2008).

Column photobioreactors are cylindrical systems with a radius of not more than 0.2 m (to avoid problems associated with the availability of light in the center of the photobioreactor) and a height limit of up to 4 m (for design reasons). Lighting can be performed both by an external light source and a source located on the cylinders' axis inside them. According to the method of mixing the aqueous medium, bubble, and gas-lift (air-lift) column, photobioreactors are distinguished. The mixing efficiency determines the amount of liquid circulation period in the reactor volume, defined as the average duration of light-dark cycles for each cell. For bubble reactors, the duration of light-dark cycles is 1–4 s, and for air-lift 10–100 s.

Of all these types of photobioreactors, the simplest solution for the cultivation of microalgae is panel reactors. They are characterized by the maximum illuminated surface and allow achieving a high density of cells. However, at their operation, it is necessary to adhere to the balance of illumination to prevent photoinhibition. The use of panel photobioreactor to cultivate microalgae Nannochloropsis 211/78 allowed obtaining 4.7 g/l biomass with a maximum biomass accumulation rate of 0.51 g/l per day and lipid content of 28.3%, in terms of the dry mass of microalgae (Hulatt et al. 2017).

5.3 Methods of Microalgae Biomass Harvesting

Extraction of microalgae biomass from the culture medium is one of the most expensive and time-consuming processes. The most widely used methods of collecting algal biomass are filtration, centrifugation, flocculation, flotation, and immobilization and various combinations of these methods.

There are different filtration processes: dead-end filtration, microfiltration, ultrafiltration, pressure filtration, and vacuum filtration. However, it should be noted that the use of these processes is associated with high running costs and time-consuming. An effective mechanical method of extracting biomass of microalgae is centrifugation because this method allows in a short time to separate more than 95% of algae cells from the environment. However, the high energy consumption of centrifugation limits its use for large-scale collection of algal biomass.

During flocculation, cell aggregates (flocs) are formed due to the adhesion of microalgae cells. Flocculation occurs spontaneously or under the action of chemical and physical factors. The surface charge of microalgae cells is usually "negative," and its intensity is a function of the type of microculture, ionic strength, and pH of the environment. The addition of chemical reagents, so-called flocculants, helps neutralize the "negative" charge on the surface of microalgae cells, which usually prevents their self-aggregation. Among inorganic flocculants, aluminum sulfate, alum-potassium, and aluminum-ammonium alum and iron chloride are most often used, and among organic – chitosan and starch. Bioflocculation is due to some species of microalgae to form exocellular polymers. This process is accompanied by changes in surface charge and cell aggregation.

Flotation is the formation of complex "particles-bubbles," the emergence of these complexes, and the removal of the foam layer formed from the surface of the treated liquid. A relatively common method of extracting microalgae biomass is cells' flocculation and bubble water with air. As a result, the floccules float to the surface, which significantly facilitates their removal.

To date, biotechnology for wastewater treatment using immobilized microalgae cells has developed intensively. The advantage of immobilized cells compared to suspended cells is the simplification of the biomass collection procedure.

5.4 Biofuel Production from Microalgae

The biofuel production scheme is shown in Fig. 5.3 (Dvoreckij et al. 2015). The cultivation stage includes the accumulation of biomass under optimal growth conditions until the stationary growth stage is reached, then the creation of stress conditions through a deficiency of nitrogen-containing substances to stimulate the accumulation of intracellular neutral lipids. The accumulated biomass with a high lipid content is concentrated. Then the cells of microalgae are destroyed for more efficient lipid extraction.

Extraction is carried out with a mixture of polar and nonpolar extractants for 2.5–3 h. The extractant is distilled off, and the mixture of lipids is separated into fractions: the fraction containing nonpolar lipids is used for further production (Callegari et al. 2019; Pfromm et al. 2011). Nonpolar lipids, after reaction with methanol in the presence of a catalyst, form fatty acid methyl esters and technical glycerin. Glycerin is used in many industries (production of detergents and

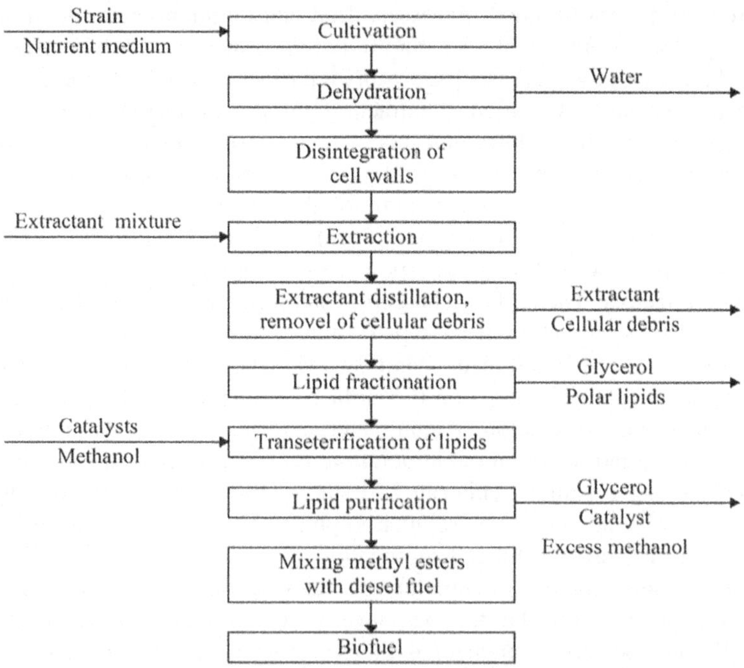

Fig. 5.3 Scheme of biofuel production from biomass of microalgae (Dvoreckij et al. 2015)

Table 5.2 Physical and chemical indicators of biodiesel fuel, obtained from the microalgae *Chlorella vulgaris*. (Reen et al. 2018)

N, no	Indicators	Characteristics of biodiesel fuel based on microalgae
1	Color	Dark brown with a greenish shade
2	Transparency	Transparent
3	Solid content	Absent
4	Density at $t = 20$ C, kg/m^3	876.0
5	Copperplate test	Withstands
6	Acid number, mg KOH/g	0.3

cosmetics, agriculture, textile, paper, and leather industries) (Trivedi et al. 2015; Reen et al. 2018).

Physicochemical characteristics of biodiesel fuel from microalgae were determined (Reen et al. 2018). Based on the results obtained, it can be concluded that biodiesel fuel meets international quality standards. The results are shown in Table 5.2.

Due to the different chemical compositions, conventional diesel fuel has a lower density, cetane number, and flash point than biodiesel (Reen et al. 2018; Yang et al. 2019). The composition of biodiesel fuel contains esters of fatty acids, and the composition of traditional diesel fuel mainly contains saturated and unsaturated hydrocarbons. An infrared spectrum of microalgae lipids and biodiesel fuel (ethyl esters of fatty acids) practically does not differ. The spectrum has an intense absorption peak at 1751 cm^{-1} for algal oil (lipids), as well as in ethyl esters of fatty acids (biodiesel fuel). It corresponds to stretching vibrations of the C-O bond. Stretching vibrations of C-O bonds appear in the form of three peaks with maxima at 1236 cm^{-1}, 1159 cm^{-1} (more potent), and 1100 cm^{-1}.

Ethyl esters of fatty acids also have these absorption bands, but with the most incredible intensity, i.e., the most vital signals are observed. The peak with the absorption maximum at 1355 cm^{-1} corresponds to the deformation vibrations of the C-H bond. In the area of 720 cm^{-1}, bending pendulum vibrations of CH_2 groups are manifested. Absorption bands with maxima at 1439 cm^{-1} and 1460 cm^{-1} are caused by shear bending vibrations of CH_2 groups. There are no absorption band maxima at 1439 cm^{-1} for ethyl esters of fatty acids.

Unsaturated hydrocarbon chains are part of triglycerides. They are detected by stretching vibrations at 3010 cm^{-1} of the CH=CH-group for cis-isomers. Trans-isomers can be detected by deformation vibrations of C-H at 765 cm^{-1} (Yadav and Sen 2017; Ahmad et al. 2011). The absorption bands in the region of 2934 cm^{-1}, 2894 cm^{-1}, 2910 cm^{-1}, 2880 cm^{-1}, and 2861 cm^{-1} correspond to stretching vibrations of CH_2 groups. In the infrared spectrum of biodiesel fuel (ethyl esters of fatty acids), there are no absorption bands at 3500–2500 cm^{-1} and 1285 cm^{-1} and there are no maxima at about 940 cm^{-1}, which are typical of acids.

Analysis of infrared spectra shows that unsaturated fatty acid fragments of the starting triglycerides are almost entirely cis-isomers. Trans-isomers are practically absent (peak at 968 cm^{-1}). In addition, after the transesterification reaction, isomerization of the double bonds of fatty acid fragments does not occur, i.e., they remain cis-isomers. Oscillations corresponding to conjugated double bonds –CH=CH-CH=CH- do not appear in the spectrum, not only in the initial lipid fraction but also after the transesterification reaction, in the reaction product – ethyl esters of fatty acids.

5.5 Biofertilizer Production from Microalgae

Algae extracts contain large amounts of cytokinins, betaine, algopolyphenols, vitamins, hormones, natural antioxidants, and mineral elements (Mg, Ca, B, Mo, and others). The contribution of these factors to the growth and development of plants (stimulate cell division, differentiation of callus tissue) and increase in soil fertility indicators were revealed (Kirsanova 2020).

The members of the genus *Anabaena* in symbiosis with the fern *Azolla* can bind up to 60 kg/ha of nitrogen per season. They are also a valuable source of organic matter that improves soil fertility. In this case, microalgae are a renewable resource, in contrast to the industrial production of nitrogen fertilizers. Along with nitrogen fixation, algae emit substances that improve plant growth. Algal biomass can be considered a good organic fertilizer and differs favorably from traditional species, as it contains no pathogenic microflora, no weed residues, or no pests. Unlike traditional mineral fertilizers, it does not harm the soil microflora. The use of *Chlorella vulgaris* is based on a very high content of biologically valuable substances (Lakaniemi et al. 2013). The dry biomass of *Chlorella vulgaris* contains more than 45% of crude protein, including essential amino acids, 30–35% of carbohydrates, and 7–10% of fat. The composition of the green cell contains essential amino acids in animal nutrition: lysine (10%), methionine (1%), tryptophan (2%), arginine (15%), histidine (3%), leucine (6%), isoleucine (3%), phenylalanine (2%), threonine (2%) valine (5%), and chlorophyll (2%).

Biogas technology allows obtaining natural biofertilizer in the shortest possible time by anaerobic fermentation (Shamanskyi and Boichenko 2015). This fertilizer contains biologically active substances and trace elements. The main advantage of biofertilizers over traditional ones is the shape, availability, and balance of all nutrients and high level of organic matter humification. Biofertilizers serve as a robust energy material for soil microorganisms. Nitrogen-fixing and other microbiological processes activate in the soil after their application. It positively effects soil fertility and improves the physical and mechanical properties of the soil.

The use of biofertilizers for growing crops will reduce the use of chemical fertilizers that adversely affect the quality and fertility of soils. Experimental data indicate the practical and economic feasibility of using the organic mass of cyanobacteria extracted from flowering spots in the waters of the Dnieper reservoirs for industrial biogas production with the subsequent use of spent substrate as a balanced organic fertilizer in forestry and agriculture.

The research results, in general, confirm the previous calculations of the average biogas yield per week in the volume of about 0.7 m^3 from 1 m^3 of the concentrated

Table 5.3 Elemental composition of cyanobacterial residues after anaerobic fermentation. (Zagirnjak et al. 2017)

Elements	C	O	Si	P	S	Cl	K	Ca	Cu
Weight, %	54.66	37.81	0.61	1.34	0.69	0.22	2.14	1.92	0.60

Table 5.4 Elemental composition of the dried spent biomass of cyanobacteria

Element	Amount, %	Element	Amount, %	Element	Amount, %
Si	4.432 ± 0.086	Ti	0.081 ± 0.019	Se	0.007 ± 0.002
P	7.160 ± 0.131	Mn	1.139 ± 0.017	Br	0.053 ± 0.002
S	11.713 ± 0.101	Fe	1.492 ± 0.015	Sr	0.029 ± 0.002
Cl	8.461 ± 0.079	Ni	0.023 ± 0.002	Zr	0.004 ± 0.002
K	20.197 ± 0.060	Cu	0.006 ± 0.001	Pd	0.008 ± 0.002
Ca	45.131 ± 0.112	Zn	0.024 ± 0.001	Sb	0.025 ± 0.004

substrate at the optimum temperature of 20–30 °C and data on the prospects of pre-treatment of the biological substrate in the cavitation field as a preparatory process for its subsequent bioconversion (Zagirnjak et al. 2017). Samples of the spent substrate remaining after the cessation of the methanogenesis process were studied separately. According to X-ray microstructural analysis, biofertilizers do not contain heavy metals (Table 5.3) and therefore can be used for feeding plants, including food crops.

The elemental composition of the used biomass is essential. It determines the efficiency of the fertilizer and the balance of macronutrients in it for plant nutrition. To establish the possibility of using spent biomass as organic fertilizers after biogas synthesis, the elemental composition of dried spent biomass was determined on an X-ray fluorescence analyzer EXPERT 3L (Table 5.4) (Zagirnjak et al. 2017). The contents of the main components of biomass are listed there. They are also compared with the limited content of heavy metals and hazardous compounds in the raw materials for the production of fertilizers, which are determined by the current Ukrainian standards.

None of the elements, the content of which in the raw materials for the production of fertilizers is limited (cadmium, lead, and arsenic), was found in the dried spent biomass. The spent substrate mainly contains calcium and sulfur (these elements are trace elements necessary for balanced plant nutrition), the application of which in the composition of fertilizers is appropriate.

The content of phosphorus and potassium – the primary plant nutrients – is at the level of the best varieties of mineral fertilizers.

The negative factor is the high chlorine content. However, it is often included in the form of chlorides of potassium fertilizers, which are widely used in agriculture, so its content in organic fertilizers from spent biomass is acceptable. In addition to a small amount of ballast silicon, the new potential fertilizer contains iron and manganese, which are necessary to ensure balanced plant development. This composition is acceptable for using spent biomass of cyanobacteria as fertilizer (Jimenez et al. 2020; Shamanskyi and Boichenko 2018).

5.6 Assessment of the Economic Effect of Microalgae Cultivating in Wastewaters for Biofuel and Fertilizer Production

The task is to quantify the economic effect and payback period of capital costs in implementing the proposed technological processes in the implementation of a new concept of drainage for a typical city with a population of 250,000 people. The daily amount of wastewater discharged by the city is $50 \cdot 10^3$ m^3/day ($18.25 \cdot 10^6$ m^3/year).

Calculation of the economic efficiency of biodiesel production from microalgae for the proposed typical city.

The annual specific productivity for biomass and lipids in the cultivation of microalgae, calculated earlier, is $P_{biomass}^{spec}$ = 11.5 kg/m^2 (57.5 kg/m^3) and P_{lipid}^{spec} = 4.1 kg/m^2 (20.5 kg/m^3).

At the introduction of technology on sewage treatment plants of the considered city, the annual productivity on biomass can be calculated by the formula:

$$P_{boimass}^{getting} = 50 \cdot 10^3 \times P_{biomass}^{spec} \times t_{PBR} \text{ kg/year.} \quad (5.1)$$

where t_{PBR} is the time spent by the sewage water in the working area of the photobioreactor (PBR) t_{PBR}= 1.25 days (30 h).

The formula is used to calculate annual lipid productivity:

$$P_{lipids}^{getting} = 50 \cdot 10^3 \times P_{lipid}^{spec} \times t_{PBR} \text{ kg/year.} \quad (5.2)$$

Thus, the productivity of biomass will be:

$$P_{biomass}^{getting} = 50 \cdot 10^3 \times 57,5 \times 1,25 = 3,59 \cdot 10^6 \text{kg/year.}$$

The annual lipid productivity will be:

$$P_{lipid}^{getting} = 50 \cdot 10^3 \times 20,5 \times 1,25 = 1,28 \cdot 10^6 \text{kg/year.}$$

If centrifugation is used, the annual amount of collected biomass, as well as the obtained lipids, will be:

$$P_{centrifuge}^{biomass} = 0,9 \times 3,59 \cdot 10^6 = 3,23 \cdot 10^6 \text{kg/year,}$$

$$P_{centrifuge}^{lipid} = 0,9 \times 1,28 \cdot 10^6 = 1,15 \cdot 10^6 \text{kg/year.}$$

The annual amount of oils obtained after their extraction from the biomass of microalgae will be:

$$P_{raw}^{oil} = 0,9 \times 1,15 \cdot 10^6 = 1,04 \cdot 10^{-6} \text{kg/year.}$$

The annual amount of residual biomass that can be used as raw material for biogas production will be:

$$P_{biom.resid.} = 3,23 \cdot 10^6 - 1,04 \cdot 10^6 = 2,20 \cdot 10^6 \text{kg/year}.$$

The annual amount of biodiesel that can be produced from the obtained oils will be:

$$P_{biodiesel} = 0,9 \times 1,04 \cdot 10^6 = 0,93 \cdot 10^6 \text{ kg/m}^2.$$

The formula is used to calculate the volume of obtained biodiesel:

$$W_{biodiesel} = P_{biodiesel} \times \frac{1}{\rho_{biodiesel}} \text{m}^3/\text{year} \quad (5.3)$$

where $\rho_{biodiesel}$ is the density of biodiesel.

The volume of biodiesel produced will be:

$$W_{biodiesel} = 0,93 \cdot 10^6 \times \frac{1}{0,86 \cdot 10^3} = 1,09 \cdot 10^3 \text{m}^3/\text{year}.$$

At 2020 prices, the price of biodiesel is $28 \cdot 10^3$ UAH/m^3 (28 UAH/l). Under such conditions, the cost of the obtained biodiesel will be:

$$C_{biodiesel} = 1,09 \cdot 10^3 \times 28 \cdot 10^3 = 30,41 \cdot 10^6 \text{ UAH/year}.$$

Specific costs for biodiesel production in 2020 prices are $15 \cdot 10^3$ UAH/m^3 (15 UAH/l). Under such conditions, the annual cost of processing oils into biodiesel will be:

$$Z_{biodiesel} = 1,09 \cdot 10^3 \times 15 \cdot 10^3 = 16,29 \cdot 10^6 \text{UAH/year}.$$

Therefore, the annual economic effect of the production of biodiesel from microalgae will be:

$$EE_{biodiesel} = 1,09 \cdot 10^6 - 16,29 \cdot 10^6 = 14,12 \cdot 10^6 \text{UAH/year}.$$

Calculation of the economic efficiency of biogas production from residual biomass of microalgae.

The raw material for biogas production is residual microalgae biomass, and the annual amount at the sewage treatment plant is $P_{biom.\ resid.} = 2,20 \cdot 10^6$ kg. When fermenting biomass according to the proposed (Shamanskyi and Boichenko 2016) four-stage technology, it is possible to obtain both methane and carbon dioxide. Their mass output will be:

methane $M_{CH_4}^{microalgae} = 2,20 \cdot 10^6 \times \frac{27,1}{100} = 0,60 \cdot 10^6$ kg,
carbon dioxide $M_{CO_2}^{microalgae} = 2,20 \cdot 10^6 \times \frac{13,9}{100} = 0,31 \cdot 10^6$ kg.

The volume of biogas components will be:

methane $W_{CH_4}^{microalgae} = 0,60 \cdot 10^6 \times \frac{1}{0,656} = 0,91 \cdot 10^6$ m³/year,
carbon dioxide $W_{CO_2}^{microalgae} = 0,31 \cdot 10^6 \times \frac{1}{1,913} = 0,16 \cdot 10^6$ m³/year.

The cost of components of biogas obtained by fermentation by new four-stage technology [37] will be (Table 5.5):

methane $C_{CH_4}^{microalgae} = 0,91 \cdot 10^6 \times 9,8 \times \frac{32,8}{34} \times 0,95 = 8,17 \cdot 10^6$ UAH/year,
carbon dioxide $C_{CO_2}^{microalgae} = 0,16 \times 10^6 \times 15 \times 0,95 = 2,28 \times 10^6$ UAH/year.

The total economic efficiency of biogas production from residual biomass will be:

$$C_{біогаз}^{microa\lg ae} = 8,17 \cdot 10^6 + 2,28 \cdot 10^6 = 10,45 \cdot 10^6 \text{UAH/year}.$$

Table 5.5 The results of the calculation of capital costs of the photobioreactor

№.	Type of costs	Sum, UAH
1	Design (project preparation, complete set of technological equipment, author's supervision)	4 500 000
2	Preparatory works and landscaping (site preparation, temporary buildings, and roads, site improvement)	5 500 000
3	Photobioreactors (base plates, cultivation tanks, bubbling systems complete with compressors and pressure valves, documentation)	51 800 000
4	Centrifuge OGSh 321 K5 – 2 pieces (433 500 UAH/piece)	867 000
5	Installation for oil extraction up to 14 t/day	450 000
6	Installation for production of biodiesel of 3 000 l/day	447 600
7	Technical and administrative buildings (pumping, buildings for electrical equipment, building for centrifuges and oil extraction plants, building for biodiesel production)	1 200 000 2 100 000
8	Equipment of technical premises (pumps, devices for mixing microalgae with sewage water, pipes, fittings, measuring devices, lightning protection, internal heating)	1 000 000
9	Costs for obtaining permits, approvals (development of a detailed plan of the territory, change of purpose of the land plot, commissioning)	3 200 000
	Together	71 064 600
	Other expenses	7 106 400
	Total	78 171 000

Calculation of the economic efficiency of phosphate fertilizer production from residual biomass of microalgae.

The annual amount of phosphates that accumulates in the biomass of microalgae during the additional treatment of the city's sewage water can be calculated by the formula:

$$M_{PO_4}^{microalgae} = W_{SW} \times \left(K_{PO_4}^{input} - K_{PO_4}^{output}\right) \times 365, \mathrm{kg/year} \quad (5.4)$$

where: W_{SW} – the daily volume of sewage water, coming for treatment, m^3/day

$K_{PO_4}^{input}$ $K_{PO_4}^{output}$ – concentration of phosphates at the input to the treatment plant and the output, respectively, kg/m^3

The economic effect of the sale of fermented biomass as a phosphate fertilizer is determined by the formula:

$$C_{PO_4}^{microalgae} = M_{PO_4}^{microalgae} \times k_{ekv} \times c_{PO_4} \quad UAH/Year \quad (5.5)$$

where: k_{ekv} – conversion factor of phosphates into phosphate fertilizers (equivalent of phosphate fertilizers, $k_{ekv} = 2$)

c_{PO_4} – the price of phosphate fertilizers, UAH

The annual amount of biomass accumulated by biomass phosphates will be:

$$M_{PO_4}^{microalgae} = 50 \cdot 10^3 \times \left(8 \cdot 10^{-3} - 0,84 \cdot 10^{-3}\right) \times 365 = 0,13 \cdot 10^6 \mathrm{kg/year}.$$

At 2018 prices, the cost of phosphate fertilizers is c_{PO_4}= 25 UAH/kg. Under these conditions, the economic effects of the sale of fermented biomass will be:

$$C_{PO_4}^{microalgae} = 0,13 \cdot 10^6 \times 2 \times 25 = 6,53 \cdot 10^6 \mathrm{UAH/year}.$$

5.7 Assessment of the Factors Influencing Atmospheric Air Emissions

The task is to calculate the factors influencing atmospheric air emissions of carbon dioxide during the cultivation of microalgae in the photobioreactors.

During the purification of sewage waters from biogenic elements, biomass is cultivated, which is used for biofuel production. It allows replacing part of traditional fossil fuels with biofuels, reducing carbon dioxide emissions, and, thus, reducing the value of factors affecting the atmosphere.

In the process of biofuel production from microalgae, electricity costs are necessary:

- Directly for the cultivation of microalgae
- For removal of microalgae from the culture medium
- For separation of oils

- For production of biodiesel from them
- For production of biogas

The cost of cultivation per square meter of the *photobioreactor's* working area is:

- For closed photobioreactors – $157.68 \cdot 10^6$ J/m^2
- For open photobioreactors – $31.54 \cdot 10^6$ J/m^2

The use of sewage water as a culture medium can reduce the cost of electricity for direct cultivation:

- For closed photobioreactors by 20%
- For open photobioreactors by 25%

Thus, provided that the photobioreactors are used to clean the sewage waters out of biogenic elements, the cost of cultivation will be:

- For closed photobioreactors – $e_{el.cult.SW}^{closed.PBR} = 126,14 \cdot 10^6 \mathrm{J/m^2}$
- For open photobioreactors – $e_{el.cult.SW}^{open.PBR} = 23,65 \cdot 10^6 \mathrm{J/m^2}$

So, the total amount of potentially consumed electricity for biofuel production will be:

- For closed photobioreactors:

$$e_{el.\cos t.biof.SW}^{closed.PBR} = (126,14 + 10,15 + 1,6 + 0,18 + 2,77) \cdot 10^6 = 140,84 \cdot 10^6 \mathrm{J/m^2}.$$

- For open photobioreactors:

$$e_{el.\cos t.biof.SW}^{open.PBR} = (23,65 + 11,59 + 0,46 + 0,05 + 0,79) \cdot 10^6 = 36,54 \cdot 10^6 \mathrm{J/m^2}.$$

The percentage of energy produced with obtained biofuel as a percentage of the energy to be consumed during the production process will be:

- For closed photobioreactors$\delta_{SW}^{closed.PBR} = \frac{(151,80+115,30)\cdot 10^6}{(140,84+19,72)\cdot 10^6} \cdot 100 = 166,36\ \%.$
- For open photobioreactors $\delta_{SW}^{open.PBR} = \frac{(43,29+33,12)\cdot 10^6}{(36,54+5,65)\cdot 10^6} \cdot 100 = 181,11\ \%.$

For comparison, provided the cultivation of microalgae without sewage water as a culture medium, the percentage of energy obtained will be 139.04% (an increase of 27.32%) and 152.58% (an increase of 28.53%), respectively.

The actual electricity costs for this process will be:

- For closed photobioreactors $e_{real.el.\cos t}^{closed.PBR} = 140,84 \cdot 10^6 \times \frac{100}{60} = 234,72 \cdot 10^6$ J/m^2.
- For open photobioreactors $e_{rael.el.\cos t}^{open.PBR} = 36,54 \cdot 10^6 \times \frac{100}{60} = 60,90 \cdot 10^6$ J/m^2.

The percentage of the actual electricity generated from biofuel produced as a percentage of the energy to be consumed during the production process will be:

- For closed photobioreactors$\delta_{SW}^{closed.PBR} = \frac{(151,80+115,30)\cdot 10^6}{(234,73+19,72)\cdot 10^6} \cdot 100 = 104,97\ \%$.
- For open photobioreactors $\delta_{SW}^{open.PBR} = \frac{(43,29+33,12)\cdot 10^6}{(60,90+5,65)\cdot 10^6} \cdot 100 = 114,82\ \%$.

For comparison, provided the cultivation of microalgae without sewage water as culture medium, the percentage of actual energy obtained will be 87.0% (an increase by 17.97%) and 95.87% (an increase by 18.95%) accordingly. Specific emissions of CO_2 during electricity production, in this case, will be:

- For closed photobioreactors $n_{el.CO_2}^{closed.PBR} = 0,107 \cdot 10^{-6} \times 140,84 \cdot 10^6 = 15,07$ kg/m^2.
- For open photobioreactors $n_{el.CO_2}^{open.PBR} = 0,107 \cdot 10^{-6} \times 36,54 \cdot 10^6 = 3,91$ kg/m^2.

The total amount of CO_2 emissions calculated in this case is:

- For closed photobioreactors $n_{CO_2}^{closed.PBR} = 15,07 + 1,42 = 16,49$ kg/m^2.
- For open photobioreactors $n_{CO_2}^{open.PBR} = 3,91 + 0,40 = 4,31$ kg/m^2.

CO_2 emissions from the processes of preparation of raw materials and production of biofuels, concerning the avoided emissions due to the substitution of traditional fuels in this case, will be:

- For closed photobioreactors $\delta_{CO_2}^{closed.PBR} = \frac{15,07+1,42}{11,02+6,41} \cdot 100 = 94,60\ \%$.
- For open photobioreactors $\delta_{CO_2}^{open.PBR} = \frac{3,91+0,40}{3,16+1,85} \cdot 100 = 86,03\ \%$.

The balance of CO_2 during cultivation in reactors of both types, in this case, is positive. CO_2 emissions are reduced for closed photobioreactors by 19.34% and for open photobioreactors by 16.76%.

There is the calculation below the quantitative values of environmental factors (atmosphere), namely, carbon dioxide emissions from the combustion of traditional fuels and third-generation biofuels from microalgae cultivated by traditional technologies and when using sewage water as a culture medium for microalgae. Specific annual CO_2 emissions per unit area of the cultivator working area when burning different types of fuel are given in Table 5.6.

The base values of emissions are specific emissions of CO_2 during the combustion of third-generation biofuels from microalgae, cultivated by traditional technologies, as they are the largest: $f_{0\ CO_2}^{closed\ PBR} = 19.86$ kg/m^3; $f_{0\ CO_2}^{open\ PBR} = 5.01$ kg/m^3.

The values of the influence factors of the traditional fuel combustion, which is replaced by biofuels, will be:

- For closed photobioreactors $F_{CO_2\ microalgae}^{closed\ PBR} = \frac{17,43}{19,86} = 0,877$.
- For open photobioreactors $F_{CO_2\ microalgae}^{open\ PBR} = \frac{5,01}{5,15} = 0,973$.

Table 5.6 Specific annual CO_2 emissions from the combustion of different fuels

№.	Type of fuel burned	Specific CO_2 emissions, kg/m^2	
		Closed PBR	Open PBR
1	Conventional fuel that is being replaced by biofuels	17,43	5,01
2	Biofuels from microalgae, cultivated by traditional technologies	19,86	5,15
3	Biofuels from microalgae cultivated in sewage water as a culture medium	16,49	4,31

The values of the influence factors during the combustion of biofuels from microalgae, cultivated by traditional technologies, will be:

- For closed photobioreactors $F^{closed\ PBR}_{CO_2\ microalgae} = \frac{19,86}{19,86} = 1,0.$
- For open photobioreactors $F^{open\ PBR}_{CO_2\ microalgae} = \frac{5,15}{5,15} = 1,0.$

The values of the factors of influence during the combustion of biofuels from microalgae cultivated in the sewage water as in the culture medium will be:

- For closed photobioreactors $F^{closed\ PBR}_{CO_2\ microalgae} = \frac{16,49}{19,86} = 0,830.$
- For open photobioreactors $F^{open\ PBR}_{CO_2\ microalgae} = \frac{4,31}{5,15} = 0,836.$

Thus, the application of microalgae cultivation technology in closed photobioreactors using sewage water as a culture medium with subsequent production of biofuels from microalgae biomass results in reducing the environmental impact factor of carbon dioxide emissions by:

$$\left[{(0,877-0,830)}/{0,877}\right] \times 100 = 5,36\%$$

5.8 Assessment of the Factors Influencing Hydrosphere Pollution

The task is to calculate the quantitative values of environmental factors (hydrosphere), namely, the discharge of biogenic elements into surface water without additional treatment of sewage water and after additional treatment using the proposed technology.

Phosphates, ammonium nitrogen, and nitrates are removed from sewage water during purification, according to the proposed technology. It allows reducing the volume of their discharge into reservoirs with treated sewage water and, thus, reducing the value of factors affecting surface water.

The base values are the concentrations of nutrients in the sewage water before additional purification: $f_{0\ PO_4} = 8.0\ \mathrm{g/m^3}$; $f_{0\ NH_4} = 1.8\ \mathrm{g/m^3}$; $f_{0\ NO_3} = 9.5\ \mathrm{g/m^3}$.

Under such conditions, the values of the factors influencing the discharge of phosphates into water bodies will be:

- Without additional treatment $F_{PO_4}^{without\ treatm} = \frac{8,0}{8,0} = 1,0.$
- With additional treatment $F_{PO_4}^{with\ treatm} = \frac{0,84}{8,0} = 0,105.$

The values of the factors influencing the discharge of ammonium nitrogen into reservoirs will be:

- Without additional treatment $F_{NH_4}^{without\ treatm} = \frac{1,8}{1,8} = 1,0.$
- With additional treatment $F_{NH_4}^{with\ treatm} = \frac{0,10}{1,8} = 0,056.$

The values of the factors influencing the discharge of nitrates into reservoirs will be:

- Without additional treatment $F_{NO_3}^{without\ treatm} = \frac{9,5}{9,5} = 1,0.$
- With additional treatment $F_{NO_3}^{with\ treatm} = \frac{2,78}{9,5} = 0,293.$

Thus, the result of applying sewage water purification technology from biogenic elements in the closed photobioreactors is a reduction of environmental impact factors: phosphates by $\left[{(1,0-0,105)}/{1,0}\right] \times 100 = 89,5\ \%$, ammonium nitrogen by $\left[{(1,0-0,056)}/{1,0}\right] \times 100 = 94,4\%$, and nitrates by $\left[{(1,0-0,293)}/{1,0}\right] \times 100 = 70,7\%$.

5.9 Conclusions

The possibility of using domestic wastewater as a culture medium for growing microalgae is considered. As a result of applying such technology, additional wastewater treatment from biogenic elements is achieved. Obtaining excess biomass is also achieved due to the metabolism of microalgae. The use of biomass for the production of biofuels and organic biofertilizers is proposed.

By replacing conventional types of liquid and gaseous fuels with biogas and biodiesel made from biomass of cultivated microalgae, the impact of carbon dioxide emissions on the atmosphere is reduced by 5.36%.

Due to the application of wastewater treatment technology in closed photobioreactors, the factors influencing the hydrosphere of discharges different biogenic elements are reduced: phosphates by 89.5%, ammonium nitrogen by 94.4%, and nitrates by 70.7%.

References

Ahmad AL, Yasin NHM, Derek CJC, Lim JK (2011) Microalgae as a sustainable energy source for biodiesel production: a review. Renew Sust Energ Rev 15:584–593

Bertildi FC, Sant'Anna E, Olivera JLB (2008) Chlorophyll content and mineral profile in the microalgae Chlorella Vulgaris cultivated in hydroponic wastewater. Cienc Rural 38:54–58

Bilgili F, Koçak E, Bulut Ü et al (2017) Can biomass energy be an efficient policy tool for sustainable development? Renew Sust Energ Rev 71:830–845

Bolognesi S, Bernardi G, Callegari A et al (2021) Biochar production from sewage sludge and microalgae mixtures: properties, sustainability and possible role in the circular economy. Biomass Conv Bioref 11:289–299

Callegari A, Bolognesi S, Cecconet D et al (2019) Production technologies, current role, and future prospects of biofuels feedstocks: a state-of-the-art review. Crit Rev Environ Sci Technol 0:1–53

Capodaglio AG, Ghilardi P, Boguniewicz-Zablocka J (2016) New paradigms in urban water management for conservation and sustainability. Water Pract Technol 11(1):176–186

Dvoreckij DS, Dvoreckij SI, Temnov MS, Peshkova EV, Akulinin EI (2015) Technologija poluchenija lipidov iz mikrovodoroslej (Technology for obtaining lipids from microalgae). Izdatelstvo FGBOU VPO "TGTU", Tambov

El-Sharony TF, El-Gioushy SF, Amin OA (2015) Effect of Foliar application with algae and plant extracts on growth, yield, and fruit quality of fruitful mango trees Cv. Fagri Kalan. J Hortic 2: 162

Ficara E, Uslenghi A, Basilico D et al (2014) Growth of microalgal biomass on supernatant from biosolid dewatering. Water Sci Technol 69(4):896–902

Gabriel F, Fernández A, Gómez-serrano C (2018) Recovery of nutrients from wastewaters using microalgae. Front Sustain Food Syst 2:1–13

Hosseinizand H, Lim CJ, Webb E et al (2017) Economic analysis of drying microalgae Chlorella in a conveyor belt dryer with recycled heat from a power plant. Appl Therm Eng 124:525–532

Hulatt CJ, Wijffels RH, Bolla S et al (2017) Production of fatty acids and protein by Nannochloropsis in flat-plate photobioreactors. PLoS One 12(1):1–17

Jimenez R, Markou G, Tayibi S et al (2020) Production of microalgal slow-release fertilizer by valorizing liquid agricultural digestate: growth experiments with tomatoes. Appl Sci 10(11): 3890

Kirsanova VV (2020) Docilnist obrobitku ta vykorystannya mikrovodorostej (chlorella) jak organichnych dobryv (Expediency of cultivation and use of microalgae (chlorella) as organic fertilizers). Ekologichni nauky (Environ Sci) 1(28):324–327

Lakaniemi AM, Tuovinen OH, Puhakka JA (2013) Anaerobic conversion of microalgal biomass to sustainable energy carriers – a review. Bioresour Technol 135:222–231

Markou G, Wang L, Ye J et al (2019) Cultivation of microalgae on anaerobically digested agro-industrial wastes and by-products. In: Gupta S, Bux F (eds) Application of microalgae in wastewater treatment. Springer, Cham

Moreno-Garcia L, Adjallé K, Barnabé S et al (2017) Microalgae biomass production for a biorefinery system: recent advances and the way towards sustainability. Renew Sust Energ Rev 76:493–506

Neczaj E, Grosser A (2018) Circular economy in wastewater treatment plant–challenges and barriers. Proceedings 2:614

Nkoa R (2014) Agricultural benefits and environmental risks of soil fertilization with anaerobic digestates: a review. Agron Sustain Dev 34:473–492

Pfromm PH, Amanor-Boadu V, Nelson R (2011) Sustainability of algae-derived biodiesel: a mass balance approach. Bioresour Technol 102:1185–1193

Reen S, Chyuan H, Wayne K et al (2018) Sustainable approaches for algae utilization in bioenergy production. Renew Energy 129:838–852

Rizzo AM, Prussi M, Bettucci L et al (2013) Characterization of microalga Chlorella as a fuel and its thermogravimetric behavior. Appl Energy 102:24–31

Salar-Garcia MJ, Gajda I, Ortiz-Martinez VM et al (2016) Microalgae as substrate in low cost terracotta-based microbial fuel cells: novel application of the catholyte produced. Bioresour Technol 209:380–385

Santos FM, Pires JCM (2018) Nutrient recovery from wastewaters by microalgae and its potential application as bio-char. Bioresour Technol 267:725–731

Shamanskyi SI, Boichenko SV (2015) Energoefektyvna ta ekologichno bezpechna technologija stabilizacii osadiv stichnyh vod aviapidpryjemstv (Energy efficient and environmentally friendly technology for stabilizing sewage sludge of aviation enterprises). Shidno-Evropejskyj Zhurnal Peredovyh Tehnologii (Eastern-European J Enterprise Technol) 5/8(77):39–45

Shamanskyi SI, Boichenko SV (2016) Rozrobka ekologichno bezpechnoi technologichnoji shemy vodovidvedennja aviapidpryjemstva (Development of environmentally safe technological water disposal scheme of aviation enterprise). Shidno-Evropejskyj Zhurnal Peredovyh Tehnologii (Eastern-European J Enterprise Technol) 6/10(84):49–57

Shamanskyi S, Boichenko S (2018) Environment-friendly technology of airport's sewerage in advances in sustainable aviation. Springer International Publishing AG, pp 161–175

Shamanskyi S, Boichenko S (2019) Environmentally friendly biogas production. IEEE 6th International Conference on Energy Smart Systems (2019 IEEE ESS), April 17–19, 2019: Kyiv: 2019 – C. 243-248.

Solovchenko A, Lukyanov A, Vasilieva S et al (2013) Possibilities of bioconversion of agricultural waste with the use of microalgae. Mosc Univ Biol Sci Bull 68(4):206–215

Sreekumar N, Giri Nandagopal MS, Vasudevan A et al (2016) Marine microalgal culturing in open pond systems for biodiesel production— Critical parameters. J Renew Sustain Energy 8(2): 023105

Trivedi J, Aila M, Bangwal DP et al (2015) Algae-based biorefinery—how to make sense? Renew Sust Energ Rev 47:295–307

Uysal O, Uysal FO, Ekinci K (2015) Evaluation of microalgae as microbial fertilizer. European J Sustain Dev 4(2):77

Viswanathan B (2017) Chapter 15—biochemical routes for energy conversion. In: Viswanathan B (ed) Energy sources. Elsevier, Amsterdam, pp 357–368

Yadav G, Sen R (2017) Microalgal green refinery concept for biosequestration of carbon-dioxide vis-à-vis wastewater remediation and bioenergy production: recent technological advances in climate research. J CO2 Util 17:188–206

Yang C, Wang C, Li R et al (2019) Pyrolysis of microalgae: a critical review. Fuel Process Technol 186:53–72

Zagirnjak MB, Nykyforov VV, Maljovanyi MS, Sameshova D, Kozlovska TF, Elizarov MO, Shtrbova E, Shlyk SV, Digtjar SV (2017) Ekologichna biotechnologija pererobky synjo-zelenych vodorostej: monografija (Ecological biotechnology of blue-green algae processing: monograph). PP Shcherbatyh O.V, Kremenchuk

Zolotareva OK, Shnyukova EI, Syvash OO, Mykhailenko NF (2008) Perspektyvy vykorystannja mikrovodorostej u biotechnologii (Prospects of the microalgae use in biotechnology). Alterpress, Kyiv

Chapter 6
Development of New Structured Honeycomb Fiber Catalysts for Hydrocarbon Conversion to Carbon-Free Fuel

Hennady Soloviov, Andryi Halstian, and Andryi Bushuev

Abbreviation

CO_2	Carbon dioxide
CH_4	Methane
CO	Carbon monoxide
H_2	Hydrogen
H_2O	Water (vapor)

6.1 Introduction

The problems of reducing greenhouse gas emissions into the air are becoming especially relevant in the modern world (Pokhodenko et al. 2010). It is planned to reduce carbon dioxide emissions by 8% annually according to the "Paris Agreement." It will guarantee the fixation of the increase in the Earth's air temperature at the achieved level plus 1.5 °C. CO_2 is one of the main air pollutants and its source is transport, including aviation, because it uses hydrocarbon fuels, which are produced mainly from oil (kerosene, gasoline, diesel fuel, etc.) (Kozin and Volkov 2006, 2002). The challenge of today is the unalterable transition to carbon-free fuels. In the

H. Soloviov (✉)
V. Dahl East Ukrainian National University, Severodonetsk, Ukraine

A. Halstian
Kyiv National University of Technologies and Design, Kyiv, Ukraine

A. Bushuev
Luhansk State Medical University, Luhansk, Ukraine

S. Boichenko et al. (eds.), *Chemmotological Aspects of Sustainable Development of Transport*, Sustainable Aviation, https://doi.org/10.1007/978-3-031-06577-4_6

last 2–3 years, publications have appeared in which ammonia figures as a "green" fuel, for example, for diesel power units of sea tankers and for tractors (Dudnik and Melakh 2007). Japanese and European car giants (Toyota, BMW, and 28 other corporations) have already declared the rapid transition of cars to "green" hydrogen as a fuel for "fuel cells," which will replace traditional internal combustion engines. Technical and economic indicators of production of "green" hydrogen and ammonia depend on the choice of raw materials and the method of its processing. The most attractive source of hydrogen is water, but modern water electrolysis technologies are energy-intensive. Nowadays hydrogen is produced from natural gas and other hydrocarbon gases and oil products. For example, ammonia produced through methane vapor conversion costs 250 $ per ton on an average, and hydrogen obtained as the result of electrolysis costs 600–650 $ per ton. The aim of our investigation is to invent such modern catalysts for the conversion of methane, n-propane, methanol, ethanol, and other hydrocarbons or alcohols, which would significantly reduce or eliminate the formation of carbon dioxide in these technologies.

6.2 Analysis of the Present State of Industrial Technologies for the Production of Hydrogen, Ammonia, and Methanol

According to "Haldor Topsoe" data all hydrogen and ammonia productions in the world use technology of coal conversion (25%), hydrocarbon gases (natural, liquefied, and others), or associated petroleum gas conversion (75%). All of them have large specific emissions of carbon dioxide into the air: from coal, 3 t/t of ammonia (17.1 t/t of hydrogen) and from natural gas, 1.6–2.0 t/t of ammonia (8.9–11.2 t/t of hydrogen). The most widespread process in the world is a multistage (more than ten stages) process based on steam conversion of methane in a tubular furnace (so-called primary reforming), followed by processing at the stage of secondary reforming and at eight more physicochemical stages, including Karsol process for the release of liquefied CO_2 from commercial gases. 270–272 gas burners are installed in tubular reactors, and the mode of endothermic reactions flow is supported by the heat of flue gases. 180 million tons of ammonia was produced in the world in 2019, and 205 billion cubic meters of natural gas was spent on its production in a year (Pokhodenko et al. 2010). 160 million tons of carbon dioxide was released into the air with rarefied flue gases from the gas burners of tubular reactors, and 240 million tons of liquefied compound gas was obtained in the concentrated form. The purpose of the planned and performed research for the past 10–20 years is to develop and implement the latest structured conversion catalysts that would be able to eliminate the combustion of natural gas in tubular reactors to reduce greenhouse gas emissions by 40%. In addition, in the second stage (at the level of search work) we are conducting investigation on the technologies that guarantee the complete exclusion

Table 6.1 The results of the study of the composition of gas samples (recalculated on "dry" ingredients) of one of the industrial tubes (inner diameter 71 mm; length 9.45 m) of a tubular furnace of the methane reforming process ($P = 3$ MPa)

Place of sampling for analysis (catalyst layer), m	Methane content in a mixture of dry gases (recalculated on "dry"), g/nm^3	The content of carbon dioxide in a mixture of dry gases (recalculated on "dry"), g/nm^3	The content of carbon monoxide in a mixture of dry gases (recalculated on "dry"), g/nm^3	The hydrogen content in the mixture of dry gases (recalculated on "dry"), g/nm^3
1.0	534.8	64.3	32.5	13.0
2.0	336.3	166.1	45.4	25.3
3.0	224.6	223.1	43.4	31.3
4.0	163.5	266.4	42.9	37.4
5.0	131.0	315.6	44.2	43.5
6.0	112.4	301.4	46.3	47.8
7.0	96.9	245.6	80.8	87.4
8.0	78.0	241.9	118.7	139.2
9.0	64.2	239.7	135.3	171.2

of carbon dioxide emissions into the air. The essence of the idea underlying our further research is that we decompose methane catalytically in one step into solid carbon (in the form of pyrolytic graphite) and hydrogen by the reaction: $CH_4 = C + 2\ H_2 + 75.3$ kJ/mol. We recommend to store graphite in the used underground (mine) layers on the period of combating with climate changes. Tables 6.1, 6.2, and 6.3 represent research results of the installation of methane steam reforming on the catalyst "ГІАП-3-6Н" (Ni/α-Al_2O_3) of industrial granulation, loaded into one of the tubes of an industrial tubular furnace operating at a pressure of 3 MPa.

The mono-endothermic reaction of methane and water takes place only on the first meter of the catalyst layer, and the reaction rate is in 2.3 times slower than on 2–3 meters of catalyst layer (see Figs. 6.1, 6.2, 6.3, and 6.4). This indicates that a significant part of the thermal energy wastes do not support the endothermic reaction but an increase in the temperature of the vapor-gas mixture, as well as the process flowing in the external diffusion region, which is controlled only by the rate of heat energy supply to the reaction zone. The rate is much higher at 2–4 meters of the catalyst layer due to the action of a sequential exothermic reaction of "water shear."

The mono-endothermic reaction of methane and water takes place only on the first meter of the catalyst layer, and the reaction rate is in 2.3 times slower than on 2–3 meters of catalyst layer (see Figs. 6.1, 6.2, 6.3, and 6.4). This indicates that a significant part of the thermal energy wastes do not support the endothermic reaction but an increase in the temperature of the vapor-gas mixture, as well as the process flowing in the external diffusion region, which is controlled only by the rate of heat energy supply to the reaction zone. The rate is much higher at 2–4 meters of the catalyst layer due to the action of a sequential exothermic reaction of "water shear."

Table 6.2 The results of tests of the catalyst "ГІАП –3-6Н" in the process of steam conversion of methane (in the industrial mode of operation of the tubular furnace)

Catalyst layer, m	Temperature of VGM (outer pipe wall), K	VGM volume, nm^3/h	Contact time, s	Rate of VGM, м/c	Methane conversion, X_{CH4},%	Methane mass, kg/h	Mass of H_2O/ dry gases, kg/h	Mass of H_2, kg/h	Mass of CO_2, kg/h	Mass of CO, kg/h
0	869	395.30	0.00	2.74	0.00	71.154	237.6/71.2	0.00	0.00	0.00
1.0	879	399.28	0.35	2.84	5.56	65.507	229.3/79.5	2.11	7.88	3.98
2.0	913	405.66	0.66	3.04	17.86	56.471	210.7/98.1	6.12	27.90	7.62
3.0	948	411.64	0.92	3.24	30.53	48.001	193.5/115.2	10.25	47.70	9.27
4.0	973	416.12	1.17	3.41	40.43	41.648	174.7/134.1	13.67	67.85	10.92
5.0	993	419.21	1.40	3.56	47.57	37.271	153.4/155.4	15.76	89.76	12.57
6.0	1021	422.00	1.61	3.74	53.13	33.318	155.1/153.6	17.26	89.32	13.72
7.0	1033	424.39	1.81	3.86	57.84	29.930	159.4/149.3	18.60	75.86	24.95
8.0	1070	427.20	1.96	4.08	63.49	25.977	142.7/166.1	20.00	80.56	39.51
9.0	1076	429.57	2.15	4.19	68.35	22.589	132.1/176.7	21.33	84.39	48.33

Table 6.3 Physicochemical characteristics of the process of steam conversion of methane (in the industrial operation mode of the tubular furnace)

Catalyst layer, m	Density of VGM, kg/m^3	Kinematic viscosity of VGM, m^2/s	Reynolds criterion	β CH_4, м/с	β H_2O, м/с	β H_2, м/с	β CO_2, м/с	β CO, м/с	Nu	α, W/m^{2}K
0	12.11	0.0016	25.63	0.033	0.01	0.04	0.01	0.014	3.19	52.6
1.0	11.37	0.0017	24.90	0.034	0.01	0.04	0.01	0.014	3.15	52.4
2.0	9.95	0.0019	23.81	0.037	0.01	0.04	0.01	0.015	3.07	51.4
3.0	8.80	0.0021	23.47	0.040	0.02	0.04	0.01	0.017	3.04	50.4
4.0	7.92	0.0023	22.70	0.042	0.02	0.04	0.01	0.017	2.99	49.0
5.0	7.23	0.0025	21.33	0.044	0.02	0.04	0.01	0.018	2.90	46.3
6.0	6.87	0.0025	22.29	0.047	0.02	0.04	0.015	0.019	2.96	46.7
7.0	6.46	0.0025	23.04	0.049	0.02	0.04	0.016	0.020	3.00	46.3
8.0	5.52	0.0029	20.60	0.051	0.02	0.04	0.017	0.021	2.85	43.7
9.0	5.16	0.0031	20.31	0.052	0.02	0.04	0.017	0.021	2.83	42.7

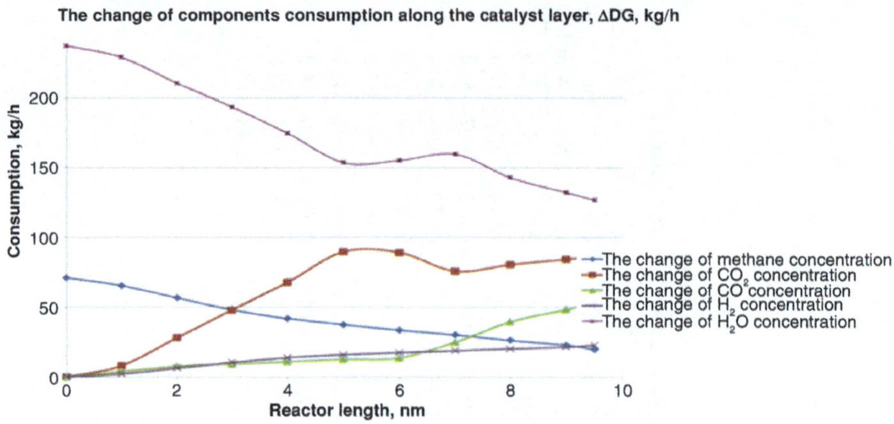

Fig. 6.1 Changes in the consumption of components in the catalyst layer

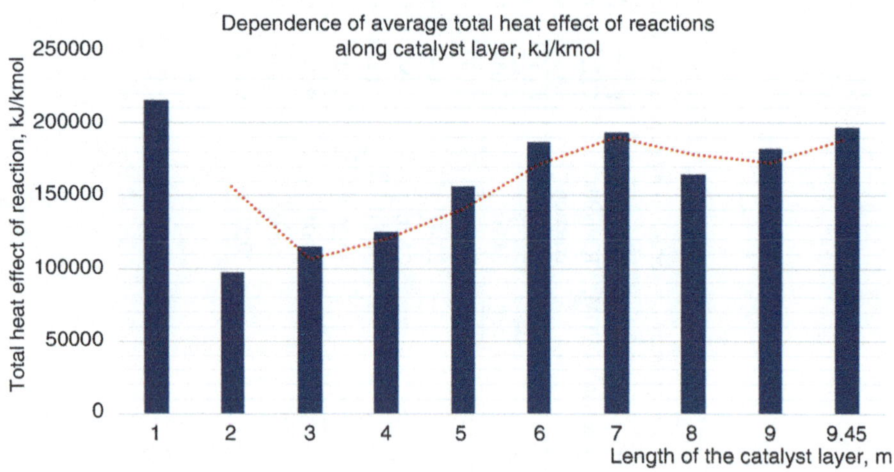

Fig. 6.2 Dependence of the average total thermal effect of reactions on the catalyst layer

The activation energy of the steam conversion process is 94 kJ/mol in the temperature range 450–600 °C, which correlates well with the data of many authors (Hou and Hughes 2001). The following equation is recommended for the internal diffusion zone:

$$r = 1,3 \cdot 10^3 \cdot A \cdot e^{-94/_{RT}} \cdot P_{CH_4} \tag{6.1}$$

where r is the rate in mol/kg•s; A contact surface, cm^2/g; and P_{CH4} methane partial pressure, atm. The process becomes completely externally diffuse at temperature higher than 600 °C. The activation energy is 10,2 kJ/mol.

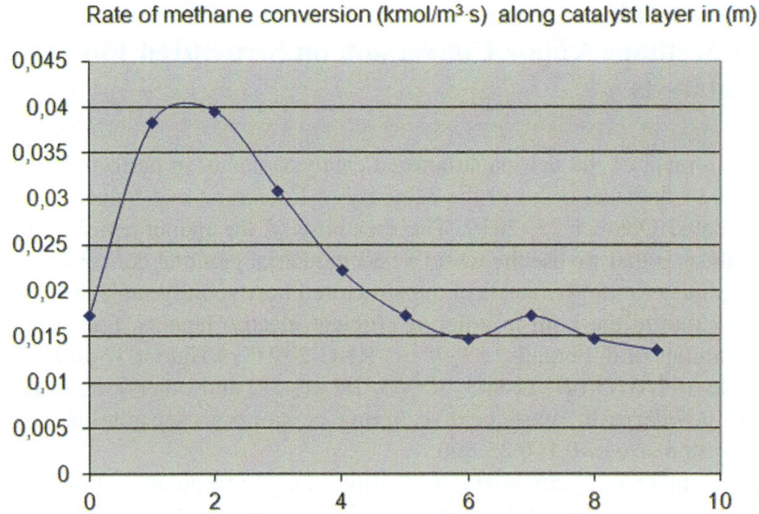

Fig. 6.3 The rate of methane conversion along the length of the layer

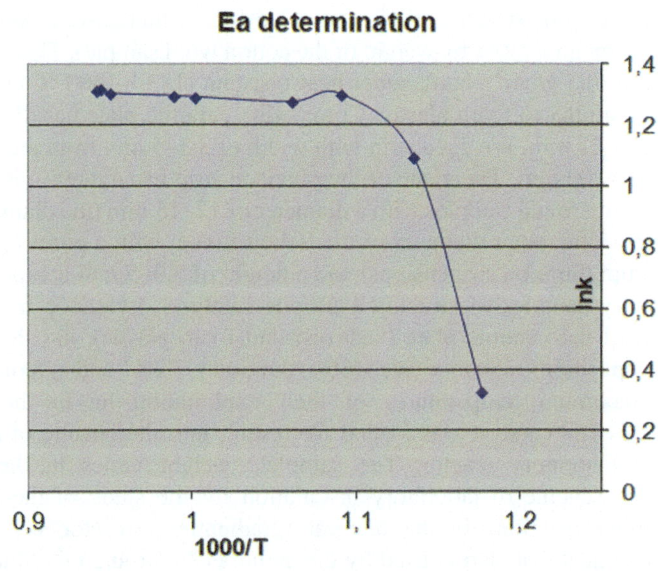

Fig. 6.4 Determination of the activation energy of the process

6.3 The Results of Studies of the Kinetics of Reactions of Methane Vapor Conversion on Structured Fibrous Catalysts

The development of the newest structured catalysts has been performed within the framework of the state order of the Ministry of Education and Science of Ukraine №ДЗ-91 dated October 25, 2019. The creativity of the manufacturing method of these catalysts is that we use the world's best industrial granular conversion catalysts as intermediates for the synthesis of our structured honeycomb catalysts, which have advantages due to the built-in units of efficient electric heaters. For example, the contact granules of ICI brands "57_4" or "RKC-2-6H" of Haldor Topsoe or domestic catalysts (ГІАП-3-6Н, Ц-905, К-905, НІАП-22, and others) are crushed and ground to powder with subsequent scattering on sieves to separate the fraction of granules with a size of 0.1–0.25 mm.

We have invented a composition of a mixture of inorganic and organic nano-adhesives at the level of "KNOW-HOW," with the help of which we fix a powdered catalyst in the structure of fibrous honeycomb material. We soak ceramic fiberboard brand "LYTX-1430B," which is made by vacuum pressing of synthetic mineral fiber (heat-resistant cotton wool) with a thread diameter of 3–3.5 microns brand "LYTX-611" by pasty aqueous suspension of catalyst powder, which also includes thermo-stable nanosilicon glue (30% by weight of the cotton wool sample). This allows us to produce a light (0.3 g/cm^3), hard, super heat-resistant (1430–1600 °C) material – a ceramic plate, similar to foam plastic. Commodity ceramic plate has dimensions of 1000 × 1200 × 25 mm. We cut a strip with width of 3,3–5 mm from the plates' end with length of 1200 mm. The plane of the obtained strip has dimensions of 1200 × 25 mm. We stamp round samples with a diameter of 12–15 mm (the diameter should be at the level of the inner diameter of the Schott funnel with a porous glass baffle used for vacuum filtration suspensions) and a height of 3.3–5 mm in a press form to make laboratory samples of structured fibrous catalysts. Synthetic mineral fiber, which is part of the ceramic plate (heat-resistant foam plastic), has the following chemical composition: silicon oxide 46%, aluminum 32%, and zirconium oxide 22%. The maximum temperatures of their exploitation lie in the range of 1430–1600 °C. The catalyst was loaded for testing into the cuvette of an integral (gradientless) laboratory reactor. The samples' weight varies in the range of 0.4–1.6 g. The scheme of laboratory installation for the study of the kinetics of steam reforming reactions in the integral (gradientless) reactor was a typical, completely identical installation used by the authors (Zaufir and Gavriilidis 2003).

The experiment was performed as follows. Laboratory samples of structured fibrous catalyst, made according to the method presented above, were loaded sequentially into the reactor cuvette. The samples' weight varied from experiment to experiment as follows: 0.516, 0.764, 1.128, 1.592, 1.972, and 2.148 grams. The catalyst was converted to the reduced state before each series of experiments by treating it with a hydrogen-argon mixture (5% hydrogen in argon) for 2 hours at a temperature of 500 °C. The same amount of reagents was added into the reactor in all experiments: methane 50 cm^3/min, hydrogen 50 cm^3/min, and water vapor 200 cm^3/min. At the outlet of the reactor the reaction mixture was cooled to room temperature in a water glass heat exchanger, and the water vapor condensate was collected in a phase separator. Next, the VGM passed through a column filled with silica gel, where it was completely dried. After the pressure control valve, the converted gases were analyzed by a gas chromatograph of brand "ЛХМ-08Д" and by an infrared gas analyzer of CO_2 of brand "Testo 330."

Gas chromatographic control of reagents and products (among which hydrogen, carbon dioxide, carbon monoxide, and methane have been fixed) was performed using thermal conductivity detectors and columns filled with 1) molecular sieves NaX, 3 m; $T_{col.} = 60\ °C$; $V_{Ar} = 40\ cm^3/min$ – H^2, CO, CH_4 and 2) activated carbon brand "СКТ," soaked with 10% solution of nickel sulfate, 1 m; $T_{col.} = 70\ °C$; $V_{Ar} = 40\ cm^3/min$ – CO_2. The results of experimental investigations to study the kinetics of vapor conversion of methane in an integral laboratory reactor are presented in Table 6.4.

Kinetic laboratory studies of the reaction of carbon dioxide with hydrogen were carried out similarly as described above. Laboratory results for the study of the kinetics of reverse reactions are summarized in Table 6.5:

$$CO_2 + H_2 \rightarrow CO + H_2 and\ CO + 3H_2 \rightarrow CH_4 + H_2O \tag{6.2}$$

The rate of chemical reactions by methane consists of the sum of reaction 6.1 and 6.3 rates in the direct way: $r_{CH4} = r_1 + r_3$, and the rate of formation and loss of CO_2 is calculated as the sum of reaction 6.4 and 6.5 rates: $r_{CO2} = r_2 + r_3$. (Tables 6.6 and 6.7) (Numbering of reactions here and further is due to Table 6.8.)

It is possible to calculate (i.e., to divide) numeral values of rate constant of reactions 6.3 and 6.5 (k_1, k_2, k_3 – rate constants, units of rate constants of reactions 6.3 and 6.5 – [kmol/kg•s•$(kPa)^{0,25}$]) and reaction 6.4 [kmol/kg•s•(kPa)] only after independent determination rate constant by reaction 6.4 (Table 6.7). The kinetic equation of reaction 6.3 for the researched sample of catalyst "structured fibrous" is shown below as a model of "Langmuir–Hinshelwood" (Eq. 6.5 is shown for comparison of kinetic data for the industrial catalyst "ГІАП-3-6Н"as a powder with particle size 0.15–0.25 mm):

$$r_1 = k_1 \cdot \frac{P_{CH_4} \cdot P_{H_2O}{}^{0,5}}{P_{H_2}{}^{1,25}} \cdot \frac{1 - P_{CO} \cdot P_{H_2}{}^{3} / K_{P1} \cdot P_{CH_4} \cdot P_{H_2O}}{Z^2} \tag{6.3}$$

$$k_{1\ fiber} = 1,150 \cdot 10^8 \cdot e^{-198,7/RT} \tag{6.4}$$

$$k_{1\ ГІАП} = 4,502 \cdot 10^8 \cdot e^{-201,42/RT} \tag{6.5}$$

Similarly according to reactions 6.4 and 6.5, we get:

Table 6.4 The results of the study of vapor conversion of methane on a structured fiber catalyst in an integral (gradientless) reactor. The composition of vapor-gas mixture at the reactor entrance is the following: methane 50 cm^3/min, hydrogen 50 cm^3/min, and water vapor 200 cm^3/min

№.	T, K	Catalyst mass, g	X_{CH4}, %	Mass frac., CH_4	Mass frac., H_2O	Mass frac., $CO \cdot 10^2$	Mass frac., $CO_2 \cdot 10^2$	Mass frac., H_2	r_{CH4}, kmol/kg·s·10^5	r_{CO2}, kmol/kg·s·10^5
1	748	0.516	2.4	0.161	0.656	0.02	0.58	0.172	0.156	0.152
2	748	0.764	3.2	0.160	0.652	0.03	0.78	0.183	0.143	0.139
3	748	1128	5.7	0.154	0.640	0.04	1.36	0.195	0.172	0.167
4	748	1532	8.7	0.148	0.627	0.07	2.05	0.204	0.194	0.188
5	748	1972	9.4	0.146	0.624	0.08	2.17	0.208	0.161	0.155
6	748	2148	10.1	0.145	0.627	0.10	2.23	0.211	0.146	0.140
7	773	0.516	3.9	0.158	0.649	0.04	0.73	0.185	0.281	0.267
8	773	0.764	7.4	0.150	0.643	0.07	1.33	0.198	0.329	0.312
9	773	1128	10.4	0.144	0.625	0.09	1.5	0.211	0.314	0.296
10	773	1532	14.9	0.135	0.607	0.14	2.1	0.229	0.331	0.310
11	773	1972	18.8	0.127	0.592	0.20	2.9	0.245	0.324	0.303
12	773	2148	21.3	0.123	0.583	0.22	3.0	0.255	0.306	0.281
13	798	0.516	5.7	0.154	0.642	0.07	0.83	0.192	0.373	0.341
14	798	0.764	9.3	0.1476	0.626	0.14	1.5	0.211	0.416	0.379
15	798	1128	14.8	0.135	0.604	0.27	2.77	0.229	0.447	0.406
16	798	1532	19.5	0.126	0.595	0.35	3.47	0.248	0.434	0.394
17	798	1972	24.8	0.116	0.564	0.37	3.63	0.269	0.427	0.387
18	798	2148	27.9	0.110	0.553	0.40	4.00	0.280	0.399	0.361
19	823	0.516	14.9	0.131	0.5996	0.25	2.09	0.230	0.974	0.870
20	823	0.764	19.4	0.126	0.5806	0.33	2.67	0.246	0.864	0.768
21	823	1128	27.7	0.110	0.5469	0.53	3.67	0.279	0.836	0.732
22	823	1532	35.5	0.096	0.5168	0.71	4.53	0.308	0.679	0.587
23	823	1972	39.4	0.089	0.5022	0.87	4.88	0.321	0.654	0.555
24	823	2148	44.3	0.0809	0.4840	1.01	5.19	0.338	0.631	0.528

Table 6.5 Results of the study kinetics of methanation reaction of CO_2 by hydrogen (reverse reaction to the "water shear" reaction) on a structured fiber catalyst in the laboratory integral (gradientless) reactor. The composition of vapor-gas mixture at the reactor entrance is the following: CO_2–190 cm^3/min and hydrogen 140 cm^3/min

№.	T, K	Catalyst mass, g	X_{CO2}, %	X_{CO}, %	X_{CH4}, %	Output CO kmol/s·10^8	Output CH_4 kmol /s·10^8	r_{CO}, kmol/kg·s·10^5	r_{CH4}, kmol/kg·s·10^5
1	598	0.422	2.7	2.00	0.65	0.2780	0.0904	0.659	0.214
2	598	0.528	2.4	1.70	0.70	0.2363	0.0973	0.448	0.184
3	598	0.633	2.16	1.40	0.76	0.1946	0.1056	0.307	0.167
4	598	0.845	2.24	1.34	0.80	0.1863	0.1223	0.221	0.145
5	598	1103	2.27	1.27	1.00	0.1765	0.1390	0.160	0.126
6	623	0.422	4.47	3.37	1.10	0.4684	0.1529	1110	0.362
7	623	0.528	4.65	3.30	1.35	0.4587	0.1877	0.869	0.355
8	623	0.633	4.46	2.89	1.57	0.4017	0.2182	0.635	0.355
9	623	0.845	4.66	2.81	1.85	0.3906	0.2572	0.462	0.304
10	623	1103	4.89	2.79	2.20	0.3878	0.3058	0.352	0.277
11	648	0.422	7.57	5.34	2.07	0.7923	0.2877	1877	0.682
12	648	0.528	7.81	5.56	2.25	0.7728	0.3128	1464	0.592
13	648	0.633	6.91	4.16	2.45	0.6199	0.3406	0.979	0.538
14	648	0.845	7.49	4.35	3.14	0.6047	0.4365	0.716	0.517
15	648	1103	8.55	4.3	4.25	0.5977	0.5908	0.542	0.536
16	673	0.422	12.6	9.50	3.10	13.265	0.4309	3120	1021
17	673	0.528	13.6	9.30	3.80	14.927	0.5282	2827	1016
18	673	0.633	14.3	9.2	5.00	12.788	0.6950	2020	1098
19	673	0.845	15.2	9.15	6.05	12.719	0.8228	1505	0.974

Table 6.6 List of reactions that are distinctive for different types and mechanisms of methane conversion processes and equations for calculating their equilibrium constants

№.	Reaction	Equilibrium constant, K_{Pi}	Equilibrium constant units, $(kPa)^v$
1	$CH_4 + H_2O = CO + 3H_2$	$1198 \times 10^{17}\exp(-26{,}830/T)$	$(kPa)^2$
2	$CO + H_2O = CO_2 + H_2$	$1767 \times 10^{-2}\exp(4400/T)$	$(kPa)^0$
3	$CH_4 + 2H_2O = CO_2 + 4H_2$	$2117 \times 10^{15}\exp(-22{,}430/T)$	$(kPa)^2$
4	$CH_4 + CO_2 = 2CO + 2H_2$	$6780 \times 10^{18}\exp(-31{,}230/T)$	$(kPa)^2$
5	$CH_4 + 3CO_2 = 4CO + 2H_2O$	$2170 \times 10^{22}\exp(-40{,}030/T)$	$(kPa)^2$
6	$CH_4 = C + 2H_2$	$4161 \times 10^{7}\exp(-10{,}614/T)$	kPa
7	$2CO = C + CO_2$	$5744 \times 10^{-12}\exp(20{,}634/T)$	$(kPa)^{-1}$

Table 6.7 The values of reaction 6.3, 6.4, and 6.5 rate constants on the "structured fibrous" catalyst and their logarithms depending on the inverse value of absolute temperature

T, K (t,° C)	1/T, 1/K·10^3	$k_1 \cdot 10^7$, kmol/ kg·s·(kPa)$^{0.25}$	$\ln k_1$	$k_2 \cdot 10^5$, kmol/ kg·s·kPa	$\ln k_2$	$k_3 \cdot 10^6$, kmol/ kg·s·kPa^{0,25}	$\ln k_3$
598 (325)	1.67	0.00055	−23.624	0.68	−0.3856	0.056	−19.001
623 (350)	1.61	0.0035	−21.773	0.78	−0.2485	0.071	−18.768
648 (375)	1.54	0.0150	−20.318	0.84	−0.1744	0.327	−17.236
673 (400)	1.49	0.0600	−18.932	0.96	−0.0408	0.659	−16.535
748 (475)	1.34	26.850	−15.130	–	–	4.65	−14.581
773 (500)	1.29	79.340	−14.047	–	–	8.67	−13.958
798 (525)	1.25	22.750	−14.994	–	–	14.08	−13.473
823 (550)	1.22	58.800	−12.044	–	–	23.38	−12.966

$$r_2 = k_2 \cdot \frac{P_{CO} \cdot P_{H_2O^{0.5}}}{P_{H_2^{0,5}}} \cdot \frac{1 - P_{CO} \cdot P_{H_2} / K_{P2} \cdot P_{CO} \cdot P_{H_2O}}{Z^2} \tag{6.6}$$

$$k_{2\ fiber} = 1,15 \cdot 10^{-4} \cdot e^{-10,8/RT} \tag{6.7}$$

$$k_{2\ \Gamma IA\Pi} = 4,58 \cdot 10^{-4} \cdot e^{-11,4/RT} \tag{6.8}$$

$$r_3 = k_3 \cdot \frac{P_{CH_4} \cdot P_{H_2O}}{P_{H_2^{1.75}}} \cdot \frac{1 - P_{CO_2} \cdot P_{H_2 4} / K_{P3} \cdot P_{CH_4} \cdot P_{H_2O^2}}{Z^2} \tag{6.9}$$

$$k_{3\ fiber} = 0,208 \cdot 10^3 \cdot e^{-92,7/RT}$$

$$k_{3\ \Gamma IA\Pi} = 0,831 \cdot 10^3 \cdot e^{-103,4/RT}$$

where:

P_i – partial pressure of the *i* component, kPa

$$Z = 1 + K_{CO,\ ad} \cdot P_{CO} + K_{H,ad} \cdot P_H + K_{H_2O,ad} \cdot \frac{P_{H_2O}}{P_{H_2}} \tag{6.10}$$

k_1, k_2, k_3 – rate constants of reactions 6.3, 6.4, and 6.5 (units of rate constants of reactions 6.3 and 6.5 are [kmol/kg·s·(kPa)^{0,25}] and of reaction 6.4 – [kmol/kg·s·(kPa)]).

Table 6.8 represents kinetic parameters of the final model for two catalyst types ("structured fibrous" and "ΓIAΠ-3-6H," powder with particle size 0.15–0.25 mm).

Table 6.8 Kinetic parameters of the final model for two catalyst types ("structured fibrous" and "ΓIAΠ-3-6H," powder with particle size 0.15–0.25 mm)

T, K	Structured fibrous $k_1 \cdot 10^7$, kmol/kg·s·(kPa)^{0,25}	ΓIAΠ-powder, $k_1 \cdot 10^7$, kmol/kg·s·(kPa)^{0,25}	Fibrous $k_3 \cdot 10^6$, kmol/kg·s·(kPa)^{0,25}	ΓIAΠ-powder, $k_3 \cdot 10^7$, kmol/kg·s·(kPa)^{0,25}	ΓIAΠ-fibrous $k_2 \cdot 10^5$, 1/kPa	$k_{ad,CO} \cdot 10^2$, 1/kPa	$k_{ad,H} \cdot 10^2$, (kPa)$^{-0,5}$	$k_{ad,H2O} \cdot 10^2$, (kPa)$^{-0,5}$
Reverse								
598	0.0006	0.0022	0.056	0.2311	2708	84.91	7800	–
623	0.0035	0.0140	0.071	0.5833	3125	29.05	4010	–
648	0.0150	0.0600	0.327	13.019	3364	9500	1972	–
673	0.0600	0.2400	0.659	26.364	3845	4013	1000	–
Steam reforming of methane								
748	2685	10.74	4648	18.59	–	–	–	0.7158
773	7934	31.73	8665	34.66	–	–	–	0.7681
798	22.750	91.10	14.08	56.30	–	–	–	0.8369
823	58.800	236.2	23.38	93.50	–	–	–	0.9014

6.4 Scientific Principles of Design of Reactors and Processes Associated with Endothermic Reactions of Conversion and Pyrolysis of Methane, N-Propane, and Methanol by Mathematical Modeling on Structured Catalysts

The task of the calculation is to determine the degree of work efficiency of a new, oxide-nickel catalyst for the wet conversion of methane and other hydrocarbons. Catalysts of such kind are highly porous metal, ceramic, honeycomb, or fibrous materials or their combinations, which have unique characteristics: high activity, heat resistance (up to 1600 °C), developed specific reaction surface (7000–9000 m^2/m^3), low aerodynamic resistance (porosity of the catalyst is 84–91%), and the ability to supply heat energy to the reaction zone due to direct heating of the heterogeneous catalyst by electric current or high frequency field. Such properties of catalysts open wide prospects of cardinal modification of existing technologies of conversion of methane, propane, and methanol with simultaneous creation of new equipment for these processes.

Replacement of traditional tubular furnaces for shaft (ball) reactors with loading of developed catalysts will allow providing the necessary volumes of production of target products on smaller and less metal-consuming equipment with specific productivity which exceeds in time the productivity of tubular furnaces which also have a complex structure.

The disadvantage of existing industrial tubular reactors of primary reforming (wet steam conversion) of methane is that the supply of heat energy (dQ+) to compensate spending on endothermic effects of reactions is provided from the outside through the metal wall of the tubes. The energy is obtained from combustion of the natural gas in ceiling gas burners and it should overcome the thermal resistance of the wall and the wall resistance of the gas film (heat supply through the point contacts of the catalyst is neglected). Subsequently, the transfer of heat energy to the reaction contact surface is carried out by heat transfer from the gas mixture to the outer surface of the catalyst (Stepanov et al. 1998).

A significant advantage of the proposed shaft reactor is that the supply of reaction energy (dQ+) is carried out directly on the reacting surface of the catalyst when loading a structured honeycomb fibrous catalyst, which is equipped with a specially built-in electric heater. It is obvious that the highly developed surface of the fibrous catalyst provides a much higher intensity of heat exchange compared to the traditional granular layer of the catalyst (Zaufir and Gavriilidis 2003; Popovich et al. 2017a, b).

The development of new technologies and equipment from laboratory research through a number of research stands to the industrial scale requires significant time spent and money and does not correspond to the current scientific and technical level. The method of mathematical modeling allows studying catalytic processes in detail in the presence of sufficiently complete initial information obtained on laboratory-experimental stands with subsequent development of their equipment design for any process on an industrial scale with a significant reduction in development time and saving money.

The bases for creating a mathematical model are the chemical kinetics of reactions taking place in the reaction volume and the chemical, thermodynamic, and thermophysical parameters of the reacting components, as well as heat-mass-exchange characteristics of interphase interactions.

The software was built with the maximum coverage of possible initial reacting components and corresponding chemical reactions, and when calculating specific processes only those components and reactions were taken into account that are directly related to this particular case.

The following gases are components of vapor-gas mixtures at the inlet of the reactor: nitrogen, oxygen, methane (n-propane, methanol, ethanol, etc.), water vapor, hydrogen, monoxide and dioxide of carbon, and other hydrocarbons – C_nH_m.

The reactions are considered in the order shown in Table 6.6 when modeling the processes of wet or mixed conversion of gas mixture.

The kinetics of the reaction of methane with water vapor at atmospheric pressure was studied in laboratory integral (gradientless) and flow reactors; the kinetic equations were obtained for the catalyst with composition: synthetic mineral wool (fiber) 35%, “nano”-silicon adhesive 10%, Al_2O_3 32.5%, MgO 15%, and NiO 2.5%, which were close to the kinetic equation obtained by foreign authors for aluminum-magnesium-nickel contact brand “57_4” company “ICI” for wet conversion of methane in a microreactor with boiling layer and grain size of 0.105–0.2 mm (Zaufir and Gavriilidis 2003).

The dependences of the reaction rates in the kinetic region are derived in the result of processing the experimental data obtained on the experimental tube of vapor-water conversion of methane operating at a pressure of 30 bar:

1.

$$W1_{CH_4}{}^{\{1C\}} = k_1{}^{\{1C\}} \cdot \frac{(C_{CH_4}{}^*)^{\alpha\{1C\}} \cdot (C_{H_2O}{}^*)^{\beta\{1C\}}}{(C_{H_2}{}^*)^{\gamma\{1C\}}} \cdot \frac{1-(C_{CO}{}^*)\cdot(C_{H_2}{}^*)^3/K_1\cdot(C_{CH_4}{}^*)\cdot(C_{H_2O}{}^*)}{Z^2}$$

$$r_1 = k_1 \cdot \frac{P_{CH_4}\cdot P_{H_2O}{}^{0.5}}{P_{H_2}{}^{1.25}} \cdot \frac{1 - P_{CO}\cdot P_{H_2}{}^3/K_{P_1}\cdot P_{CH_4}\cdot P_{H_2O}}{Z^2}$$

$$k_{1\,fiber} = 1,150 \cdot 10^8 \cdot e^{-198700/RT}$$

2.
$$W2_{CO}{}^{\{2C\}} = k_2{}^{\{2C\}} \cdot \frac{(C_{CO}{}^*)^{\alpha\{2C\}} \cdot (C_{H_2O}{}^*)^{\beta\{2C\}}}{(C_{H_2}{}^*)^{\gamma\{2C\}}} \cdot \frac{1 - (C_{CO_2}{}^*) \cdot (C_{H_2}{}^*)/K_2 \cdot (C_{CO}{}^*) \cdot (C_{H_2O}{}^*)}{Z^2}$$

$$r_2 = k_2 \cdot \frac{P_{CO} \cdot P_{H_2O^{0,5}}}{P_{H_2{}^{0,5}}} \cdot \frac{1 - P_{CO_2} \cdot P_{H_2}/K_{p_2} \cdot P_{CO} \cdot P_{H_2O}}{Z^2}$$

$$k_{2\,fiber} = 1,15 \cdot 10^{-4} \cdot e^{-10800/RT}$$

3.
$$W3_{CH_4}{}^{\{3C\}} = k_3{}^{\{3C\}} \cdot \frac{(C_{CH_4}{}^*)^{\alpha\{3C\}} \cdot (C_{H_2O}{}^*)^{\beta\{3C\}}}{(C_{H_2}{}^*)^{\gamma\{3C\}}} \cdot \frac{1 - (C_{CO}{}^*) \cdot (C_{H_2}{}^*)^4/K_3 \cdot (C_{CH_4}{}^*) \cdot (C_{H_2O}{}^*)}{Z^2}$$

$$r_3 = k_3 \cdot \frac{P_{CH_4} \cdot P_{H_2O}}{P_{H_2{}^{1,75}}} \cdot \frac{1 - P_{CO_2} \cdot P_{H_2{}^4}/K_{p_3} \cdot P_{CH_4} \cdot P_{H_2O^2}}{Z^2}$$

$$k_{3\,fiber} = 0,208 \cdot 10^3 \cdot e^{-92700/RT}$$

4.
$$W4_{C_3H_8}{}^{\{4C\}} = k_4{}^{\{4C\}} \cdot (C_{C_3H_8}{}^*)^{\alpha\{4C\}} \cdot (C_{H_2O}{}^*)^{\beta\{4C\}}$$

5.
$$W5_{CH_3OH}{}^{\{4V\}} = k_5{}^{\{5C\}} \cdot (C_{CH_3OH}{}^*)^{\alpha\{5C\}} \cdot (C_{H_2O}{}^*)^{\beta\{4V\}}$$

6.
$$W6_{CH_4}{}^{\{6C\}} = k_6{}^{\{6C\}} \cdot \frac{(C_{CH_4}{}^*)^{\alpha\{6C\}} \cdot (C_{CO_2}{}^*)^{\beta\{6C\}}}{(C_{H_2}{}^*)^{\gamma\{6C\}}} \cdot \frac{1 - (C_{CO}{}^*)^2 \cdot (C_{H_2}{}^*)^2/K_6 \cdot (C_{CH_4}{}^*) \cdot (C_{H_2O}{}^*)}{Z^2}$$

7.
$$W7_{CH_4}{}^{\{7C\}} = k_7{}^{\{7C\}} \cdot \frac{(C_{CH_4}{}^*)^{\alpha\{7C\}}}{(C_{H_2}{}^*)^{\gamma\{7C\}}} \cdot \frac{1 - (C_{H_2}{}^*)^2/K_7 \cdot (C_{CH_4}{}^*)}{Z^2}$$

Wx {nm} – reaction rate, kmol/(kg·s), where:

x – main component of the reaction.
n – sequence number of the reaction.
m – place of reaction flowing (in the catalyst or in the free volume).
κ_{1-7} – rate constants of reactions 6.1, 6.2, 6.3, 6.4, 6.5, 6.6, and 6.7.
k_1, k_2, k_3 – rate constants of reactions 6.3, 6.4, and 6.5 (units of rate constants of reactions 6.3 and 6.5 are [kmol/kg·s·(kPa)^{0,25}] and of reaction 6.4 – [kmol/kg·s·(kPa)]).
C^* – concentration on the surface of heterogeneous contact (kmol/nm^3).
P – as partial pressure of the component in a mixture (kPa).
K_{1-7} – equilibrium constants of reactions 6.1, 6.2, 6.3, 6.4, 6.5, 6.6, and 6.7.
Z – coefficient in the formula of "Langmuir-Hinshelwood," which takes into account the slowing down of the reaction due to adsorption of derivative reagents and reaction products:

$$Z = 1 + K_{CO,\ ad} \cdot P_{CO} + K_{H,ad} \cdot P_H + K_{H_2O,ad} \cdot \frac{P_{H_2O}}{P_{H_2}}$$

where:

$K_{CO,\ ad}$, $K_{H,\ ad}$, $K_{H_2O,ad}$ – adsorption constants of CO, H (atomic hydrogen), H_2O, and others.
P_{CO}, P_H, P_{H_2}, P_{H_2O} – partial pressure of ingredients in kPa.

Structure of Arrhenius formula was used to approximate experiment results on the study of kinetics or reaction flowing:

$$k_{N_p} = K_{N_p} \cdot e^{-E_{N_p}/R \cdot (t+273)},$$

where K_{N_p} – preexponential multiplier.

E_{N_p} – activation energy of reaction N, kJ/mol.
$R = 8{,}14$ kJ/(kmol·K) – gas constant.
t – temperature (θ – for reactions on the catalyst, T – for reactions in the volume), °C.

The system of equations that describes the model process of hydrocarbon conversion has mixed characteristics (includes differential and algebraic equations) and has the following view.

For the gas stream core.

Material balance.

$$\frac{dG_{CH_4}}{dx} = -S_c \cdot W^{1c}{}_{CH4} - S_c \cdot W^{3c}{}_{CH4} - S_c \cdot W^{6c}{}_{CH4} - S_c \cdot W^{7c}{}_{CH4} \qquad (6.11)$$

$$\frac{dG_{CH_3OH}}{dx} = -S_c \cdot W^{5c} \quad (6.12)$$

$$\frac{dG_{CH_3H_8}}{dx} = -S_c \cdot W^{5c}{}_{C_3H_8} \quad (6.13)$$

$$S_c = F \cdot (1 - \varepsilon_c), \; S_v = F \cdot \varepsilon_c, \; F = \pi \cdot R_c{}^2. \quad (6.14)$$

Equations for the mass transfer between the core of the gas stream and the catalyst surface have the following view taking into account stoichiometry:

$$g_{CH_3OH} \cdot S_s = S_s \cdot \beta_{CH_3OH} \cdot (C_{CH_3OH} - C^*{}_{CH_3OH}) = W^{4c}{}_{CH_3OH},$$

$$g_{CO} \cdot S_s = S_s \cdot \beta_{CO} \cdot (C_{CO} - C^*{}_{CO}) = W^{2c}{}_{CO},$$

$$g_{CH_4} \cdot S_s = S_s \cdot \beta_{CH_4} \cdot (C_{CH_4} - C^*{}_{CH_4}) = W^{3c}{}_{CH_4} + W^6{}_{CH_4} + W^7{}_{CH_4},$$

$$g_{C_3H_8} \cdot S_s = S_s \cdot \beta_{C_3H_8} \cdot (C_{C_3H_8} - C^*{}_{C_3H_8}) = W^{3c}{}_{C_3H_8}.$$

Here: $g_i = \beta_i \cdot (C_i - C^*{}_i)$ [kmol/m^2·s] – density of the diffusion stream.
β_i [m/c] – mass exchange coefficient.
Energy balance:

$$\Sigma_i \cdot G_i \cdot C_{P_i} \cdot \frac{dT}{dx} = F \cdot (1 - \varepsilon_c) \cdot S_s \cdot [\alpha_T \cdot (\theta - T) + \Sigma_i g_i \cdot (I^*{}_i - I_i(T))]$$
$$+ F \cdot \varepsilon_c \cdot \left[-W^{2v}{}_{CO} \cdot Q_2 + W^{4v}{}_{CH_3OH} \cdot Q_4 - dQ_+\right] \quad (6.15)$$

where: θ, T – temperature of catalyst and gas stream.
α_T [W/m^2·degree] – coefficient of heat release for interphase surface.
I_i [kJ/kmol] – enthalpy of gas stream components

$$I^*{}_i = \begin{cases} I_i(\theta), & g_i > 0, \\ I_i(T), & g_i < 0. \end{cases}$$

dQ_+ [kJ/s] – forced supply of heat energy externally built into the structured honeycomb block by an electric heater.

6.5 Catalyst

6.5.1 Energy Balance

We have the condition of continuity of a heat stream through a surface interphase and take into account losses through an external wall:

$$\rho_c \cdot c_c \cdot \frac{\partial \theta}{\partial \tau} = S_s \cdot \left(\alpha_T \cdot (T - \theta) + \lambda_c \frac{\partial \theta}{\partial x^2} - \Sigma_i g_i I_i \right) + W^{4c}{}_{CH_3OH} \cdot Q_4$$
$$+ \qquad W^{2c}{}_{CO} \cdot Q_2 + W^{3c}{}_{C_3H_8} \cdot Q_3$$
$$- 2k_{T_e} \cdot \frac{(\theta - T_e)}{R_c \cdot (1 - \varepsilon_c)} \tag{6.16}$$

where k_{T_e} [kW/m^2·degree], T_e – the heat transfer coefficient from the catalyst to the environment and the temperature of the environment, respectively. In this case, we can neglect heat loss through the end surfaces, i.e., these are following ratios:

$$\lambda_c \cdot \frac{\partial \theta(0)}{\partial x} = \lambda_c \cdot \frac{\partial \theta(L_1)}{\partial x} = \lambda_c \cdot \frac{\partial \theta(L_1 + L_2)}{\partial x} = \lambda_c \cdot \frac{\partial \theta(L_1 + L_2 + L_3)}{\partial x} = 0 \tag{6.17}$$

Criteria dependences were used to determine the values of mass transfer and heat transfer coefficients:

$$Nu = \begin{cases} 0.725 \cdot Re^{0.47} \cdot Pr^{0.43}, & Re = 2 \div 30, \\ & Pr = 0.6 \div 3600, \\ 0.395 \cdot Re^{0.64} \cdot Pr^{0.33}, & Re = 30 \div 8 \cdot 10^4, \end{cases}$$

$$Sh = \begin{cases} 0.725 \cdot Re^{0.47} \cdot Sc^{0.43}, & Re = 2 \div 30, \\ & Sc = 0.6 \div 3600, \\ 0.395 \cdot Re^{0.64} \cdot Sc^{0.33}, & Re = 30 \div 8 \cdot 10^4, \end{cases}$$

$$Re = w \cdot \frac{d_{eq}}{\nu_g}; \quad Pr = \frac{\nu_g \cdot \rho_g \cdot c^p{}_g}{M_g \cdot \lambda_g}; \quad Sc = \frac{\nu_g}{D_i}$$

$$w = 22.4G \cdot \frac{T + 273.15}{273.15 \cdot F \cdot \varepsilon_c}$$

$$G = \Sigma_i \cdot G_i; \quad d_{eq} = \frac{4 \cdot \varepsilon_c}{(1 - \varepsilon_c) \cdot S_s}; \quad \alpha_T = \frac{Nu \cdot \lambda_g}{d_{eq}}; \quad \beta_I = \frac{Sh \cdot D_i}{d_{eq}} \quad i \sim$$

where:

G – the total consumption of all components of the reaction mixture.
w[m/s] – the average rate of the reaction gas movement in the porous catalyst.

The obtained system of Eqs. (6.11, 6.12, 6.13, 6.14, 6.15, and 6.16) with limits (6.17) is inhomogeneous in its structure, as it includes linear and nonlinear equations of algebra, ordinary differential equations, and equations in partial derivatives. A special mathematical software was created to solve it. Calculations were performed, using the PC program developed by us, on mathematical modeling of an industrial

Table 6.9 Indicators of a typical series of structured fibrous contacts

№. of typical series	Dimensions, a·b·h, mm	Area, m^2	Volume, l/m^3	Mass of the block, kg	Power of the heater, kW
1	71 × 71 × 100	0.005	0.5/0.0005	0.25	2.5
2	100 × 100 × 100	0.01	1.0/0.001	0.5	5.0
3	200 × 200 × 200	0.04	8.0/0.008	4.0	40.0

shaft reactor for wet conversion of methane, which is designed for a capacity of 500,000 t/year of ammonia, loaded with a new structured fibrous catalyst equipped with electric heaters. It is planned in 2021 to introduce in Ukraine experimental and industrial production of new structured catalysts with rectangular honeycomb blocks with the dimensions of the standard series given in Table 6.9.

The following parameters of the shaft reactor are recommended for adequate replacement of tubular reactors in the current technologies of steam conversion of methane:

- Inner diameter of the reactor, 2200 mm; the area occupied by the catalyst, 4 m^2; the height of the reactor, 4000 mm; the height of the catalyst layer, 2800 mm.
- Volume of the catalyst 11,2 m^3; weight 5600 kg (1400 honeycomb blocks of a typical row №3, each of them has volume of 8 l, weight of 4 kg, bulk specific weight of 470–500 kg/m^3, porosity – 86%).

6.6 Isothermal Conversion of Natural gas for the Production of Derivative Gas for Methanol Production

Preparation of the derivative synthesis gas for methanol production is carried out by the method of vapor-oxygen-carbon dioxide conversion of natural gas in most industrial technologies used in Ukraine. For example, the isothermal principle of methane conversion in a shaft reactor on a nickel-alpha-Al_2O_3 catalyst under atmospheric pressure is used in the current production of methanol with a capacity of 100,000 tons/year at public company "Severodonetsk Nitrogen." The temperature mode of the process is maintained at the level of 870 ± 50 °C. This is ensured by the catalytic combustion of part of the methane or commercial hydrogen produced from it. This does not seem very rational to us, because all combustion products (CO_2 and H_2O) remain in the commercial mixture, which is fed to the methanol synthesis reactor. According to the technological regulations, the optimal yield of methanol takes place in compliance with the parameter ratio, mentioned below, in the range of 2.0–2.07, but not less than 1.7:

$$\frac{C_{H_2} + C_{CO}}{C_{CO} + C_{CO_2}} \geq 1,7 \div 2,07 \tag{6.18}$$

Besides it is more important that residual carbon dioxide which can go out to the air is produced. That is, the production of methanol itself is not carbon-free and it is rational to improve this technology in order to reduce the formation of carbon dioxide. In addition, it will allow increasing the yield of commercial hydrogen, as well as crude methanol by 30–40% if we are to exclude the combustion of hydrocarbons and hydrogen to heat a mixture of reacting gases.

We have developed the newest technology for the production of structured fibrous honeycomb catalysts equipped with electric heaters, the implementation of which will solve these problems. Laboratory and experimental-industrial tests of the developed catalysts in isothermal conversion are carried out (Tables 6.10 and 6.11). The examination of the working scheme of derivative synthesis gas production for methanol production is performed; the technical and economic comparison of these processes is carried out.

The comparison of new and existing technologies is performed below in the form of material balances. Regulations data of the working production of methanol are presented in Tables 6.12 and 6.13.

The comparison showed that the implementation of a structured fibrous honeycomb catalyst with direct heating of the contact with "green" electric current reduces oxygen consumption by 3876 kg/g and increases the yield of hydrogen from 1090.6 kg/g to 1553 kg/g (on 462.4 kg/g).

6.7 Conclusions

Methods of manufacturing the newest structured fibrous catalysts with built-in honeycomb blocks and electric heaters have been developed and experimental samples of catalysts were manufactured with the execution of the state order of the Ministry of Education and Science of Ukraine №ДЗ-91 dated October 25, 2019, for the period 2019–2020. Experimental and analytical works were performed on the examination of the operation modes of an industrial tubular furnace and a shaft reactor for isothermal methane conversion. Investigations of the kinetic parameters of the developed catalysts have been performed in the integral laboratory reactor for the reactions of wet methane conversion and reduction of carbon dioxide with hydrogen to carbon monoxide (2CO) and to methane (reverse process); the values of the rate constants were obtained, the kinetic equations of the investigated reactions were recommended, and the necessary macrokinetic parameters were calculated. Drawings were developed; molds, stamp forms, and other equipment were manufactured for the organization of experimental-industrial production of catalysts. "IT" programs were developed for the design of processes and reactors using mathematical modeling methods, and calculations were performed and recommendations were issued for industrial implementation of efficient technologies that reduce greenhouse gas CO_2 emissions to air by 40% and increase hydrogen yield at the stage of isothermal methane conversion in methanol production with a capacity of 100,000 tons/year for 11 tons/day (from 1090.6 kg/h to 1553 kg/h).

Table 6.10 Analysis of the process of isothermal vapor-oxygen-carbon conversion of natural gas in an industrial shaft reactor for the methanol production with a capacity of 100,000 tons/year

Catalyst layer, m	Contact time, s	Temperature of VGM, °C	Methane conversion, XCH4,%	Oxygen conversion, XO2,%	Concentration of CH4, vol. %	Concentration of O2, vol. %	Concentration of H2O, vol. %	Concentration of H2, vol. %	Concentration of CO2, vol. %	Concentration of CO, vol. %
0.0	0	440	0	0	35.71	21.43	38.10	0	4.76	0
0.2	0.043	606	8.56	10.06	32.65	16.15	40.30	4.50	5.10	1.30
0.4	0.123	640	27.86	30.94	27.03	12.32	38.48	16.37	5.40	2.50
0.6	0.150	675	46.14	53.26	21.73	6.98	37.30	25.40	5.50	3.80
0.8	0.173	700	53.48	82.65	16.61	2.57	36.71	31.61	5.60	6.90
1.0	0.192	720	61.62	87.85	13.21	1.10	36.34	35.10	5.65	8.60
1.2	0.215	748	83.17	93.04	6.01	0.63	35.66	39.30	5.80	12.6
1.4	0.195	760	97.94	96.35	0.736	0.33	34.36	40.87	6.10	17.6
1.8	0.215	831	99.02	98.34	0.350	0.15	33.80	41.07	6.34	18.29
2.2	0.315	871	99.35	98.56	0.232	0.13	33.31	41.6	6.40	18.33
2.6	0.311	905	99.43	98.79	0.204	0.11	32.49	42.4	6.45	18.35

Table 6.11 The results of tests of the structured fibrous catalyst on the experimental installation (flow rector with internal diameter 100 mm) in the process of vapor-carbon dioxide-oxygen conversion of methane (replacement of 65% oxygen by electric heating)

Catalyst layer, m	Contact time, s	Temperature of VGM, °C	Methane conversion, X_{CH4},%	Oxygen conversion, X_{O2},%	Concentration of CH_4, vol. %	Concentration of O_2, vol. %	Concentration of H_2O, vol. %	Concentration of H_2, vol. %	Concentration of CO_2, vol. %	Concentration of CO, vol. %
0.0	0	400	0	0	35.71	9.05	46.76	0	8.48	0
0.1	0.043	569	8.36	10.06	32.33	8.14	9300	10.50	7.43	2.30
0.2	0.123	608	25.96	30.94	29.35	6.25	34.38	18.26	7.26	4.50
0.3	0.150	644	45.14	53.26	25.76	4.23	31.30	24.76	7.15	6.80
0.4	0.173	670	51.50	82.65	21.23	1.57	27.82	34.62	6.86	7.90
0.5	0.192	725	60.63	87.84	16.61	1.10	23.39	42.10	6.70	10.10
0.6	0.215	738	84.17	93.04	13.70	0.63	15.60	49.80	6.47	13.8
0.8	0.195	779	98.84	96.35	6.00	0.33	10.02	57.87	6.17	19.6
1.0	0.315	789	99.25	98.53	0.349	0.13	9071	61.60	5.75	23.1
1.2	0.311	792	99.33	98.73	0.232	0.11	8358	62.40	5.48	23.2

Table 6.12 Material balance of methanol production with a capacity of 100,000 tons/year at public company "Severodonetsk Nitrogen" (methane conversion stage)

№.	Name	Volume, m^3/g	Mass, kg/g	Quantity, kmol/g	Concentration, vol. %
Loaded					
1	Methane (CH_4)	7500	5355	334.8	35.71
2	Oxygen tech. (O_2)	4500	6408	200.9	21.43
3	Water (vapor) (H_2O)	8000	6432	351.1	38.1
4	Carbon dioxide (CO_2)	1000	1960	44.7	4.76
5	Carbon monoxide (CO)	–	–	–	–
6	Hydrogen (H_2)	–	–	–	–
7	Nitrogen and others (N_2)	–	–	–	–
Obtained					
1	Methane (CH_4)	101.7	5355	4.81	0.36
2	Oxygen tech. (O_2)	43.1	6408	1.90	0.15
3	Water (vapor) (H_2O)	9497.1	6432	424	31.74
4	Carbon dioxide (CO_2)	1884.7	1960	81.14	6.34
5	Carbon monoxide (CO)	5438.6	–	242.8	18.29
6	Hydrogen (H_2)	12.214	1090.6	545.3	41.04
7	Nitrogen and others (N_2)	430.8	–	–	0.96

Table 6.13 Material balance of methanol production with a capacity of 130,000 tons/year when a new structured fibrous catalyst with an electric heater ("green" electricity – 15.7 MW·h) was implemented

№.	Name	Volume, m^3/g	Mass, kg/g	Quantity, kmol/g	Concentration, vol. %
Loaded					
1	Methane (CH_4)	7500	5355	334.8	35.71
2	Oxygen tech. (O_2)	1771.1	2522	71.9	8.71
3	Water (vapor) (H_2O)	9820	7895.2	438.4	46.76
4	Carbon dioxide (CO_2)	1808.9	3545.4	80.75	8.82
5	Carbon monoxide (CO)	–	–	–	–
6	Hydrogen (H_2)	–	–	–	–
7	Nitrogen and others (N_2)	–	–	–	–
Obtained					
1	Methane (CH_4)	107.7	76.9	4.8	0.31
2	Oxygen tech. (O_2)	43.48	61.4	1.9	0.126
3	Water (vapor) (H_2O)	7608.7	6117.4	339.7	22.06
4	Carbon dioxide (CO_2)	1052.3	2009.8	47	3.05
5	Carbon monoxide (CO)	7838.4	9798	350	22.73
6	Hydrogen (H_2)	17.408.6	1553	777.2	50.47
7	Nitrogen and others (N_2)	430.8	538.5	–	0.96

References

Dudnik AN, Melakh VG (2007) Hydrogen filling stations. Ecotechnolog Resour Saving 4:3–12

Hou K, Hughes R (2001) The kinetics of methane steam reforming over a Ni/α-Al2O3 catalyst. Chem Eng J 82:311–328

Kozin L, Volkov S (2002) Hydrogen energy and ecology. Naukova Dumka, Kiev

Kozin LF, Volkov SV (2006) Modern energy and ecology. Naukova Dumka, Kiev

Pokhodenko VD, Skorokhod EV, Solonina Yu M (2010) Fundamental problems of water energy. Naukova Dumka, Kiev

Popovich A, Soloviev H, Orlyk V et al (2017a) Development of mathematical model of methane oxidation on fibrous catalyst. East-Euro J Enterprise Technolog. 6/6(90):33–40

Popovich A, Soloviev H, Suvorin A (2017b) Research into methane oxidation catalysts of the applied type. East-Euro J Enterprise Technolog 4/6(88):29–34

Stepanov AV, Sulzhik NI, Salo VP (1998) Resource-saving technologies for the production of ammonia. Energy Technolog Resour Saving 3:21–25

Zaufir M, Gavriilidis J (2003) Catalytic combustion assisted steam reforming in a catalytic plate reactor. Chem Eng Sci 58:3947–3960

Chapter 7
Providing Wasteless Manufacturing of Aviation Biofuels by Using Camelina Seed Residues for Producing Functional Bread

Iryna Korniienko, Andrii Anatskyi, and Mykhailo Baranovskyi

Nomenclature

FAO	Food and Agricultural Organization
OECD	Organisation for Economic Co-operation and Development
WHO	World Health Organization
RR	Redox reactions
DRP	Determination of redox potential

7.1 Introduction

One of the priority areas in biotechnology is the bioconversion of plant raw materials. This process can be implemented by fermenting the substrate with a specific consortium of microorganisms. Biofuels, bioethanol, biobutanol and biohydrogen are considered to be valuable products' bioconversion of technical plants (rapeseed and camelina). Nowadays biofuels have been recognized as an excellent alternative to traditional aviation fuels. According to the concept of the target complex research programme of the NAS of Ukraine "Biological resources and newest technologies of bioenergy conversion" for 2013–2017 (order of the Presidium of the NAS of Ukraine from 20.03.2013, № 189), there is a threat of depletion of fossil energy sources and, accordingly, their significant rise in price in the near future. Another

I. Korniienko (✉) · A. Anatskyi · M. Baranovskyi
National Aviation University, Kyiv, Ukraine

Dniprovsk State Technical University, Dnipro, Ukraine

S. Boichenko et al. (eds.), *Chemmotological Aspects of Sustainable Development of Transport*, Sustainable Aviation, https://doi.org/10.1007/978-3-031-06577-4_7

stimulus for the development of the latest bioenergy conversion technologies is the potential to reduce carbon dioxide emissions into the atmosphere through the expansion of bioethanol and biodiesel consumption, improving fuel properties due to biological impurities.

According to FAO (Food and Agricultural Organization) and the OECD (Organisation for Economic Co-operation and Development), the world's bioethanol production exceeded 100 billion litres (80 million tons) in 2012. Leaders in fuel bioethanol production – Brazil, the USA, China and France – process sugar and starch into bioethanol, which are obtained from sugar cane, sugar beet and cereals: corn, wheat, etc. These substances belong to the first generation, so it is advisable to look for cost-effective technologies for processing lignocellulose – a second-generation raw material and a third-generation raw material – into biofuels, the sources of which are almost inexhaustible. Such sources should include non-food crops, perennial herbs and waste from agriculture, food and forestry (straw, corn stalks, stems and husks of sunflower, sawdust, camelina, rapeseed, soybeans, etc.). Biodiesel production from camelina and rapeseed is aimed at developing modern approaches to fatty acid esterification, the use of modern catalysts and the use of biodiesel by-products, primarily glycerine. An important aspect of biodiesel production is the use of oils with an improved fatty acid composition of raw materials, increasing its productivity.

In their scientific work (Boichenko et al. 2017), NAU specialists have developed recommendations on expanding the raw material potential of camelina. The current state and prospects of aviation biofuel production in Ukraine have been analysed. The potential of using camelina as a raw material for obtaining alternative aviation fuels has been determined and the technological scheme of its processing has been given. Camelina is an unpretentious crop and has a small vegetation period, so it does not need any additional components (Seal and Baranowski 2000).

As a result of the proposed processing technology, camelina forms crist, a fat-free flour with a high protein content and a full composition of amino acids, which are considered useful components for the human body. Camelina is not only valuable in terms of bioenergy conversion, it can be considered a functional product that has certain therapeutic and prophylactic properties in cases of cardiovascular disease (due to its high omega fatty acid content) and gastrointestinal disease (due to its high content of mucous substances). Camelina is recommended for use in the complex treatment of diabetes (it enhances the synthesis of pancreatic hormone – insulin) and as a source of essential amino acids.

This work is the first to explore modern biotechnological approaches to the use of camelina as the main care of bioenergy conversion, in the production of functional bread with a high content of lactic acid bacteria.

In accordance with the requirements of the International Conference on Nutrition organized by the WHO/FAO (December 1992), representatives of 159 countries, including Ukraine, unanimously adopted the "World Declaration and Programme of Action on Nutrition". To implement the provisions of this Declaration, most countries, including Ukraine, have already formulated and are implementing national programmes for healthy nutrition at the state level. They are extremely necessary, as the ratio of the impact of various factors on the health of the nation, according to WHO experts, looks as follows:

- Environmental conditions affect human health by 20–25%.
- Genetic factors – by 18–20%.
- The health system – by 8–12%.
- Lifestyles by 52–55%, with diet and nutrition structure being the most significant with the following factors.

Therefore, the top priority problem for Ukraine and its food industry is the creation of fundamentally new technologies for deep integrated processing of agricultural raw materials into high-quality products that have a health impact on the human body, provide prevention of alimentary conditions and diseases and help eliminate shortages of vitamins, macro- and micronutrients and other essential substances. All these requirements are met by health products – functional food products and functional ingredients, biologically active additives to food. Ukraine has all the necessary prerequisites for the creation of a national healthy food industry.

According to the presented research results (Neposhyvaylenko and Kornienko 2020), an intensive growth of morbidity among the younger generation in the gastrointestinal tract as a result of the consumption of poor quality food products with an excess in the diet of a significant proportion of simple carbohydrates has been established. To address this issue, the authors proposed the development of technology for the production of functional foods: bread, fruit pastila and dairy products, which are enriched with nutrients and live cultures of lactic acid bacteria.

Bread is one of the most important products for daily consumption. The relevance of this work therefore lies in the need to create an improved formula for functional bread, in which 5% of the flour is recommended to be replaced with crist/camelina. Regular consumption of functional bread guarantees the elimination of the consequences of daily consumption of unhealthy fast-digesting carbohydrates, replenishes the body with the necessary components and helps to reduce the toxic effects of elements contained in products or formed in the body by increasing the mass fraction of insoluble fibre, the source of which are camelina seeds.

7.2 Main Part

The aim is to develop a modern, energy-efficient biotechnology to produce functional bread from the flour-fermented symbiosis of pure cultures of lactic acid bacteria to enrich bakery products with nutrients from crist camelina, which is a waste of bioenergy conversion. The objective of the study is to determine the impact of camelina and additional energy sources (ATP, folic acid vitamins and succinic acid) on the fermentation process and the increase in the biomass of lactic acid bacteria in sourdough for baking.

The object of the study is a biotechnology for producing ferment for bakery needs.

Subject of the study – qualitative indicators of ferment, technological process parameters, energy efficiency.

Scientific novelty of the obtained results:

- For the first time, it has been established to use camelina seeds in the cycle of sourdough breeding and dough preparation, based on the functional and growth-stimulating factors of lactic acid bacteria biomass.
- The increase in the biomass of lactic acid bacteria in yeast with the addition of camelina seeds has been scientifically confirmed.
- For the first time, the use of adenosine triphosphate has been proven to be effective in bakery practice (ATP), thus reducing the fermentation time of the flour component of the lactic acid bacteria at the production stage of yeast production and providing a significant energy saving effect.
- Experiments have been found to increase the gas-forming ability of the dough in the presence of camelina and ATP, which improves the rheological properties of the dough.
- Developed and improved technological scheme for the production of sourdough and functional bread, taking into account the needs and characteristics of lactic acid bacteria in sourdough in accordance with the test for osmosensitivity.
- Based on the analysis of technical and technological indicators of technological equipment for the production of functional bread, a set of energy-efficient technical measures is developed aimed at modernizing individual units of equipment, the implementation of which will reduce electricity consumption by 28% per production cycle.

Practical significance of the obtained results. An improved formula for functional bread has been broken down, which is recommended for daily consumption as the main source of useful carbohydrates, fibre, nutrients and lactic acid bacteria. An energy-efficient technology has been developed for producing functional bread.

Efficiency (economic, ecological). Biotechnological solutions are economically justified due to the introduction of fermentation process biostimulants – camelina and ATP, which reduce fermentation time and temperature limits for this process. The eco-fermentation of this product is confirmed by the bread biosafety indicators which have been achieved by increasing the titre of lactic acid bacteria which, as a result of fermentation of the flour component, synthesize lactic acid, a natural preservative which ensures the extension of the shelf-life of finished products to ensure their compliance with the established microbiological biosafety indicators. The proposed production technology makes bakery products environmentally safe with therapeutic and prophylactic effect, which is especially necessary for maintaining the human gastrointestinal tract healthy.

7.2.1 Experimental Part

In order to obtain a biologically active sourdough for baking with a permanent consortium of lactic acid bacteria, it was proposed to use a commercial biopreparation of pure cultures of Vivo (Ukraine) brand, the quality of which is confirmed by certificates of the International Organization for Standardization (ISO) 9001: 2008, as well as ISO 22000: 2005, which is a type of European quality standards. Biopreparation Vivo is a symbiotic complex of pure cultures: *Streptococcus thermophilus*, *Lactobacillus delbrueckii* ssp. *bulgaricus*, *Lactobacillus acidophilus*, *Bifidobacterium lactis*, *Lactobacillus casei*, *Lactobacillus rhamnosus*, *Lactobacillus paracasei* and *Bifidobacterium infantis*. The sourdough technology is based on fermentation of flour by a permanent consortium of lactic acid bacteria of Vivo bio-preparation in the amount of 1 g of the preparation per 1 kg of flour at 20–22 °C.

To provide functional features of sourdough and bread and to improve the rheological properties, it is recommended to use crist or camelina seeds. Crist is a protein low-fat flour from which soluble non-protein components are isolated, which have highly active functional characteristics. It affects the structure, technological and nutritional properties of the finished product. It has a high moisture-binding effect, plasticity, adhesion and sorption. Crist or crushed camelina seeds are characterized by a high content of nitrogenous components; sugars; dextrins; tannins; fibre; carotene; vitamins C, B1, B2, B3, B5 and B6; vegetable fats; and mineral salts (about 19) – potassium, iron, manganese, phosphorus as well as flavonoids and caffeoylquinic acids. Camelina seeds contain a large amount of B vitamins – thiamine and pyridoxine in the amount of 11 and 46 mg/kg of flour. It is known that these vitamins are key enzymes of carbohydrate metabolism in the cells of living organisms, which in this case is relevant for accelerating the fermentation process of the flour component. Crist contains nicotinic acid in the amount of 45–51 mg/kg of flour, which is involved in redox processes. As stimulators of lactic acid bacteria growth, as well as to improve their immobilization during fermentation, it is recommended to use crist or crushed camelina seeds (genus *Camelina*, species *Camelina glabrata* or *Camelina sativa*), which contains 40% more fatty oils (omega-3 and omega-6). The peculiarity and advantage of camelina seeds compared to flax seeds, sesame seeds, milk thistle meal and even sturgeon species are too high content of omega-3 and omega-6 fatty acids, which allows this type of seed to be classified as a functional food (Peshchuk and Novesnko 2008).

In order to create the necessary cultivation conditions for lactic acid bacteria and to intensify the process of flour component fermentation, it has been proposed to introduce 5% mashed potatoes into the dough. This is justified by the chemical composition of potatoes – the maximum dry matter content is 36.8%, starch 29.4%, protein 4.6% and vitamins C, B1, B2, B6, PP and E. Potatoes contain enzymes of hydrolase class – amylase (α and β) oxidoreductase – polyphenoloxidase (tyrosinases), peroxidase and catalase.

This technology provides a recipe for functional bread with low gluten content due to the use of such types of flour – chickpea, whole grain spelt and wheat. Due to fermentation process there is a significant reduction in gluten content in low-gluten types of flour, which allows the use of this bread for patients with celiac disease.

Research methods and techniques. The method of the experiment involves studying the qualitative indicators of sourdough during the production cycle of breeding in the main physical, chemical and microbiological ostensibly to study the effect of the introduced stimulants on the quantity and fermentation activity of lactic acid bacteria.

Physico-chemical methods of research: lifting power behind the ball, titratable acidity, pH and rH2 were conducted according to recommendations (Drobot 2006; Auerman 2005). Microbiological research to identify lactic acid bacteria, *Escherichia coli*, was carried out by sowing sourdough and bread samples on elective nutrient media – bifidoagar, lactoagar and Endo environment. Organoleptic evaluation of functional bread was carried out in accordance with DSTU4583: 2006.

7.2.2 The Results of the Experiment

In order to accelerate the process of sourdough removal and reduce the duration of the production stage of increasing its biomass, the influence of stimulating factors – crist (crushed seeds) of camelina, ATP long, B vitamins with folic acid and succinic acid – was studied.

The main stage in baking is considered to be the biotechnological stage of sourdough extraction based on a low-gluten mixture of chickpea and spelt flour, which was used in a ratio of 1:3. In order to optimize the process of its increase, in terms of achieving optimal acidity and lifting power in the shortest possible time, it is recommended to carry out cultivation at a cultivation temperature of 20–22 °C, which is lower than the classical 30 °C. Previous studies have found it impractical to increase the cultivation temperature to 40–42 °C, as the increase in temperature led to an increase in titratable acidity and an increase in the lifting time for the ball to 40–45 min, which is unacceptable in this technology. In order to create optimal conditions for yeast excretion, it is recommended to apply the following growth stimulants – ATP long (5 mg/100 g flour), rye meal/seeds (5%), lactulose (4%), a complex of B vitamins with folic acid (0.01%) and succinic acid (0.1%) and propolis (1%). The application of all growth stimulants is calculated according to the applied weight of flour.

At the end of the production stage of obtaining the sourdough, measurements of the concentration of hydrogen ions and redox potential (ORP) were performed. The results are presented in Fig. 7.1.

Redox reactions (RR) are one of the main parameters of microorganism cultivation, especially anaerobic (Sogomonian et al. 2011). RRs play a key role in metabolic processes and energy production. The oxidation reaction is not separated from the reduction reaction, they are in inseparable unity. Determination of redox potential (DRP) of cells makes it possible to assess the impact of these processes on cell morphology, reproduction process, the direction of metabolic processes and the synthesis of metabolites. In electrochemistry, the redox potential is determined by Eh (millivolts), and to assess biochemical transformations, this figure is expressed in conventional units. The value of the redox potential of rH2 for reducing and oxidizing media is in the range of 0–42.6. In open biological systems, in the

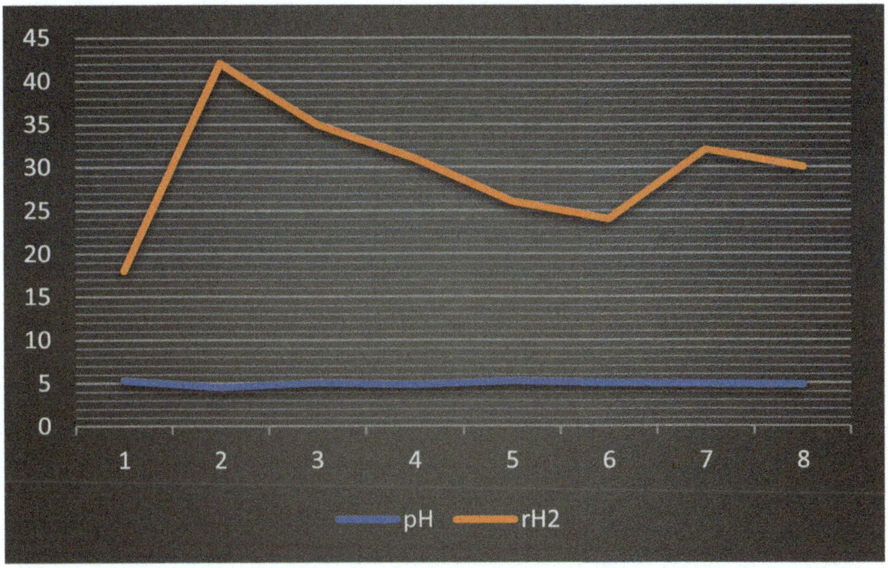

Fig. 7.1 The result of determining the concentration of hydrogen ions (pH) and redox potential (rH2) of sourdough

implementation of the biotechnological process, the maximum rate of redox reactions is related to the magnitude of the oxidation potential. As a result, at a certain period of time a steady state is established, which is characterized by a magnitude of the potential. The change in the magnitude of the potential in biological systems leads to changes in the direction of biochemical transformations and the inclusion or exclusion of enzyme systems due to changes in DRP. In this work, the redox potential (electronic nature of the process, biochemical transformations direction) of the nutrient medium in which cultured lactic acid bacteria was determined. Based on the obtained results of redox potential studies, it is possible to conclude that only camelina changes the direction of the biochemical reaction process towards a complete oxidative process (rH2 = 42); ATP without the use of impurities has a pronounced reducing potential (rH2 = 18). In this case, the redox potential shows the direction and depth, as well as the intensity of biochemical processes that are catalyzed by enzymes consisting of cofactors and coenzymes, so they can act as oxidants or reducing agents. Systems with low redox potential donate electrons, and those with high potential receive them. Introduction into the medium of propolis with the prebiotic lactulose brings the state of the medium by redox potential (rH2 = 26) to equilibrium. Despite the closest redox potential to equilibrium in samples with a complex of B vitamins with folic acid and succinic acid, unfortunately, there was no significant increase in the number of lactic acid bacteria in the environment, due to their synthetic origin and lack of need for such stimulants biochemical transformations (Table 7.1). Lactic acid bacteria are most likely to have a preference for natural substances (camelina seeds), which contain not only these

Table 7.1 Quality indicators of spelt chickpea sourdough at the end of the production cycle

Qualitative indicators of sourdough	Test samples							
	1	2	3	4	5	6	7	8
Acidity, degrees	10.3	14.8	13.1	12.3	10.0	9.1	9.6	8.4
Lifting force, minutes	15	17	16	16	17	45	39	56
Fermentation activity, minutes	9	12	11	10	12	41	37	65
The number of bifidobacteria, cells/cm^3	600×10^8	1500×10^9	900×10^9	900×10^8	1200×10^9	600×10^4	600×10^3	300×10^3
The number of lactobacilli, CFU/cm^3	300×10^8	1200×10^8	900×10^8	600×10^8	900×10^8	150×10^4	300×10^3	150×10^3

components but also enzymes and fatty acids, which they use to build cell membranes. Succinic acid is more often used as an intermediate of the Krebs cycle (after such a cycle there is an increase in yeast biomass), in which it affects the rapid resynthesis by ATP cells, because it is a powerful stimulator of energy production. In lactic acid bacteria, the fermentation process does not occur according to the Krebs cycle, but by glycolysis and pentose phosphate, so probably there was no significant increase in their biomass.

When assessing the quality of sourdough for baking, pay attention to the following indicators: lifting force per ball (not less than 25 min), acidity of 10–14 degrees (depending on the type of flour), the number of lactic acid bacteria (lacto-, bifido), sensitivity fermentation microflora and fermentation activity (10–25 min). These indicators were evaluated during the removal of yeast; the results are shown in Table 7.1. For the studied samples the following combinations of components are used: 1 ATP, 2 lactulose + camelina, 3 ATP + camelina, 4 ATP + lactulose, 5 propolis + lactulose, 6 complex of B vitamins + folic acid, 7 succinic acid and 8 control sample without the addition of growth stimulants. The initial value of the acidity of the tested samples before fermentation was 5.1 degrees.

The indicator of fermentation activity for samples with fermentation stimulants (ATP, lactulose, camelina, propolis) is within 9–12 min, which indicates an intensive process of substrate oxidation. Analysing the results of research, it can be argued that the introduction of an additional energy source is a technologically sound solution, as evidenced by the indicators shown in Table 7.1. The introduction of additional energy in the form of ATP molecules immediately releases energy to trigger energy-intensive chemical reactions (exergonic process). In cells, it is an endergonic process that requires additional introduction of energy sources – substrates (in our case, cellulose, camelina, chickpea flour). Adding an additional source of ATP in the process of removing sourdough is an appropriate solution, given the amount of excretory ATP during anaerobic fermentation (fermentation – glycolysis and oxidative phosphorylation – pentose phosphate). As a result of glycolysis, only two molecules of ATP are synthesized, and as a result of oxidative phosphorylation, one molecule of ATP is obtained by converting one molecule of glucose. At a time when each cycle of beta-oxidation of fatty acids gives 14 molecules of ATP, and in the respiratory process of oxidation – 28 molecules of ATP. Such types of fermentation (a set of anaerobic redox reactions in which organic compounds are electron donors and acceptors) are characteristic of homofermentative fermentation by representatives of lactobacilli – *L. bulgaricus*, *L. lactis* – and milk streptococci *Streptococcus cremoris* level. As a substrate it is used as mono- and disaccharides.

$$C_6H_{12}O_6 + 2FDF + 2F_H = 2CO_3 - COOH + 2ATP \tag{7.1}$$

What cannot be said about heterofermentative fermentation (7.2), which results in the formation of various components – lactic and succinic acid, ethyl alcohol, carbon dioxide and glycerin Eq. (7.2). This type of fermentation occurs by pentose phosphate and is characteristic of bifidobacteria *B. bifidum*, as well as lactobacilli species *L. brevis* and *L. lactobacterium*. From the energy point of view, this path of catabolism is two times less efficient than the glycolytic path of glucose oxidation, because the oxidation of one molecule of glucose produces one molecule of ATP. But its value lies in the fact that it provides bacteria with pentoses (ribulose-6-phosphate), which are precursors of nucleotides and nucleic acids. In addition, in this cycle, two molecules of NADP * H2 are formed, which are necessary to restore biosynthesis reactions. ATP is synthesized according to energy and constructive types of metabolism (these types of metabolism are characteristic of lacto- and bifidobacteria); they occur simultaneously in bacteria due to successive enzymatic reactions as a result of substrate phosphorylation.

$$\begin{aligned} 2C_6H_{12}O_6 = {} & CH_2CHOH\text{-}COOH + COOH\text{-}CH_2\text{-}CH_2COOH \\ & + CH_3COOH + CH_3CH_2OH + CO^2 + H_2 + X_{kcal} \end{aligned} \quad (7.2)$$

Microbiological studies have shown that ATP, lactulose camelina and propolis can be considered a stimulating factor in the environment for the biomass of lactic acid bacteria. Using a specific substrate – lactulose – the cultivation of lactic acid bacteria is directed, as evidenced by the results of microscopy (no yeast in the tested samples).

After counting the number of colonies and cells on solid/semi-liquid electrical nutrient media, microscopic studies were conducted (Figs. 7.2 and 7.3). The results of the microscopy revealed that all samples except for the ATP sample contain sufficient starch grains. This is due to the fact that starch belongs to the first-order polysaccharides used when there are no easily digestible carbohydrates (glucose, fructose). When ATP is applied, single grains of starch are recorded in the field of view, indicating that fermentation of almost the entire flour component is activated. In the presence of other additional growth stimulants for lactic acid bacteria, their moderate presence was noted, which indicates an incomplete process of hydrolysis of the flour component. It is known that for fermentation processes, it is assumed that there is no complete oxidation of the substrate, as this process is considered to be energy-poor, as a result of which fermenting microorganisms digest a huge amount of the substrate to produce incomplete ATP energy. Fermented fibres and filler

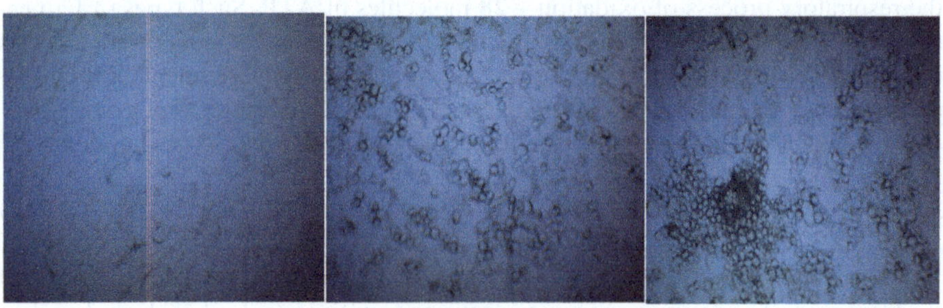

1 2 3

Fig. 7.2 Results of microscopy of tested samples of starchy grains (without dyestuffs): (**a**) sourdough with ATP, (**b**) sourdough with camelina and lactulose, (**c**) sourdough (control sample)

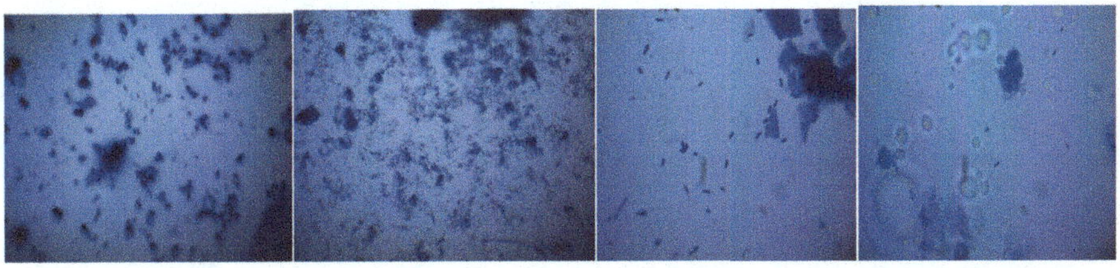

Fig. 7.3 Result of microscopy (dyed cells with methylene blue) of tested samples of starter: (**a**) ATP + lactulose, (**b**) camelina + lactulose, (**c**) propolis + lactulose, (**d**) control

substrates are subject to further fermentation in the human intestine, which will result in the biosynthesis of fatty acids in the human intestinal microflora. These acids in the human intestine stimulate the processes of cell proliferation and differentiation; form mucus; promote the absorption of water, potassium and magnesium salts; regulate the acid-alkaline balance; necessary material for the synthesis of lipids and cell membranes; and promote cell regeneration.

Analysing the results of microscopy, it becomes obvious that the addition of ATP, lactulose and camelina helps maintain the maximum number of lactic acid bacteria that meet morphological characteristics: rod-shaped, Gy +, immobile, do not form spores and capsules, facultative anaerobes that do not have the function of catalase performs the enzyme - lactoperoxidase, energy is obtained exclusively through the fermentation process (homofermentative and heteroenzymatic). increase in the number of lactic acid bacteria, especially bifidobacteria). Thanks to the products of lactic acid bacteria biosynthesis (bacteriocins, lactic acid), there is a full provision of appropriate microbiological requirements for the quality of the finished product (absence of *Escherichia coli*, other pathogens and pathogens of bread); also, lactic acid bacteria contribute to coenoanabiosis creating the necessary conditions for their development and reproduction.

Salt is an abiotic factor that is associated with osmotic activity, especially for lactic acid bacteria. According to the results of experiments, it was found that lactic acid bacteria are sensitive to salt, even at a concentration of 1.0%. And with the addition of 2% salt to the dough, plasmolysis (loss of water by the cell when placing it in a hypertonic solution) is complete. The addition of salt to the paste or other semi-finished products reduces the activity of amylase. Also, it reduces the attack of starch by amylase and increases the temperature of the beginning of its gelatinization. Therefore, in this technology it is recommended to introduce salt at the end of the dough formation process, so that the fermentation process proceeds in full, while maintaining the best rheological properties of the dough. Experiments have shown that the concentration of salt 1.0% by weight of flour significantly affects the intensity of reproduction of lactic acid bacteria (the number of lactic acid bacteria is reduced by several orders of magnitude). The increase in salt and decrease in the biomass of lactic acid bacteria leads to a decrease in gas formation in the pre-ferment

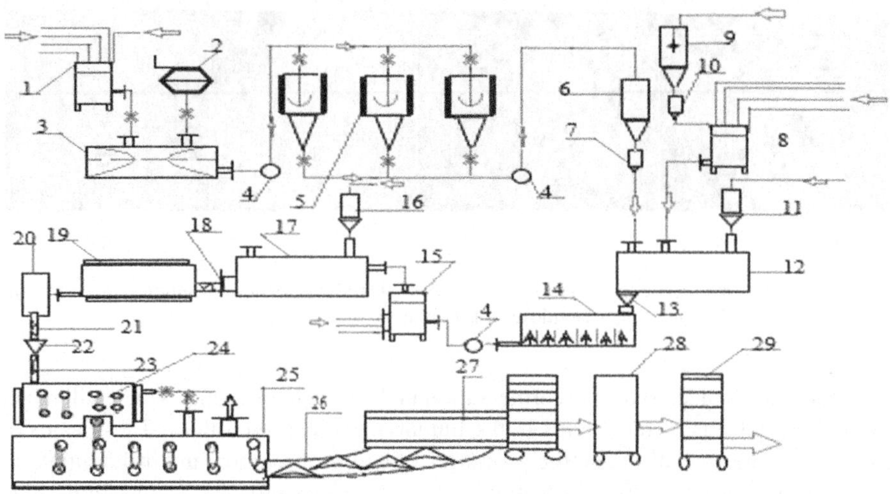

Fig. 7.4 Biotechnological scheme for the production of functional bread: 1, 8, 15 dosing station; 2 flour dispensers; 3 brewing machine; 4 dosing pump; 5 fermentation vat with stirring and heating; 6 expendable vat; 7, 10 scoop-type dispenser; 9 shredder; 11, 16 bulk component dispenser; 12, 17 kneading machines for pre-ferment and paste; 13 vane pump dispenser; 14 trough-shaped capacity six sections; 18, 21 conveyor pipe; 19 fermentation hopper; 20 dough receiver; 22 dough stacker; 23 loader; 24 stand; 25 rotary oven; 26 conveyor; 27 bread machine; 28 distribution carriage; 29 bread trolley

or dough by 30–42% and also reduces the intensity of acid accumulation and looseness of the dough.

Taking into account the obtained research results and certain regularities of the influence of various factors, an improved technological scheme for the production of functional bread using lactic acid bacteria growth stimulants has been developed (Fig. 7.4). Optimized biotechnological stages of the process – obtaining sourdough, pre-ferment and dough, taking into account the peculiarities of the cultivation of lactic acid bacteria, as well as the main technological stages of the process – proofing of dough pieces, baking bread and packaging.

A technology has been developed for preparing dough on thick sourdough with brewing. This method is based on an analysis of the advantages and disadvantages of well-known brewing methods – the Yugon method (Japanese technology for brewing flour) and the technology for obtaining brewing proposed by Soviet scientist L. Y. Auerman.

Through the dosing station of liquid components (1) (up to five components), water, a solution of lactulose, ATP, sourdough and propolis are fed by gravity from the collectors to the brewing machine (3). The flour is fed through the flour dispenser (2) in the required amount (not more than 10% of the total mass for dough preparation). Brewing of flour takes place by supplying water with a temperature of 80–85 °C, while the semi-finished product – brewing –should have a temperature of not more than 50 °C. Reasonable compliance with temperature conditions is due to the fact that enzymes – biocatalysts of protein nature and their denaturation begins with excess temperature. When the brewing temperature rises above 50–55°C, functional disorders of gluten proteins can occur, "Alpha phenomenon", which is associated with gelatinization of starch and denaturation of gluten. The sugaring time of sourdough and water-based brewing lasts from 60 to 120 min (depending on the type of flour: flour with low gluten is subject to saccharification for 60–90 min, and strong flour with developed gluten is subject to saccharification for 120 min, for each type of flour experiments by setting the saccharification time); in this case, the saccharification process lasts 60 min until the temperature drops within 28–300 °C. At the end of the saccharification process, solutions of lactulose, propolis, ATP and sourdough are added to the pre-fermentation. The sourdough on the pre-ferment with a high content of lactic acid bacteria is sent by means of a pump (4) for pumping the nutrient mixture to the fermentation vats (5), which are equipped with stirrers and steam jackets. The process of increasing the biomass of the sourdough lasts 3.0 h (for classic options, without the use of stimulants, the production sourdough ferments for 5–6 h). The technological process of increasing the sourdough (production stage) is constantly monitored by indicators as acidity and lifting force per ball (acidity 9–12 degrees depending on the selected type of flour, lifting force per ball – not less than 25 min). Fermentation temperature of the sourdough when lactulose and energy sources (ATP) are added may not exceed 20–22 °C. All other fermentation processes (obtaining pre-ferment and dough) are recommended to be carried out without changes of the classical temperature of 28–30 °C. Increasing the fermentation temperature of the sourdough to 40–42 °C (according to previous studies) is impractical. This is due to the deterioration of technological indicators of sourdough quality – an increase in acidity and lifting time for the ball, which is not a reasonable technological solution. Next, the obtained sourdough is sent by means of the dosing pump (4) to the expendable vat (6), by means of which the sourdough is sent to the kneading machines (12) through a ladle-type dispenser (7); the mixing time lasts 10 min. In this technology, it is recommended to obtain a thick paste, because during its removal about 70% of the total flour is used. Warm water, lactulose solution, ATP and, if necessary, a solution of propolis (depending on the acidity of the sourdough) in the amount of 30–50% of the first application to the kneading machine are supplied through the dosing station 8 (it contains rotary pumps and heats the liquid components). From the shredder (9), in which the grinding of camelina seeds and potato roots takes place, the mixture is sent to the dosing station (8) by means of a scoop-type dispenser (10). The flour is fed to the kneading machine for the dough (12) and through the screw dispenser for bulk components (11). The obtained steam with the help of a vane pump dispenser (13), which is located at the bottom of the kneading machine for steam (12), enters the trough-shaped tank (14), which is divided into six partitions (six section units); in each section there are individual propeller stirrers to homogenize the medium, and the paste passes from the previous section to each subsequent. This tank has a water jacket, which is supplied with steam to maintain the temperature at 28–30 °C. Due to the introduction of additional substrates (growth stimulants for lactic acid bacteria – lactulose, mashed potatoes, camelina seeds and ATP), the paste ferments for an average of 1.5 h in contrast to the classical standards (3 to 4 h). After fermentation, the pre-ferment by means of a pump (4) from the trough-shaped container (14) is fed to the dosing station of liquid components (15) (water, salt, pre-ferment). The flour is metered into the dispenser for bulk components (16) and together with other liquid components of the dosing station (15) is fed to the kneading machines (17); the mixing process lasts 10 min. After kneading, the dough is sent to the fermentation hopper (19) through the conveyor pipe (18). The fermentation hopper has a water jacket (with its help we adjust the fermentation time and acidity) to maintain the fermentation temperature at 28–30 °C; the fermentation time of the dough lasts 1.5 h. At the end of the fermentation process, the dough is sent to the dough receiver (20), and then through the dough conveyor pipe (21) is sent to the dough divider-stacker (22), in which the dough is divided into separate pieces, rolled in a dough rounder and loader (23); the dough products are sent to the stand (24), in which the steam is fed in order to maintain the required humidity of the dough pieces and the stand temperature of 30–38 °C.

After fermentation in the moulds (1.2 h), the dough pieces will be sent for baking to a rotary oven (25) (according to the classical scheme, the aging time lasts 100–120 min). At the end of the baking process (baking temperature 200 °C, baking time 45–50 min) from the rotary oven (25), the baked loaves are knocked out of the moulds and conveyor (26) or product line and sent to the bread machine (27), which is loaded with trays with baked goods, then the bread is sent to the bread trolley (29). Pre-baked products are marked and formed along separate lines using a distribution carriage (28). To control compliance with the recipe, on the dispensers are placed solenoid valves for sampling at any time of the kneading machine. This technology at the stage of obtaining production sourdough creates improved conditions for autolysis. Autolysis (Greek autos – self + lysis – decomposition, decomposition) – self-digestion, decomposition of substances under the influence of hydrolytic/proteolytic enzymes without the participation of living organisms.

Taking into account the effect obtained from the use of growth stimulants of lactic acid bacteria, which made it possible to reduce the fermentation time at the stages of production and fermentation of the dough, as well as to reduce the temperature conditions of fermentation at the stage of production sourdough (20–22 °C), technical solutions were proposed (Siganova 2001; Pashchuk 2011). The production of bakery products is characterized by high costs of heat and electricity required for the implementation of technological processes and equipment operation. That is why one of the necessary prerequisites for the successful operation of the enterprise for the production of bread, including functional, is to increase the energy efficiency of the production process. As a rule, manufacturers use standard technological equipment and energy saving measures are considered only in terms of reducing energy losses to the environment (insulation of steam and pipelines, reuse of water vapour condensate from heaters, proofer cabinets, fermentation vats and other equipment, which is heated by steam). However, in our opinion, the efficient use and distribution of energy in the production process also requires technical upgrades to existing equipment to improve its energy efficiency.

The purpose of this work stage is to develop technical and technological measures to reduce energy consumption for the production of functional bread. To achieve this goal, the following tasks are formulated:

- To analyse the proposed technological scheme in terms of energy use and consumption
- To propose measures to modernize the most energy-intensive technological equipment
- To assess the possible technical and economic effect of the implementation of the proposed energy-efficient solutions.

In our proposed technological scheme (Fig. 7.4), we can distinguish the main equipment (brewing machine, fermentation vat (three units), shredder, kneading machines, dough divider-stacker, stand cabinet, rotary oven, bread machine) and auxiliary (dispensers, augers, pumps, conveyors, etc.), intended for dosing, transportation and temporary storage of semi-finished products.

At the first stage of research according to the technological scheme, the list of the equipment for production of functional bread is made and the quantity of units of each type of the equipment and its capacity and working time a day are specified (Table 7.2). The data in Table (7.2) are given based on the capacity of a typical bakery – 6 tons per day. The operating time of the equipment is determined according to the technological regulations of bread production, and electricity costs were calculated by the formula:

$$N_i = n_i \cdot N_{1i} \cdot \tau_i \tag{7.3}$$

n_i – number of equipment units of this (i-th) type, pcs
N_{1i} – electricity consumption per unit of equipment of the i-th type, kW
τ_i – operation duration of the equipment unit of the i-th type, hours per day

Calculations were made for one production cycle, the duration of which is 12 h. Two production cycles are realized per day.

The results of the calculations show that the highest energy consumption is characterized by transportation technological processes of raw materials and semi-finished products and fermentation of sourdough in the vat and kneader.

In order to reduce the energy consumption of process equipment, we have proposed the following technical solutions aimed at modernizing the equipment.

7.2.3 Belt Conveyor Modernization

Belt conveyors with drum drives are traditionally used in the production of bread. In order to reduce their power consumption, the drive is modernized. The technical effect is achieved due to the fact that the traction force of the conveyor belt is

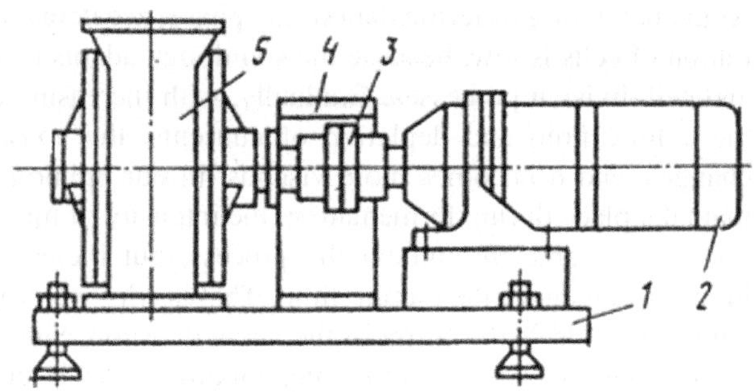

Fig. 7.5 Dough mixer I8-XTA-12/3: 1 plate; 2 drive; 3 coupling; 4 pump; 5 fencing

7.2.5 Modernization of the Kneading Machine

The most common brand of kneading machines used in bakeries is I8-XTA-12/3 (Fig. 7.5). In this machine, the dough enters the top of the loading hopper and fills the inside of the pump housing and is captured by the gate valve and injected into the outlet.

The kneading machine has a central shaft with blades, with which the process of kneading the dough is actually performed. In order to intensify this process and reduce its duration, it is proposed to place three additional shafts with blades parallel to the central shaft, at equal distances from its axis, with the possibility of planetary movement around the central axis. The blades on the shafts are fixed in a spiral line.

The installation of the additional shafts will allow turbulent flows to be created in the masses and thus intensify the process of mixing components, swelling of gluten and fringing.

The dough kneading machine works as follows: flour and liquid components are continuously dosed through a feed hopper into a trough-shaped container, where the dough mass is continuously mixed and moved along with the blades as the shaft rotates. The wire transfers the movement from the engine to the central shaft, with which the gears rotate. As the central shaft rotates, the additional shafts rotate both around the central shaft and around its own axis, i.e. they make planetary movement, which guarantees high-quality intensive dough mixing. The partition rotates together with the additional shafts.

After intensive kneading, the dough mass exits the outlet opening, creating enough pressure to move the mass through the pipeline for further processing.

The increase in productivity is due to the fact that the process of mixing components is reduced as the planetary movement of the working elements.

It is known from experience with dough mixing machines that the power required to knead dishes is a variable: it is minimal at the beginning of the process, and when mixing components it gradually increases, reaching its maximum value at the end of the process.

The power balance per one operating cycle of the machine looks like (formula 7.7):

$$N = N_1 + N_2 + N_3 + N_4 \tag{7.7}$$

N – the power required for kneading the pre-ferment, to kW

N_1 – drive power, which is spent on mixing the mass, kW
N_2 – drive power consumed to move the blades of the machine, kW
N_3 – drive power consumed to heat the steam and tangential metal parts of the machine, kW
N_4 – drive power spent on changing the structure of the pre-ferment

The axial component of the equivalent resistance forces is determined by formula 7.8:

$$\mathrm{P}_0 = F \cdot \left[r \cdot \rho_o \cdot tg^2\left(45^0 + \frac{\varphi}{2}\right) + 2 \cdot C \cdot tg\left(45^0 + \frac{\varphi}{2}\right)\right] \cdot (\sin\alpha - \mu \cdot \cos\alpha) \quad (7.8)$$

F – the area of the blade immersed in the pre-ferment, m^2, $F = 0.006$
r – the radius to the application point of equivalent resistance forces, m
φ – angle of internal friction of the pre-ferment, deg; $\varphi = 45^\circ$
C – specific adhesion of the pre-ferment with the material of the blade, PA; $C = 5000$
α – the inclination angle of the blade to the rotation axis, deg; $\alpha = 30°$
μ – the coefficient of friction of the pre-ferment with the blade, $\mu = 1$
ρ_o – pre-ferment density, kg/m^3; $\rho_o = 1080$ kg/m^3

The radius to the application point of resistance equal forces is determined by formula 7.9:

$$r = R_1 + \frac{2}{3} \cdot b \quad (7.9)$$

R_1 – is the distance from the axis of the shaft to the blade, m; $R_1 = 0.1$ m
b – height of the blade, m; $b = 0.63$ m

$$r = 0.1 + \frac{2}{3} \cdot 0.63 = 0.52\,м;$$

$$P_0 = 0.006 \cdot \left[0.52 \cdot 1080 \cdot tg^2\left(45 + \frac{45}{2}\right) + 2 \cdot 5000 \cdot tg\left(45 + \frac{45}{2}\right)\right] \cdot (\sin 30 - 1 \cdot \cos 30) = -66.2\ \mathrm{H}, \quad (7.10)$$

The radial component of the equivalent resistance forces according to formula (7.11) is:

$$P_p = F \cdot \left[r \cdot \rho_o \cdot tg^2\left(45^0 + \frac{\varphi}{2}\right) + 2 \cdot C \cdot tg\left(45^0 + \frac{\varphi}{2}\right)\right] \cdot (\cos \propto + \mu \cdot \sin\alpha);$$

$$P_p = 231.7\,H$$

The power required for mixing the pre-ferment (7.11) is:

$$N_1 = \frac{\sum[(P_0 \cdot V_0 + P_p \cdot V_p) \cdot K]}{1000 \cdot \eta}, \tag{7.11}$$

V_0 – axial speed of the refractive point movement is equal to the resistance forces acting on the blade, m/s:

$$V_0 = V_\text{p} \cdot \cos\alpha \cdot \sin\alpha,$$

V_p – circle speed of the refractive point of the equivalent resistance forces acting on the blade, m/s (formula 12):

$$V_\text{p} = \frac{r^2 \cdot n}{30}, \tag{7.12}$$

n – the number of kneading blade turns, rpm; $n = 56.3$ rpm
K – number of blades on one kneading machine shaft
η – drive efficiency factor; $\eta = 0.85$

$$V_\text{p} = \frac{0.52^2 \cdot 56.3}{30} = 0.5\frac{\text{m}}{\text{sec}};$$

$$V_0 = 0.5 \cdot \cos 30 \cdot \sin 30 = 0.2\frac{\text{m}}{\text{sec}};$$

$$N_1 = \frac{\sum[(-66.2 \cdot 0.2 + 231.7 \cdot 0.5) \cdot 11]}{1000 \cdot 0.85} = 1.3\ \text{KW}$$

Work, spent on the rotation of the kneading blades, is:

$$A_2 = \frac{2}{3} \cdot K \cdot b \cdot \sigma \cdot \rho_o \cdot \pi^2 \cdot n^2 \cdot (R_2^3 - R_1^3),$$

where

σ – blade thickness, m; $\sigma = 0.01$ m
R_2 – blade rotation radius, m; $R_2 = 0.16$ m

$$A_2 = \frac{2}{3} \cdot 11 \cdot 0.063 \cdot 0.01 \cdot 1080 \cdot 3.14^2 \cdot 56.3^2 \cdot \left(0.16^3 - 0.1^3\right) = 482.8 \text{ J/r}$$

$$A_2 = \frac{2}{3} \cdot 11 \cdot 0.063 \cdot 0.01 \cdot 1080 \cdot 3.14^2 \cdot 56.3^2 \cdot \left(0.16^3 - 0.1^3\right) = 482.8 \text{ J/r}$$

The work is spent on heating pre-fermentation and metal structures of the kneading machine in one blade revolution (formula (7.13):

$$A_3 = \frac{(t_1 - t_2) \cdot \left(m_p \cdot a_p + m_m \cdot a_s\right)}{n \cdot \tau}, \tag{7.13}$$

t_1 – mass temperature at the end of the batch, °C; $t_1 = 0$ °C
t_2 – mass temperature at the beginning of the kneading, °C; $t_2 = 25$ °C
m_p– pre-fermentation mass in kneading machine, kg
a_p–average heat capacity of pre-fermentation at 30 °C, J/(kg – K); $a_p = 2500$ J/(kg – K)
m_m= mass of metal structures' kneading machine, heated during dough mixing, kg; $m_m = 100$ kg
a_s – average heat capacity of stainless steel, J/(kg – K); $a_s = 500$ J/(kg – K)
τ – kneading duration, s; $\tau = 180$ min
R_2 – blade rotation radius, m; $R_2 = 0.16$ m

The total volume of the kneading tank (trough) is 0.27 m^3. With a trough filling factor of 0.8, its useful volume is 0.216 m^3. At a density of 1080 kg/m^3, the trough will weigh 235 kg.

$$A_3 = \frac{(30 - 20)(235 \cdot 2500 + 100 \cdot 500)}{56.3 \cdot 180} = 629.07 \text{ J/r}$$

The energy consumption per kneading unit cycle per 1 kg of pre-fermentation is:

$$A_{n(2.3)} = \frac{\sum A_{2.3}}{m_p-} = \frac{482.8 + 629.07}{235} = 4.73 \text{ J/kg}$$

Drive power to rotate the machine blades and heat up the pre-fermentation is:

$$N_{2.3} = \frac{\sum A_{2.3}}{1000 \cdot \eta} = \frac{482.8 + 629.07}{100 \cdot 0.85} = 1.31 \text{ kW}$$

Since the structural changes in the mass of the pre-fermentation depend on the mixing intensity and are proportional to the mixing operation, we take the power required to change the structure at 10% of the power required to stir the pre-fermentation:

$$N_4 = 0.1 \cdot N_1 = 0.1 \cdot 1.3 = 0.13 \text{ kW}$$

$$N = 1.3 + 1.31 + 0.13 = 2.74 \text{ kW}$$

Performance of the kneading machine is:

$$P = \frac{\varphi \cdot \rho_o \cdot \pi \cdot D^2}{4 \cdot 60 \cdot S \cdot n},$$

where φ – feed factor, which depends on the design of the blades and their placement on the shaft; $\varphi = 0.5$ m

D-blade diameter, m; $D = 0.32$ m
S-blade pitch, m; $S = 0.055$ m

$$P = \frac{0.5 \cdot 1080 \cdot 3.14 \cdot 0.32^2}{4 \cdot 0.155 \cdot 56.3} = 5 \frac{\text{kg}}{\text{min}}$$

Thus, the installation of additional shafts in the kneading machine makes it possible to increase its capacity from the nominal value of 2–5 kg/min with the total drive capacity of the kneading machine not exceeding the capacity of the electric motor installed on the machine (Table 7.3). Thus, the time required for preparation of pre-fermentation is reduced by 2 times and the electric power consumption of the kneading machine is reduced (from 4.5 to 2.25 kW).

7.2.6 *Replacement of the Stove for Baking Bread*

Cyclothermic hearth stoves, in which the electric energy is converted into heat, as well as belt and conveyor gas ovens, are traditionally used for baking bread at

Table 7.3 List of energy efficiency measures in the production of functional bread

Name of equipment	The content of the energy efficiency measure	Electricity consumption per production cycle	
		Basic scheme	Improved scheme
Loaders	Changing the design of the conveyor drum drive	10.8	5.4
Vat for fermentation	Installation of the frequency converter for the mixer drive	8.1	5
Kneading machine	Installation of two additional shafts for mixing the pre-ferment	4.5	2.25
Stove	Replacement of a cyclothermic hearth furnace with a rotary one running on wood fuel	12.9	2.3

Fig. 7.6 Rotary stove for baking bread LIDER140

medium- and low-capacity enterprises. Given the instability of the natural gas and electricity market in recent years and significant fluctuations in their value for industrial enterprises, which affects the cost of the finished products, manufacturers are looking for equipment capable of working on alternative fuel sources – solid (coal, wood), liquid (diesel fuel, fuel oil), etc. The most common practice today is the technical modernization of existing stoves in order to convert them to other fuels. In our opinion, in this type of equipment, which carries out the key process of all bread production, it is necessary to provide the ability to work on different fuels, so that, depending on the current energy market in the country, promptly, taking into account the economic component, them without stopping production and additional modernization of the furnace. To this end, as part of this work, we proposed to replace the electric hearth oven for baking bread with a rotary stove LIDER140 from the manufacturer "Kumkaya" (Fig. 7.6). The operation principle of the rotary stove is as follows: blank from the dough; the cart clings to the hook of the movable mechanism and begins to rotate; the workpiece is evenly baked, rotating around its axis until ready; the baked product is rolled out on a rack for cooling. Burning during cooking is excluded. The result is a bakery product that has a crispy crust on the outside and a delicious crumb inside. In addition, the stove is equipped with a steam generator that injects water on the dough blanks in a timely manner to prevent them from drying out and burning by maintaining the required level of humidity in the chamber.

Another undeniable advantage of this type of stove is the ability to use as an energy source of solid wood fuel (pallets), diesel fuel, natural gas and electricity. This allows the furnace to operate in the most economical way, based on the cost of a particular type of fuel (energy source). In particular, taking into account the next increase in tariffs for electricity and gas supply for industrial enterprises, which took

Table 7.4 Evaluation of the technical and economic impact of the proposed technical solutions

	Electricity consumption per production cycle, KW	
Indicator	Basic scheme	Improved scheme
Electricity consumption per operating cycle, kW	76.24	54.89
Energy savings per production cycle: kW %	– –	21.35 28.0

place in October 2020, it is proposed to organize the operation of a rotary kiln on solid wood fuel (pallets). This will eliminate the cost of natural gas and electricity for the operation of the furnace.

Thus, summarizing the above technical measures aimed at improving the energy efficiency of technological equipment for the production of functional bread, we can present them as follows (Table 7.3).

Thus, based on the technological scheme analysis of functional bread production in terms of energy consumption, it is established that the highest energy consumption is characterized by equipment for transportation of raw materials and semi-finished products (conveyors), vat for sourdough fermentation and dough mixer.

In addition, in the classic technological schemes, conveyor (gas) and hearth (electric) stove for baking bread are installed. The floor electric furnace consumes much more electric energy than our proposed type of rotary furnace, which can operate in combined modes – on gas, pallets, diesel and electricity. In order to increase the energy efficiency of the equipment and reduce the cost of electricity for the production cycle, a number of technological solutions aimed at equipment modernization or its replacement have been proposed (Table 7.4).

In particular, it is proposed to change the design of the conveyor drum drives, to install a frequency converter to drive the fermentation mixer and to replace the gas furnace with a rotary one using solid wood fuel. The implementation of the proposed solutions will reduce by 28% of the energy consumption of the technological unit for functional bread production and make a rapid transition between different energy sources during the operation of the rotary kiln depending on the cost of energy in the Ukrainian market.

7.3 Conclusions

Among all countries in the world, Japan has the largest number of functional food items registered, taking into account different age and individual characteristics of a person. The development of such food groups in the twenty-first century is closely linked to the use of various ingredients (food additives, spices, BAR, complex natural enhancers, nutraceuticals and microorganisms). The consumption of functional foods in this country is 2–4% higher than the production of other food products.

Providing the population of Ukraine with health products is extremely of important social and political significance, creating the necessary conditions for food security and preservation of the nation's gene pool. The food industry can increase the production of traditional foods enriched with essential micronutrients. Especially available are bread; bakery; flour confectionery with supplements of vitamins B, A and E, calcium, iron, iodine and selenium; and milk and dairy products with multivitamin complexes and lactic acid bacteria; it is necessary to increase the range of low-calorie fat-containing products with functional ingredients and soft drinks with extracts of medicinal plants.

The proposed functional bread is enriched with fermentation products of lactic acid bacteria, which are endowed with a powerful synthesis of antioxidants: enzymes – catalase and oxidoreductase, amino acids (methionine and cysteine) and vitamins (niacin, PP, C, K). The most pronounced feature is endowed with bifidobacteria. Probiotic strains have proven antioxidant and antimutagenic activity, which allows them to be considered as the most promising tool in the fight against free radicals (oxidants, oxidants) to prevent the processes of free radical oxidation and oxidized compound reduction.

The technological effect (increase of acidity and reduction of lifting time behind the ball; fermentation activity; porosity and shape stability of finished products, improvement of organoleptic characteristics of bread; increase in the number of lactic acid bacteria) from the use of experimentally established concentrations of process stimulants – ATP, lactulose and camelina seeds.

The technological scheme of obtaining functional bread with high content of lactic acid bacteria and vitamins and amino acids has been improved. The optimal technological conditions for the biotechnology of sourdough cultivation (production cycle) and obtaining high-quality unleavened bread without the use of baking sourdough with improved organoleptic characteristics have been established.

The implementation of improved technical solutions offered by us will allow to reduce biotechnology energy consumption of functional bread with useful nutrient addition, an additional source of energy in the form of ATP and camelina by 28%. It is recommended to organize this technology of functional bread preparation in medical and preventive establishments, schools, higher educational establishments, sanatoria, hospitals and dispensaries.

References

Auerman L (2005) Tekhnologiia khlebopekrskogo proizvodstva. Professiia, Sankt-Petersburg

Boichenko S, Leida K, Yakovleva A, Vovk O, Kuzhevskii K (2017) Influence of Rapeseed Oil Ester Additives on Fuel Quality Index for Air Jet Engines. Chemistry and Technology of Fuels and Oils Chemistry and Technology of Fuels and Oils. View Correspondence (jump link) 53(3): 308–317

Drobot V (2006) Laboratornyi praktykum z tekhnolohii khlibopekarskoho ta makarannoho vyrobnytstva. Tsentr navchalnoi literatury, Kyiv

Neposhyvaylenko N, Kornienko I (2020) Current problems of individual health of adolescents and the use of modern food biotechnology to solve them. In: Actual problems of natural sciences: modern scientific discussions, collective мonograph. University of life sciences in Lublin, Lublin, 391–409

Pashchuk Z (2011) Tekhnologiia proizvodstva khlebobulochnykh izdelii. Giord, Sankt-Petersburg

Peshchuk L, Novesnko T (2008) Biokhimiia ta tekhnolohiia oliie-zhyrovoi syrovyny. NUKhT, Kyiv

Seal K, Baranowski M (2000) Communication to the Editor Economic use of pesticides in the Ukraine. Pest Manag Sci:475–476

Siganova T (2001) Tekhnologiia khlebopekarnogo proizvodstva. Pishchevaia promyshlennost, Moscow

Sogomonian D, Akopian K, Prchunian A (2011) Izmenenie pH i okislitelno-vosstanovitelnogo potentsiala sredy v protsesse rosta molochnokislykh bakterii vliianie okislitelei i vosstanovitelei. Prikladnaia biokhimiia i mikrobiologiia1(4): 33–39

Chapter 8
Technologies for Restoring Soil from Oil Pollution

Nadiia Adamchuk-Chala, Sergii Ponomarenko, Liubov Yankiv-Vitkovska, and Yelyzaveta Chala

Nomenclature

PA	Precision agriculture
TM	Technical material
EC	Electroconductive
DEM	Digital elevation models
NDVI	Normalized difference vegetation index
GSP	Global System Position

8.1 Introduction

Nowadays, in the practical absence of technologies for the utilization of oil waste, there is systemic soil contamination, which causes the inhibition of biological processes in soil ecosystem by toxic elements and vapors of light hydrocarbons; violation of the natural ratio of carbon and nitrogen in the soil due to oil

N. Adamchuk-Chala (✉)
Institute of Agroecology and Environmental Management, National Aviation University, Kyiv, Ukraine

S. Ponomarenko
Igor Sikorsky Kyiv Polytechnic Institute, National Technical University of Ukraine, Kyiv, Ukraine

L. Yankiv-Vitkovska
Department of Higher Geodesy and Astronomy, Lviv Polytechnic National University, Lviv, Ukraine

Y. Chala
Kyiv National Linguistic University, Kyiv, Ukraine

S. Boichenko et al. (eds.), *Chemmotological Aspects of Sustainable Development of Transport*, Sustainable Aviation, https://doi.org/10.1007/978-3-031-06577-4_8

hydrocarbons; stable hydrophobization of the soil surface with heavy oil hydrocarbons (resins, asphaltenes); reduction of soil moisture, which blocks the availability of nutrient medium; soil salinization with monovalent cations of concomitant salts; etc. (Pikovsky et al. 2003).

From oil-contaminated soils, oil-oxidizing bacteria identified as *Rhodococcus erythropolis* EK-1 and *Acinetobacter calcoaceticus* K-4 (registration numbers in the Depository of the Institute of Microbiology and Virology of the National Academy of Sciences of Ukraine IIB Ac-5017 *al* and 24, respectively) were *isolated.* K-8 and their ability to synthesize surfactants on hydrophilic (ethanol) and hydrophobic (hexadecane) substrates, is as the industrial waste (petroleum, oil and fat, biodiesel production) (Terek and Velichko 2009; Pyrog 2012).

Of the plants, only some species are able to survive in oil-contaminated soil. Methods of monitoring so areas are not sufficiently developed (Terek and Velichko 2009).

Underdeveloped methods of monitoring such areas can be improved by using soybean and alfalfa plants, since legume-rhizobium symbiosis with nitrogen fixation has the ability to degrade oil hydrocarbons (Terek and Velichko 2009). Also, alfalfa hop plants (*Medicago lupulina* L.) are resistant to drought and salinity due to a well-developed taproot system (Velychko and Terek 2008; Jura et al. 2006).

The research was conducted on model sites. Crude oil with a density of 0.96 g/ml was added to the experimental soil of 50 ml of oil per 1 kg of soil, which corresponds to 48 g/kg, and 100 ml of oil per 1 kg of soil, which corresponds to 96 g/kg. Soil without oil was taken as a control. The formation of the plant-microbial system was carried out from planting plants for 3 weeks after oil was brought in the soil (the required time for weathering of light toxic petroleum products).

The survival of plants in oil-contaminated soil is possible due to adaptive restructuring at all levels of the organization. In particular, adaptive reactions aimed at optimizing water absorption, which consists in phenotypic changes in the structure of the root system and an increase in the suction power of root cells, have been identified. A vivid example of modification of the structure of root systems in the conditions of oil-contaminated soil has been established for hop alfalfa plants (Terek and Velichko 2009).

The formation of long thin lateral roots provided an increase in the area of the absorbing surface and also promoted the movement between the particles of densely bituminous oil-contaminated soil. Such changes in the morphological structure of the roots, along with such adaptive rearrangements as crusting and sovereignty of the exoderm, accelerated cell differentiation or earlier formation of Casparian bands in the cells of the root epidermis and so on.

On the background of various negative consequences of multicomponent oil pollution, the occurrence of drought conditions is especially dangerous given the independence of water potential and cell turgor; the intensity of physiological processes and biochemical reactions (photosynthesis, respiration, etc.), which underlie the formation of cell structures; and the formation of cells themselves and biomass in general. The lack of nutrients available in the soil is caused by moisture deficiency. Adaptation to extreme conditions of water supply, in addition to adaptive conditions at the level of changes in morphological structure (Terek and Velichko 2009; Crowley et al.1996), happens due to the growth of suction power of root cells.

Regulatory responses of water supply plants contribute to their adaptation in conditions of soil contamination by oil. There are changes in the functioning of the respiratory apparatus of the leaves, aimed at reducing the intensity of transpiration, namely – changes in the number, structure, and aperture of the stomata of cells of sedge plants. Studies revealed the accumulation of rough sedge and hop alfalfa in plant organs under the action of oil pollution of low molecular weight antioxidants – ascorbate and carotene. The increase in carotene content can be considered as an adaptive response of the plant in response to stress. Under the action of high concentrations of oil in the soil (96 g/kg) in the leaves of experimental plants, carotene content decreases (Terek and Velichko 2009), which according to the authors indicates a weakening of adaptive capacity and reduced tolerance of plants to high concentrations of oil in soil.

These plants can be used in technologies for cleaning and restoration of oil-contaminated soil, as physical and chemical properties of oil-contaminated soil have already been improved after growing of plants and growth in the soil of basic nutrients – phosphorus, potassium, magnesium, and iron – the main ecological and physiological groups of microorganisms. The positive influence of plants on the sorption properties of soil and optimization of soil conditions by pH level is revealed. It is shown that growing plants promotes oil degradation, it leads to a decrease in cellulose-destroying bacteria, and at the same time, the cultivation of plants ensures the restoration of their numbers. The increase in the number of nitrogen-fixing bacteria in the contaminated soil after growing the plants, as well as aerobic microorganisms, some of which are destructors of oil hydrocarbons, has been proved. Also in the rhizosphere of plants due to root secretions, the solubility of nutrients present in oil-contaminated soil increases, which leads to the transition of hard-to-reach forms of mineral elements in more accessible forms and promotes root nutrition and viability of soil microbiota (Korovetska et al. 2008; Moroz et al. 2006).

Plants for their growth and development require a certain physical and mechanical condition of the soil and the presence of many different nutrients. Analysis of the state of the soil contaminated with petroleum products in particular allows you to control the potential of plants for growth and development in specific soil conditions (Adamchuk et al. 2015).

To control the dynamics of main soil remediation processes – physical, chemical, biological, and others – diagnostics and analysis of soil condition for management of the expanded reproduction of their fertility are carried out (Aniskevich et al. 2019). For a reliable assessment of soils, and, especially, the forecast of their fertility, information about the climate, soil processes, groundwater status, quantity and quality of crop products, etc. is needed. The main information for assessing the condition of soils is the change in soil structure and the rate of change of basic properties: humus, pH, physical, water, air and nutrient regimes, biological activity of soils, their pollution, assessment of the intensity of erosion, irrigation water quality, groundwater level and mineralization, soil salinity in general and aeration zones; secondary salinization, and finally, assessment of effective soil fertility.

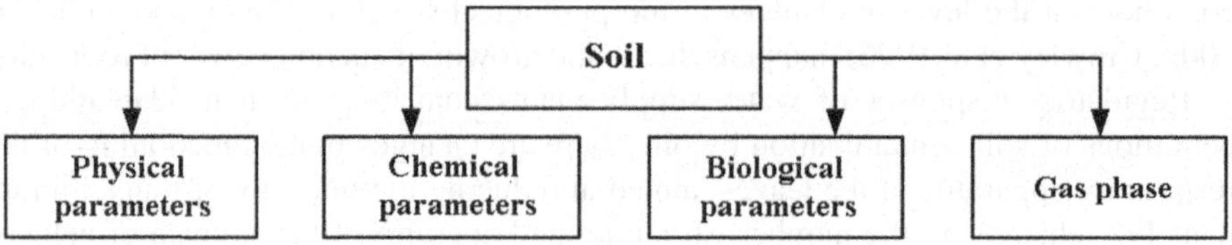

Fig. 8.1 Parameters that characterize the soil

The soil consists of three phases – solid, liquid, and gaseous – as well as biological part: soil biota (soil flora, fauna, bacteria, fungi, etc.). The solid phase is the basis of the soil, which consists of mineral and organic parts. The mineral content of most soils is 80–90% of the solid phase, which was part of the parent rock. The largest mineral particles in the soil, or so-called mechanical elements, are divided into fractions (Aniskevich et al. 2019; Adamchuk et al. 2017).

Soil, as the main means of crop production, is characterized by a number of parameters (Fig. 8.1), which can be classified according to their physical and mechanical, agrochemical, biological, etc. indicators.

For example, the physical parameters of the soil are understood as its hardness, solid phase density, built-up density (bulk density), porosity (porosity), plasticity, stickiness, moisture, etc. The relief of the surface and the contours of the field are important for the performance of mechanized works in crop production.

Chemical parameters mean the presence of nutrients in the soil (primarily nitrogen, phosphorus, and potassium), heavy metals, and other substances. For example, soil acidity determines the availability of nutrients for plants, creating the conditions of growth and development of plants.

8.2 Areas of Local Management

Soil properties vary in size fields, resulting in registration of variability during certain manufacturing operations which is one of the central tasks of the precision agriculture (PA). To identify areas of local management, it is necessary to collect data on soil conditions, analyze them, and make recommendations on variable rates of technical material (TM) (Adamchuk et al. 2017; Gebbers and Adamchuk 2010).

In essence, zone management is the division into subsections of the field with the same approach to the economic efficiency of growing products. This approach allows us to identify the following factors that determine the formation of management zones:

Sampling Density the more samples are taken from locally defined data sets, the less problems there will be with the interpolation operation and the more accurate the mapping of their distribution over the field area. The method of sampling and the density of sampling determine the accuracy of interpolation. There is a necessity to consider the comprehensive selection of elementary sections at large fields, for example, using modern methods of remote sensing, determination of electrically conductive properties of the soil, building digital elevation model (DEM) fields, etc.

Cost of Information First of all you should try to use local, available for free or at low cost, data. They include maps of the survey and electrically conductive properties of the soil, photos and some close monitoring of remote images which are available online.

Locally Defined Data of High Stability and Reliability The use of data that do not change significantly over time, including topography, electrical conductivity, soil color, and other physical properties of the soil. The advantage of this type of information is that it is stable over time and only needs to be measured once.

Scale of Locally Defined Data Spatial data must be collected at a scale at least close to the scale of the control area. For example, when planning a strategy for applying a high-quality distributor with a width of 16 m, the scale of management zones should preferably be no more than an elementary section of 16 × 16 m.

There are three methods for assessing the correctness of the definition of management areas:

- *Evolutionary:* to conduct a historical analysis of the rate of growth of profitability or income level by constructing appropriate maps for consecutive years of production.
- *Indirect method:* a detailed study of existing locally defined data on the state of the field to assess the most influential factors. To do this, we use multifactor regression analysis.
- *Direct method:* involves a direct comparison of the proposed and basic methods of local management in one field. The field yield map will show the advantages and disadvantages of each method. The difference in comparison of the cartograms can be easily recalculated by using simple spatial analysis tools within the appropriate software.

All three assessment methods have their advantages and disadvantages. The evolutionary method is simple and essentially free. The application of the indirect method may have limited value with low reliability of the initial data for regression analysis.

Direct comparison of economic efficiency of two management approaches in one field is spatially reliable and does not require any specialized equipment. The development of strategies for building management zones must be continuous.

8.3 Acoustic Sensors

At present, sound and ultrasonic frequency waves are actively used by many companies for the production of equipment for vehicle technology (Aniskevich et al. 2019).

Interference sensors based on thin-film magnetic sensors are mainly used to measure the level of ultrasonic and sound signals. Such systems can also be used to measure temperature, level of environmental pollution, pressure, humidity, chemical concentration, etc.

The sensor has two microphones: main and compensation. The main one registers "useful" noises when moving the soil on the working surface of the sensor, and the compensating one – noises that are characteristic of the day surface of the field (mainly – noises of the engine and the support system). The signals from both microphones go to a special device, where the signal from the compensation microphone is removed from the signal of the main microphone and the result is a "pure useful" signal, which is subject to spectral analysis. The speed of the sensor must be constant and correspond to the speed at which the exemplary base of reference noise is created for different types of soils of a given humidity.

Acoustic sensors provide high performance and are extremely effective for detecting soil texture and its density (density) and measuring the depth of the subsoil.

8.4 Electrochemical Sensors

Electrochemical methods are widely used to analyze the state of substances and TM (Aniskevich et al. 2019; Adamchuk et al. 2017). They are based on the use of processes occurring in the investigated solution of the medium and on the surface of the electrodes (phase boundary), which are in contact with the solutions. One of the parameters serves as a measuring signal that is functionally related to the composition and concentration of the substance: potential at the electrodes, current strength, electrical resistance, specific conductivity, mass, amount of electricity, etc.

Depending on the type of membrane, ion-selective electrodes are divided into the following types:

- Solid phase and electrodes. The membrane of such electrodes is made of mono- or polycrystals of sparingly water-soluble salts. Solid membrane electrodes are used as sensors to detect those ions that are part of the membrane, as well as those ions that are able to interact with the active components of the membrane material. The advantage of solid-phase electrodes is a long service life. However, the number of solid ionic crystalline compounds having ionic selective conductivity is limited.
- Electrodes with a liquid membrane. These electrodes have a diaphragm of soluble ion exchangers, such as liquid cation exchangers or anion exchangers. By means of porous partitions sample solutions are separated from the analyzed solutions; thus, there is a potential difference.
- Glass electrodes are for structures that are intermediate between liquid and solid membranes. Glass electrodes were the first ion-selective devices. Recently, there has been an active search for ways to obtain new grades of glass to create a new quality membrane. A large number of glass samples of different composition

have been developed, which are characterized by selectivity for potassium, sodium, phosphate ions, etc.

For glass pH electrodes, characteristics such as measuring range, temperature range, and electrical resistance are closely related and determined by the brand of electrode glass. Electrical resistance pH-electrodes at 20 °C in the vast majority are in the range of 5–1000 mOm.

The choice of electrodes to perform a specific task must be made taking into account the following factors:

Parameters of the studied environment:

- Range of ion concentrations, the analysis of which is performed
- Temperature and measuring range
- Presence of ions interfering with the measurement
- Presence of aggressive substances to the electrode material
- Sample volume restrictions
- Physical condition of the studied material

Methods of analysis:

- Method (direct potentiometry, potentiometric titration, or additive method)
- Measurement mode (continuous or periodic)
- Measurement accuracy

Restrictions on measuring instruments and other equipment used in the experiment:

- Overall dimensions of electrodes
- Electrode configuration and membrane sensitivity
- Length of connecting cable and type of socket

The operating range is taken to be such an area of the electrode function in which the deviation from linearity does not exceed some predetermined value, such as pH 0.2. Ideally, the measurement range of the electrode should cover the range of possible concentrations of the test solutions. If it is necessary to analyze more concentrated solutions than the electrode allows, the sample can be diluted.

The next most important parameter is the temperature range of the electrode. For pH electrodes, the measuring range is in the range from 0 °C to 150 °C. For ion-selective electrodes modifications with different temperature ranges are almost nonexistent. If it is not possible to select electrodes with the desired temperature range, the problem can be solved by heating or cooling the test solution. It is possible to arrange even in case of continuous measurements.

The electrode function also depends on the temperature. As the temperature increases, the slope of the electrode characteristic increases.

The concentration of the studied ions at which the electrode potential does not depend on temperature is called the isopotential point. Modern measuring devices allow you to automatically take into account the temperature changes of the electrode characteristics (thermal compensation); for this you need to enter the

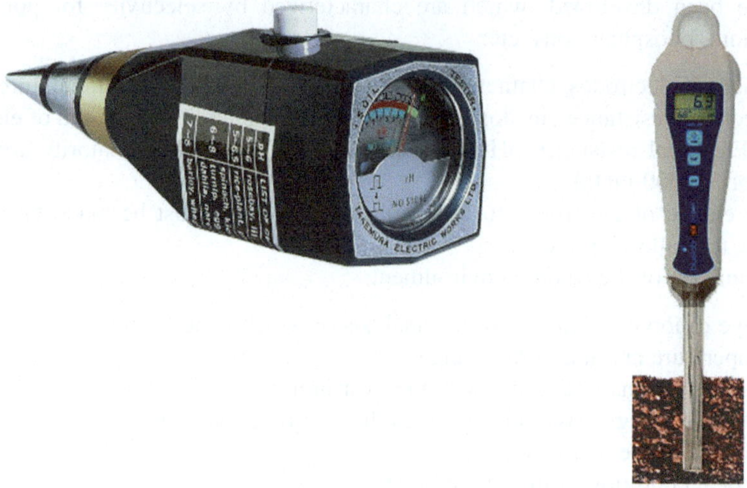

Fig 8.2 pH meter for field conditions by STELZNER Agrar-Fachberatungs-GmbH, Germany (left), and Bluelab Corporation, New Zealand

coordinates of the isopotential point and the current temperature. The latter can be entered either manually or using a temperature sensor connected to the device.

For measurements, it is desirable to choose an electrode in which the isopotential point lies near the average concentration of the investigated solutions. This will reduce the error of value associated with the measurement of temperature solutions.

It is necessary to carefully study the presence of interfering ions and substances that are aggressive to the electrode materials in the test solution. If such a situation occurs, the removal of the latter from the solution may require the use of additional methods. Under these conditions, conductivity without continuous measurements will most likely be impossible.

The electrodes are characterized by a certain photosensitivity, so their operation and storage under direct sunlight are not recommended. The presence of strong oxidants or reducing agents in the test solutions is also undesirable. In the field, cylindrical or cone-shaped devices are used to measure soil acidity. These are, first of all, electrodes that are able to adsorb hydrogen and aluminum ions. Determining the amount of adsorbed ions over a period of time assesses the acidity of the soil. In Fig. 8.2 the following devices are shown.

The devices record the pH and temperature of soils and environments, which allows you to quickly monitor and make recommendations for adjusting the pH level to optimize plant health and productivity.

The devices have the following characteristics:

- Allow direct measurements of pH and temperature in soils
- Have a display with backlight support
- Battery charge indicator

- Waterproof
- Have automatic temperature compensation

8.5 Sensors of Electrically Conductive Properties of Soil

Research in the world is intensively and widely conducted in the direction of finding opportunities to obtain reliable information about the state of soil pollution. One of these areas is the use of indirect information about the state of the soil with a reliable algorithm for converting such information into objectively necessary data – such as the level of nutrients, humus, moisture, trace elements, etc. in the soil. Effective indirect data on soil condition can be indicators of soil electrical conductivity, dielectric constant, magnetic properties, etc.

The need to determine a wide range of soil parameters necessitates the use of a wide range of highly specialized devices and equipment. Classical methods have disadvantages, such as cost, or require postoperative processing, or the duration of the analysis significantly exceeds the appropriate time limits. This requires the search for complex high-tech methods for determining soil properties. It is necessary to apply certain integrated indicators of the state of the soil environment.

Determination of electroconductive (EC) and electromagnetic properties of the soil is one of the areas in improving the monitoring of soil condition. The relative simplicity and rapidity of determining soil EC-properties combined with a high level of accuracy allows the creation of a base of indicators of the dependence of conductivity with agrobiological soil parameters in the field (Ehlert et al. 2008; Gebbers and Adamchuk 2010).

At different soil depths, electrophysical parameters can change very quickly. The intensity of electrochemical processes depends not only on the concentration of chemical ions in groundwater but also on the humidity of the analyzed medium. When the chemical composition of the analyzed medium fluctuates, the measurement error increases significantly: the change in the concentration of ions observed in practice in the soil solution leads to the fact that in some cases the intensity of electrochemical processes is determined not only by fluctuations in humidity but also by salinity, which, in turn, increases the error to 100%. In connection with the latter, the increase in the accuracy of soil monitoring is achieved by combining field measurements with the analysis of selected samples in special laboratories.

8.6 Soil Doctor Conductivity Measurement System

Soil Doctor (USA) is another commercial system for recording EC soil properties, which works on approximately the same principle as Veris. Representatives of this company for the first time introduced the technology of registration of locally

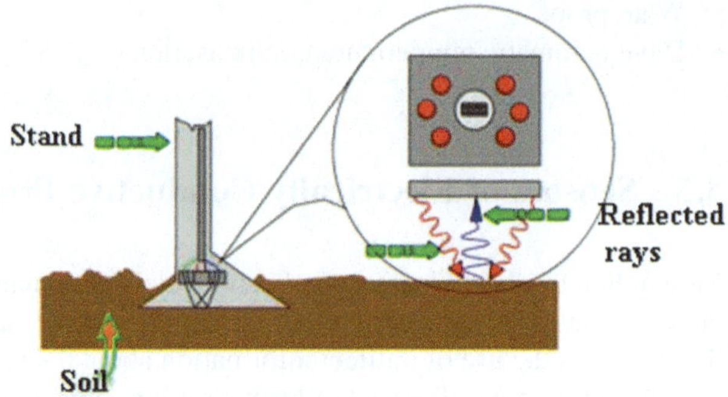

Fig. 8.3 Scheme of an optical sensor for soil type identification

defined soil parameters in real time (On-The-Go). The equipment allows to carry out sounding of soil on the move (Fig. 8.3).

Since the conductivity measurement is almost instantaneous, the data from this device can be used to control the seeding rate and the depth of tillage in the mode of sensor technology.

8.7 Optical Sensors

Optical sensors are used in the study of the optical properties of the object (the ability to transmit electromagnetic waves, spectral analysis, polarization of objects, etc.) or in determining the state of the studied objects by appearance (Adamchuk et al. 2017). Sensors are divided into sensors of passive and active type. Passive type sensors can only function in the presence of external lighting. Unlike passive light-sensitive sensors, the operation of active sensors does not depend on lighting due to specialized technologies of the light source and receiver. Such sensors can operate both day and night.

Optical sensors are used to identify the type of soil with its color. The black color is associated with fertility of soil – a measure of the content of organic carbon. Depending on the order, as the content of organic carbon decreases, the color of the soil is lighter and reflects the mineral composition. Thus, the red color of soils, for example, is an indicator of high iron concentrations.

In Fig. 8.4 the scheme of functioning of the optical sensor for identification of type of soil is presented. The sensor is installed in the subpaw space of the cultivator paw at a distance of 5–10 cm from the soil surface formed by the blades of the paw. The sensor itself has a light source that irradiates the soil surface and a light-sensitive matrix that receives light reflected from the ground. The signal from the sensor is sent to the spectrometer, which determines the type of soil.

With optical sensors measure the vegetative index (NDVI – normalized difference vegetation index) – index relative normalized vegetation (Daughtry et al. 2000).

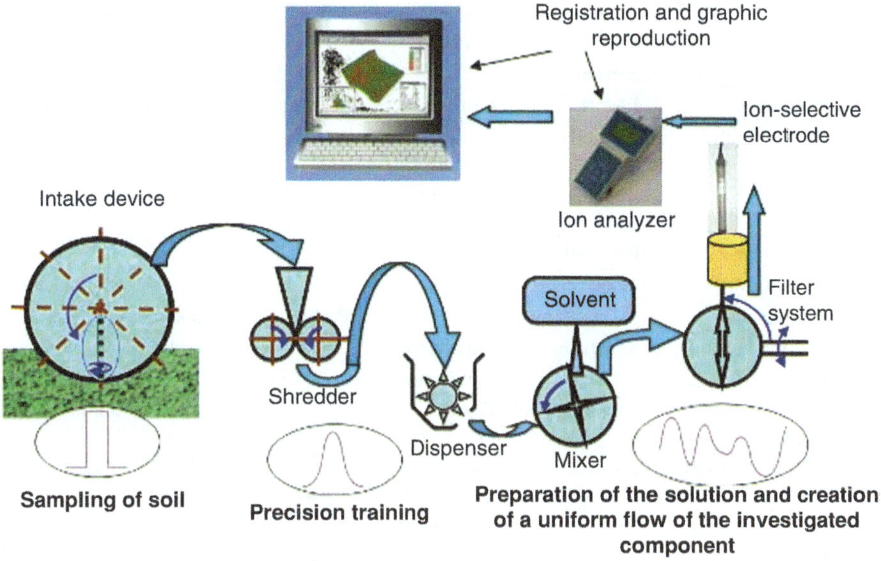

Fig. 8.4 The scheme of operation of the technological module for measuring soil pH

This is an indicator of the amount of photosynthetically active biomass and one of the most common indices, which uses quantitative estimates of vegetation to solve problems of variable rates of application of technological materials. Infrared illuminates the surface of plants. Chlorophyll in the leaf-stem mass of plants actively reflects radiation invisible to human eyes. To determine the activity of plant growth by the vegetative index using, for example, the system "GreenSeeker's" main function is to identify areas with a low NDVI index to have information about the areas of the field where there is contamination. The GreenSeeker system has its own light source and therefore can be used around the clock.

8.8 Systems of Sensors of Registration of a Condition of a Ground Cover

These types of sensors for measuring the physical and mechanical properties of soil are used in various integration options for devices with a wide range of applied registration methods.

In Fig. 8.4 the scheme of functioning of the technological module for measurement of locally defined parameters of soil is presented.

The technological module for measuring the pH of the soil performs several operations:

Table 8.1 Evaluation of different types of sensors for measuring physical and mechanical parameters of soil

Soil properties	Sensor type				
	Electric	Optical	Mechanical	Acoustic	Ion selective
Texture (sand-silt-clay)	Perfectly	Good			
Organic matter (humus)		Perfectly			
Soil moisture	Perfectly	Good			
Salinity (sodium)	Good	Satisfactorily			Satisfactorily
Density (density)			Perfectly	Good	
Depth of the subsoil	Satisfactorily	Satisfactorily	Good	Good	
Soil pH		Satisfactorily			Perfectly
Nitrogen (nitrate or complete)	Satisfactorily	Good			Good
Nutrients (potassium, phosphorus)		Satisfactorily			Good
Capacity of exchange cations	Good	Perfectly			

- Soil sampling
- Conducting precision preparation of the sample for measurement
- Preparation of solution and creation of a uniform stream of the investigated component
- Measurement and registration of the level of the studied parameter

Model mobile diagnostic unit MDA-30 school to gather information about the parameters of the fertility of the soil and condition of the plants during their growing season is shown in Fig. 8.4. The diagnostic unit is equipped with a GSP coordinate sensor and has adapters for detecting relatively homogeneous areas of the field by estimating the electrical conductivity of the soil and the spatial variability of the vegetative mass of plants.

In Table 8.1 is given a scale of "satisfactory," "good," and "excellent" assessment of different types of sensors to measure certain soil parameters.

Although many soil properties can be determined through traditional methods (laboratory analysis), automated methods for collecting locally defined information are becoming more widespread. Some parameters, such as pH and soil moisture, can change so dramatically over the field area that the use of conventional methods for collecting locally defined information becomes ineffective. In addition, there are parameters which vary less sharply such as the content of organic matter, the depth of the fertile soil layer, etc., and therefore, it is necessary in each case to conduct staging studies to determine the spatial variability of the factor and possible methods of analysis, and these operations are costly.

The use of the diagnostic unit allows to reduce labor costs of monitoring works for the construction of electronic maps for optimum application by means of chemicalization and carrying out sowing; increase the payback of fertilizers by

1.5–1.7 times; reduce the chemical load on the environment; and improve the quality of the received products.

Skillful analysis of currently existing large market instruments and equipment for recording physical and mechanical properties of soil by the criterion of quality and price, taking into account potential long-term use and improvement of equipment, opens up broad prospects for the implementation of one of the most common elements of the system vehicle – a high-performance collection local information about the state of the field and the construction of cartogram tasks to perform technological operations.

References

Adamchuk VI, Allred B, Doolittle J, Grote KK, Viscarra Rossel RA (2015) Tools for proximal soil sensing. In: Ditzler C, West L (eds) Soil survey manual. USDA, Washington, DC

Adamchuk V, Reumont F, Kaur J et al (2017) Proximal sensing of soil biological activity for precision agricultural. In Advances in animal biosciences: precision agriculture (ECPA), vol 8, no 2, pp 406–411

Aniskevich LV, Voytiuk DG, Vygera SM et al (2019) Precision phytotechnologies in the agro-industrial complex of Ukraine: monograph. NULES of Ukraine, Kiev, p 799

Crowley D, Brennerova M, Irwin C et al (1996) Rhizosphere effects on biodegradation of 2,5-dichlorobenzoate by a bioluminescent strain of root colonizing Pseudomonas fluorescens. Microbiol Ecol. 20:79–89

Daughtry CST, Walthall MS, Kim E (2000) Estimating corn leaf chlorophyll concentration from leaf and canopy reflectance. Remote Sens Environ:229–239. https://doi.org/10.1016/S0034-4257(00)00113-9

Ehlert D, Horn HJ, Adamek R (2008) Measuring crop biomass density by laser triangulation. Comput. Electron. Agric. 61:117–125. https://doi.org/10.1016/j.compag.2007.09.013

Gebbers R, Adamchuk V (2010) Precision agriculture and food security. Science 125:828–831. https://doi.org/10.1126/science.1183899

Jura N, Romanyuk O, Gonsior J et al (2006) Use of plants for reclamation of soils contaminated with oil and oil products. Ecol Noospherol 17(1-2):55–60

Korovetska G, Sohanchak R, Jura N et al (2008) Terek State apparatus stomata leaf plant *Carex hirta* L. under the influence of oil pollution. Visn. Lviv National University. Biol. Ser. 47:166–171

Moroz OM, Jura NM, Beznosko GY et al (2006) Influence of *Carex hirta* plants on the microflora of oil-contaminated soils. Scientific Bulletin of Uzhhorod University. Ser Bio 19:149–154

Pikovsky YI, Gennadiev AN, Chernyansky SS et al (2003) The problem diagnosis and normalization of pollution by oil and petroleum products in soil. Pochvovedeniye 9:1132–1140

Pyrog TP (2012) Physiological bases of intensification of synthesis of microbial surfactants. Paper presented at Microbial biotechnology topicality and future. Kyiv, 19–22 November 2012

Terek OI, Velichko OI (2009) Jura NM Physiological aspects of plant adaptation to oil- contaminated soil. In: Morgun VV (ed) Plant physiology: problems and prospects of development, vol 2. Logos, Kyiv, pp 217–226

Velychko O, Terek O (2008) Influence of oil pollution of the soil on the content of ascorbic acid and ascorbate oxidase activity in the organs of hop alfalfa plants. Nauk. Spring Lviv. Nat Acad Vet Med Biotechnol 10:29–32

Chapter 9
Investigation for a Sustainable Use of Fossil Coal Through the Dynamics of Interaction of Smokeless Solid Fuel with Oxygen and the Possibilities of Its Practical Application

Yevgen Zbykovskyy and Iryna Shvets

Nomenclature

NO_x Nitrogen oxides
SO_2 Sulfur oxides
CO Carbon monoxide
CO_2 Carbon dioxide

9.1 Introduction

One of the ways to solve the problem of resource conservation and environmental protection in the fuel, energy, and metallurgical sectors of the economy is the introduction of new complex technological processes, which can be implemented by the principle of energy chemical and technological procession of carbon raw materials (Zbykovskyy 2018, 2019; Zbykovskyy et al. 2008, 2011; Ilyshov et al. 2009). It involves the extraction of valuable chemical components and compounds in the form of coke oven gas (combustion heat ~17700 kJ/m^3), benzene hydrocarbons, hydrogen sulfide, coal tar, ammonia, and other chemical compounds, as well as the production of a solid product with new consumer properties.

Y. Zbykovskyy (✉) · I. Shvets
Donetsk National Technical University, Pokrovsk, Ukraine

S. Boichenko et al. (eds.), *Chemmotological Aspects of Sustainable Development of Transport*, Sustainable Aviation, https://doi.org/10.1007/978-3-031-06577-4_9

9.2 Technology for Smokeless Solid Fuel Production

Smokeless solid fuel can be compared with anthracites in terms of its technological and technical characteristics (Korchevoi et al. 2004). In the future, it may serve as an alternative to their use. At the same time, this fuel can be an alternative to nut coke in ferroalloy production due to its high reactivity.

Smokeless solid fuel is a product of high-temperature coking of a preheated coal charge, which consists of low-metamorphosed low-sulfur coal (80–100%) and coal waste (up to 20%) (Zbykovskyy 2008; Alvarez et al. 1998).

The main technical characteristics of solid smokeless fuel are as follows:

- Total sulfur content of S_t^d <1%
- The yield of volatile substances $V^{daf} = 1–3\%$
- Ash content $A^d = 10–20\%$
- Lower combustion heat ≥26,000 kJ/kg
- Reactivity CRI ≥ 65%
- Maximum heat dispersion rate at T > 860 °C

The technology of obtaining carbon smokeless solid energy fuel was developed in three stages (Gordienko et al. 2007, 2009):

- 1st stage – preparation of the initial coal mixture
- 2nd stage – thermal preparation of coal mixture
- 3rd stage – coking of preheated mixture

At the first stage, the preparation of the initial mixture is carried out. As a raw material, it is proposed to use a mixture of enriched low-sulfur of low-metamorphosed coal (65–85%) and sludge – coal beneficiation waste (15–35%) – which has an ash content of 30–50%. The recommended sulfur content in coal should not exceed 0.7%. Coal must be pre-milled to a size class less than 3 mm (100%) and thoroughly mixed with coal beneficiation wastes, which have high ash content.

The percentage of components in the mixture can be adjusted in a wide range, depending on the requirements of consumers and the ash content of the components. The percentage of coal in the mixture (the optimal ash content of the finished mixture is 25%) is calculated by the formula:

$$P_c = \frac{A_m\text{-}25}{A_m\text{-}A_c} \tag{9.1}$$

P_c – the percentage of coal in the mixture, %
A_m – ash content of mixture, %
A_c – ash content of coal, %

Preparation of the initial mixture must comply with the following basic requirements:

- Dosing of components according to a percentage of the mixture
- Thorough mixing of the components of the mixture before further processing
- Compliance with the requirements for the level of grinding of the initial mixture

The use of this scheme for preparing the initial mixture provides the following advantages of the developed technology:

- Use as raw materials of relatively cheap low-sulfur coal and coal beneficiation waste, which provides a low cost of the finished product, which is competitive compared to traditional thermal coal
- The low-sulfur content of the initial mixture (up to 0.7%) provides high environmental performance of the finished product
- Reduction of raw material dependence on scarce high-grade coal with high cost

In the second stage, the thermal preparation of the initial mixture is implemented. The initial mixture of this composition has a low level of caking and coking: low caking (Roga Index RI less than 50 mm), small thickness of the plastic layer ($Y = 8$–10 mm), high ash content (A^d up to 30%), and high yield of volatile substances (V^{daf} up to 40%). Coking of the obtained initial mixture is possible only under the condition of its preliminary heating by the inert gaseous heat carrier to the set temperatures (150–250 °C) by the method of high-speed heating within 3–5 s.

The preliminary heat treatment of the initial mixture with low caking and coking affects the process of coke formation. As a result, the heating rate of the mixture increases to the temperature of the transition of the organic mass of coal into the plastic state. The heating time in the plastic state and the thickness of the plastic layer increase. At the stage of formation of semi-coke, heat-reducing processes occur with a smaller temperature gradient, which leads to a decrease in the shrinkage of the carbon residue. The thermal conductivity of the heated mixture of coal and coal beneficiation waste does not decrease, because of the density of the heated mixture increases. The heat content of a load of the coke oven chamber and the speed of the cooking process is increased. This process enables a porous carbon product with sufficient mechanical strength from a low-caking mixture of coal and coal beneficiation waste. Therefore, thermal preparation of the initial mixture enables producing solid smokeless fuel in traditional coke ovens.

In the third stage, cooking of the thermally prepared initial mixture is carried out. It must be performed in accelerated coking periods of 10–14 h. Particular attention should be paid to the readiness of the coke pie for delivery.

Coking can be finished when the temperature reaches the axis of the coking chamber 750–1050 °C. In this way, the intensification of the coking of the thermally prepared mixture of coal and coal beneficiation waste is provided. The content of lump coke with a particle size of more than 40 mm in the solid smokeless fuel is 60%, and with a size of 25–40 mm, 22.5%.

The solid smokeless fuel obtained by the proposed technology has many advantages over other types of solid lump fuel:

- Increased water resistance, which is provided by a homogeneous structure that does not collapse under the influence of water in contrast to briquettes.
- High mechanical strength, which can withstand significant mechanical loads on crushing and abrasion, as well as during overloads and transportation.

- Technological process of fuel production is carried out without binders and briquetting operations.
- This type of fuel enables refusal from energy brands of coal with a high cost.
- Emissions of NO_x, SO_2, and CO are reduced, and environmental performance of boilers is improved when using solid smokeless fuel in boilers.
- The use of the proposed technology of production of solid smokeless fuel provides utilization of coal beneficiation waste.
- The possibility of obtaining solid smokeless fuel with different quality indicators depending on the requirements of the consumer (ash content, volatile matter yield).

In order to study the kinetics of the interaction of the carbonaceous substance of the fuel with the gas reagents at a pressure of 0.1 MPa and a temperature up to 1200 °C, a laboratory unit was used. The laboratory unit was a differential impulse reactor operating in modes close to perfect extrusion by the gas phase and perfect mixing solid phase. The laboratory unit also enabled to maintenance of small temperature gradients (less than 50 °C) and pressures (less than 100 Pa) in the reaction zone over the height of the fuel layer while conducting the experiments.

Intense mixing of the fuel particles in the fluidized zone provided a mode of perfect mixing of the solid phase, as well as the almost homogeneous concentration of the reagent gas over the layer of the analyzed solid fuel. The current nature of the gas movement in the reactor provided a mode of perfect extrusion by the pulse of the gas reagent.

The experiments and processing of the obtained data were carried out following a specially developed methodology. The order of measurements was determined following the kinetic experiment:

- Study of the dependence of the reaction rate on the conversion level at a given temperature ($R_m = f(X)$ at $T = const$)
- Study of the dependence of the reactivity rate on the temperature ($R_m = f(T)$)

In order to obtain the dependence of the burning rate on the conversion level of the fuel sample, the experiment was conducted as follows. A pre-dried and weighed sample of fuel weighing 0.1 g was placed on a reactor's perforated grate in a cold furnace. The reactor was then purged with inert gas (argon or helium) while heating the furnace at a speed of 8–10 °C/min to the desired temperature. After the temperature stabilization, the impulses of the reagent gas were sent to the reactor. Under the impulse conditions, a predetermined portion of the reagent gas from the calibrated loop was injected into the stream of inert gas by a six-way valve, continuously fed into the lower part of the reactor. The reagent gas interacts with the fuel sample and reaction products are formed. The component composition of the reaction products was determined at the reactor's output by the gas analyzer. Thus, the experiment involves the sequential injection of impulses of the reagent gas and the measurement of the formed products' concentrations.

The duration of the impulse response at the reactor's output was measured by a chronometer according to the measurements of the rheometer/rotometer. The jump-alike change of the density while transiting from inert gas to the reagent gas and back led to a jump-alike change in the measurements of the rheometer. That enabled to determine the moment of occurrence of the impulse response. The composition of combustion products (CO, CO_2, O_2) was measured by the gas analyzer.

The experiments were done in order to determine the dependence of the reaction rate on the temperature was also carried out using the impulse supply of reagent gas. The sample of solid fuel was placed in the reactor. After that, the reactor was continuously heated up to a predetermined temperature with simultaneous purging of the inert gas to prevent the fuel sample from interacting with the oxygen of the air. The component composition of the gaseous products was determined by the gas analyzer at the reactor's output.

For the given temperature level, the impulse input occurred no more than one or two times, depending on the sample conversion rate. After that, it went to a higher temperature level (the temperature step is 25–50 °C). The stepwise temperature increase, followed by the analysis of the gaseous products of the impulse response, was carried out until the maximum required temperature was reached.

In all the experiments, the impulse value of the reagent gas was chosen in such a way that the mass reduction of the solid fuel sample during the impulse would be much less than its initial value. This impulse value also was chosen, depending on the limiting sensitivity of the used gas analyzer. During the experiments, the impulse value of the reagent gas was chosen experimentally.

9.3 Study of Properties of Solid Smokeless Fuel

The coal samples used for the research were as follows: anthracite, highly metamorphosed coal, low-metamorphosed coal, and solid smokeless fuel (Table 9.1). For the kinetic studies, the coke residues of the presented solid fuels were prepared at the

Table 9.1 Technical and elemental analysis of the researched types of solid fuel

Type of solid fuel	Technical analysis			
	W^a, %	V^{daf}, %	A^d, %	S^d_t, %
Anthracite	2.3	2.00	4.39	0.75
Highly-metamorphic coal	1.8	11.10	10.90	1.54
Low-metamorphic coal	3.3	41.60	6.20	0.46
Solid smokeless fuel	1.9	2.70	11.80	0.74
Type of solid fuel	**Elemental analysis**			
	C^{daf}, %	H^{daf}, %	N^{daf}, %	O^{daf}, %
Anthracite	96.27	2.20	0.92	0.61
Highly-metamorphic coal	91.00	3.98	2.28	2.74
Low-metamorphic coal	81.62	5.95	1.61	10.82
Solid smokeless fuel	97.89	0.68	0.73	0.70

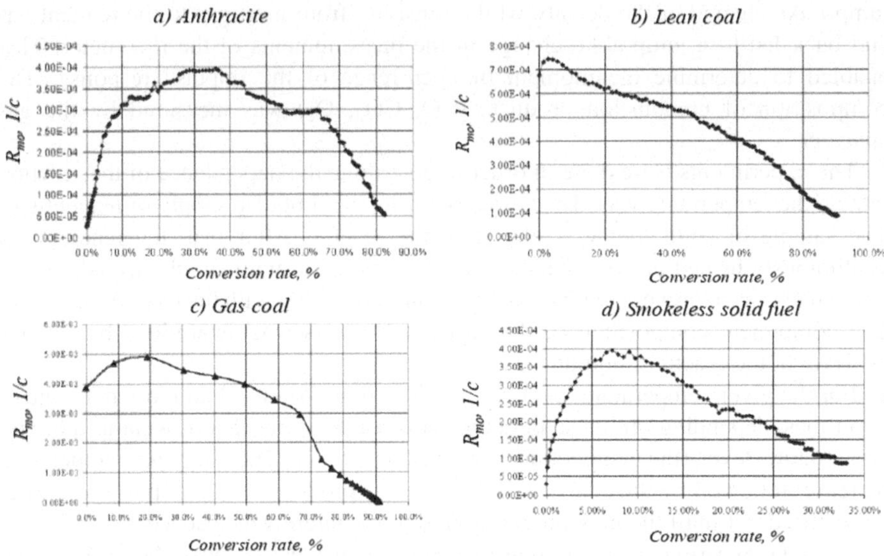

Fig. 9.1 Dependence of the interaction rate of energy fuel with oxygen on the conversion rate. (**a**) Anthracite, (**b**) Lean coal, (**c**) Gas coal, (**d**) Smokeless solid fuel

temperature of 850 °C for 30 min in an inert gas stream. Working with coke residues of fuel during kinetic studies was a prerequisite. It enabled determination of the reactivity of coke fuel residue, which was responsible for the longest stage of combustion of fuel particles, without the specific properties of volatile substances.

As a result of kinetic experiments, the dependence of the interaction rate of anthracite coke residue with oxygen at a constant temperature $T = 560$ °C was determined. The dependence of the anthracite sample burnout rate (R_{m0}), which was normalized to the initial mass of fixed carbon in the sample, was given (Fig. 9.1a). The kinetic experiments on samples with a mass of about 0.1 g with fuel particles with a diameter of 0.1–0.16 mm were performed.

The total flow rate of the reagent gas through the reactor at atmospheric pressure was 1.2–1.6 dm^3/min. The speed of burnout was high. It varied during the conversion of the sample. The speed increased to the maximum value ($R_{m0} \cong 4 \times 10^{-4}$, 1/s at $X = 25–40\%$) and decreased in the final stage of burnout. The maximum conversion level of the sample of anthracite coke residue at a temperature $T = 560$ °C was 80–82%.

The dependence of the interaction rate between the lean coal coke residue and oxygen on the conversion level ($T = 560$ °C, $d = 0.1-0.16$ mm) was investigated (Fig. 9.1b). Similar to anthracite, the dependence of the burnout rate during the conversion was non-stationary. While passing through the maximum value ($R_{m0} \cong 7.4 \times 10^{-4}$, 1/s) at the very beginning of the conversion curve ($X = 3–5\%$), at the final stage of burnout, the speed decreased monotonically.

The burnout rate of coke residue of highly metamorphosed coal at the temperature $T = 560$ °C and the range of conversion level were higher than the rate of burnout of anthracite coke residue. The maximum conversion level of the coke residue sample of highly metamorphosed coal at a temperature $T = 560$ °C was 90–92%.

During the kinetic experiments, the dependence of the interaction rate of the coke residue of low-metamorphosed coal with oxygen at a constant temperature $T = 560$ °C was established (Fig. 9.1c). The kinetic experiments on samples weighing 0.1 g for the particles' diameter $d = 0.1$–0.16 mm within other equal parameters were performed.

As in the previous cases, the dependence of the burnout rate during the conversion of the sample changed. It passed through the maximum value ($R_{m0} \cong 4.8 \times 10^{-3}$, 1/s) in the region of the conversion level ($X = 20\%$) and decreased at its final stage. The burnout rate of coke residue of low-metamorphosed coal at the range of conversion level was higher than the burnout rate of both high-metamorphosed coal and anthracite. This rate was characterized by the highest reactivity (burnout rate). The maximum conversion level of coking coal residue at the temperature $T = 560$ °C was 92%.

The dependence of the interaction rate of the coke-residue of solid smokeless fuel with oxygen with the conversion level ($T = 560$ °C, $d = 0.1$–0.16 mm) was investigated (Fig. 9.1d). As with the previous types of fuel, the burnout rate depended on the conversion level of the sample. It passed through the maximum value ($R_{m0} \cong 4.2 \times 10^{-4}$, 1/s) at the very beginning of the conversion curve ($X = 6$–8%) and monotonically decreased at the final stage of burnout. The smokeless coke residue rate at the temperature $T = 560$ °C at the beginning of conversion exceeded the anthracite burnout rate. At the same time, it decreased at a lower burnout rate of low-metamorphosed and highly metamorphosed coal.

The maximum conversion level of coke residue of solid smokeless fuel at the temperature of $T = 560$ °C was 33–35%. This fact can be explained by the significant deactivation of the active surface of coal during the previous high-temperature carbonization.

The summary graphics of the results of the experiments enabled us to compare the dependences of the interaction rate for different types of energy fuel with oxygen and conversion rate (Figs. 9.2 and 9.3).

In the whole range of conversion, the burnout rate of coke residue of anthracite particles was the lowest compared to the burnout rate of high-metamorphosed and low-metamorphosed coal. The temperature dependence of the burnout rate of solid smokeless fuel was intermediate. This dependence was characterized by low reactivity, which resulted from the deactivation of the activated carbon surface in the process of preliminary high-temperature heating without oxygen access.

In the experimental studies of the burnout dynamics, the samples of the original fuels' (anthracite, high-metamorphosed, low-metamorphosed coal, solid smokeless fuel) particles were used. The experiments on the burnout of carbon particles were carried out at starting temperatures of approximately 860 °C. In order to ensure an accurate comparison of the results for all coal samples, identical experimental conditions (initial temperatures, sample masses, airflow rates, particle sizes, etc.)

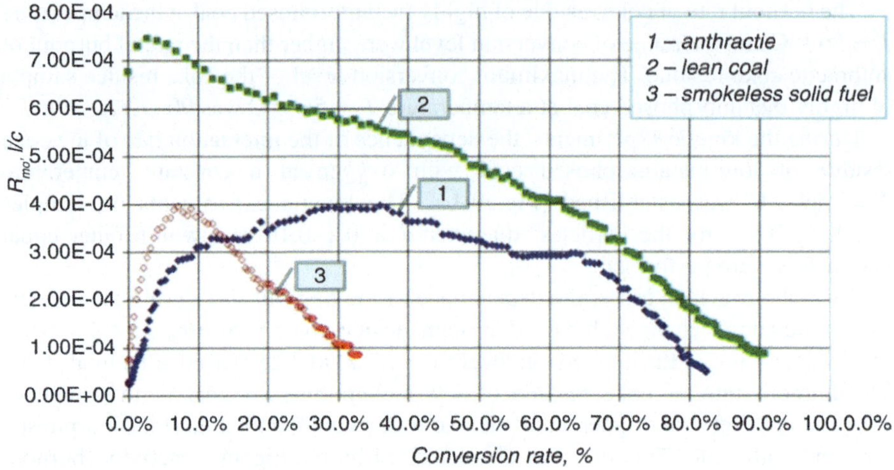

Fig. 9.2 The relation of the energy fuel interaction rate with oxygen and conversion rate

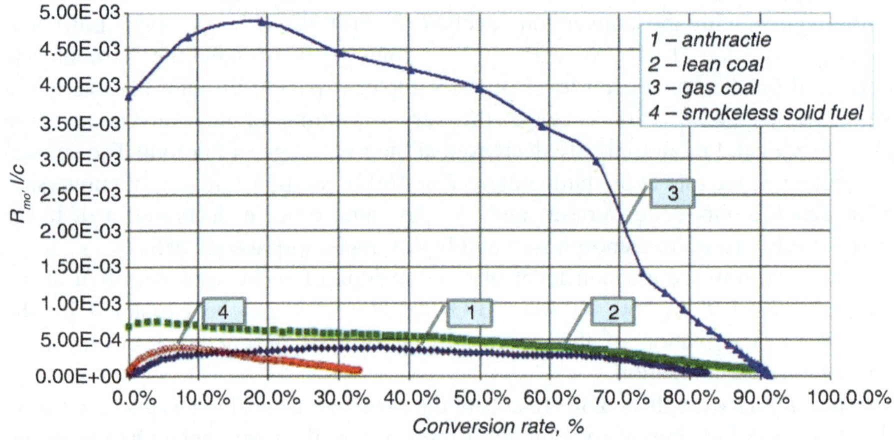

Fig. 9.3 The relation of the energy fuel interaction rate with oxygen and conversion rate compared to gas coal

were maintained. The registration of the temperature measurements in the reactor was stopped after reaching the minimum values of concentration (limit of measurement) of the reaction products (CO and CO_2) according to the gas analyzer (for CO, 0.001% vol., for CO_2, 0.01% vol.). Thus, the burnout time corresponded to the full burnout time of the coal particles. This time was determined equally for all tested coals samples. The experimental studies on the dynamics of burnout simulated the natural conditions of particles' burnout in a furnace/boiler.

The dynamics of the temperature change in the reactor during the burnout of discrete anthracite samples with diameter $d = 0.1–0.16$ mm sea-level air pressure

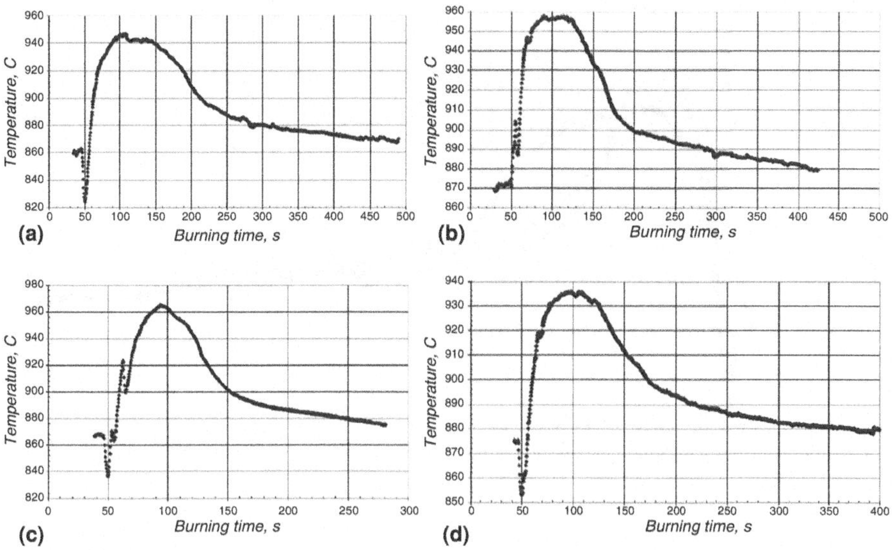

Fig. 9.4 Dynamics of coal particles combustion of anthracite (**a**), lean coal (**b**), gas coal (**c**), and solid smokeless fuel (**d**) by reaction with oxygen and at atmospheric pressure

was investigated (Fig. 9.4a). The weight of the sample in experiments was the same (≅1 g) with a coal fraction of 0.3 g and ash at 0.7 g, taking into account the constancy of carbon in samples. The linear speed of the gas stream was calculated, taking into account the prevention of removal from the reactor of minimum size fuel particles ($Ug = 0.38$ m/s, $T = 900$ °C, $P = 0.1$ MPa).

During the burnout of the anthracite sample, a dramatic change of the heat release in the reactor was detected. The temperature rose simultaneously from initial (≅860 °C) to 945–950 °C. The temperature curve of the anthracite burning corresponded to the burning of carbon of the coke residue. This curve did not contain the initial area, which characteristic of the release and burnout of volatile substances. The adequate burnout time of anthracite particles for a given high temperature (≅860 °C) was about 490 s. That was the highest value for all studied fuels types.

A similar burning curve was obtained for high-metamorphosed coal particles with equal experimental parameters (Fig. 9.4b). Its peculiarity was the existence of two extreme areas. The first extreme area had a maximum that corresponded to the burnout of volatile substances over a short time. The second extreme area corresponded to the burning of the coke residue over a longer time. The burnout duration for high-metamorphosed coal particles at given high temperatures (≅860 °C) was about 425 s.

The dynamics of burnout of low-metamorphosed coal discrete samples were characterized by the presence of two distinctive peaks. The first peak was short. It corresponded to the burnout of volatile substances (Fig. 9.4c). The second peak was more durable, and it corresponded to the burnout of coke residue. The burnout

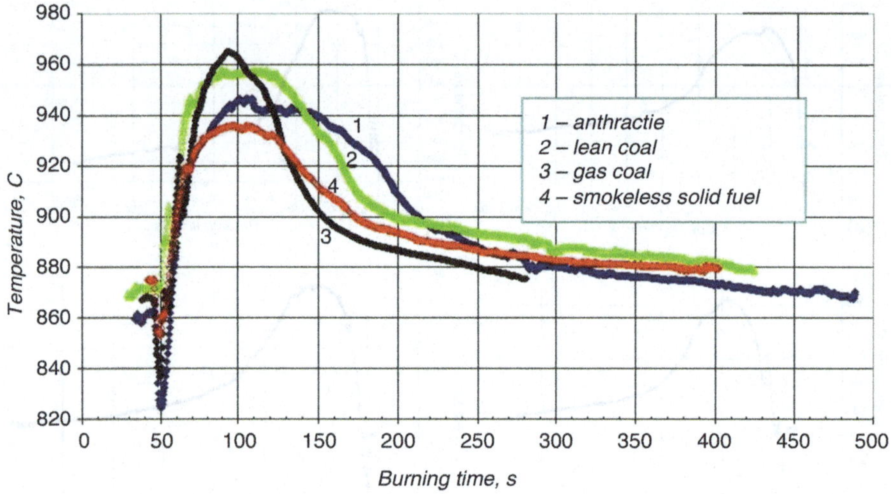

Fig. 9.5 Dynamics of combustion of energy fuels

duration of low-metamorphosed coal particles at given high temperatures (≅860 °C) was the lowest (280 s).

Although the solid smokeless fuel was derived from low-metamorphosed coal (with the content of volatiles $V^{daf} = 41.6\%$ vol.), the temperature curve of the discrete smokeless solid fuel burnout did not contain a peak that corresponded to the yield and burnout of volatiles. The temperature curve corresponded to burnout of the coke residue (Fig. 9.4d). The burnout duration of the smokeless solid fuel particles at given high temperatures (≅860 °C) was about 400 s.

The comparative characteristics of dynamic curves of burnout of the test fuels (Fig. 9.5) showed that low-metamorphosed coal had the shortest burnout duration. Solid smokeless fuel and high-metamorphosed coal had medium burnout duration. At the same time anthracite had the most significant burnout duration. The obtained temperature curves enabled us to analyze the level of maximum heat release during the burnout of the test fuels, the duration of the pyrolysis stages and the burnout of coke residues, the complete burnout time, and the reactivity of the solid fuels.

Yield and burnout of volatile matters of the fuel contribute to its ignition and further burnout in the furnace. The discrepancy of the characteristics between the used fuel and the projected one, in particular the absence of volatiles, can lead to an increase of the induction time, its late ignition, and, as a consequence, an increase of the mechanical under burn. This situation leads to a deterioration of the technical and economic indicators of the power-generation unit (Korchevoi et al. 2004).

The possible use of solid carbon fuel is the production of ferroalloys in electric furnaces. Nut coke (10–25 mm), which is the product of sorting metallurgical coke by its size, is used as a reducing agent in the production of ferroalloys. There are specific requirements for reducing agents. These requirements are due to the technological features of the ferroalloys production in powerful recovery ore furnaces. However, such requirements as electrical resistance and reactivity to difficult-to-recover oxides are practically impossible using nut coke (10–25 mm).

In some ferroalloy processes, such types of special reducing agents as petroleum coke, pitch coke, semi-coke, hard coal, and charcoal (in a mixture with other types of reducing agents) are used in a limited number.

The main requirements for the quality of carbon-reducing agents differ while producing various ferroalloys groups (siliceous, manganese, chromium, etc.) and demand the control of the following indicators:

- The content of solid (nonvolatile) carbon, ash, volatile substances, moisture, and sulfur
- Chemical composition of ash
- Electrical resistivity
- Porosity
- Physical and mechanical properties (particle size distribution, strength characteristics)
- Reducing ability regarding oxides of a specific element, which is determined by the ratio of CO/CO_2

The most essential characteristics of reducing agents used in ore reduction electric furnaces are electrical resistance and reactivity. The use of reducing agents with high electrical resistance is necessary. It is due to a decrease in charge conductivity and an increase in the proportion of electric current that passes through the arc – a high-temperature concentrated heat source. Both of these properties of coke depend on its chemical activity and structure, i.e., the degree of molecular order and density of the carbon.

The increase in the chemical activity of coke depends on the following factors:

- Use of low-metamorphosed, low-coking, and petrographically inhomogeneous coal
- Reducing the state of coke readiness
- Reducing the density of coal loading
- Increase in the number of mineral impurities

The use of such type of coal causes difficulties for horizontal coke ovens, which produce blast furnace coke from well-agglomerating coal while pushing coke. Such difficulties such as slow running and drilling of coke ovens lead to rapid wear and destruction of refractory masonry. This situation can be avoided by coking the preheated coal in the coke ovens. This coal charge should consist of high-volatile and low-metamorphosed coal without deteriorating the formation conditions of the "coke pie" in the coke oven.

Depending on the brand of smelted alloys and the type of new reducing agents, the productivity of ferroalloy furnaces can be increased by 5–10% with a proportional decrease in specific electricity consumption by 5–7% and an increase in the share of essential elements that are converted into ferroalloys.

To test the efficiency of solid smokeless fuel, its experimental samples were produced on the coke battery № 3 of the Yasinivsky Coke Plant (Ukraine). In the production of solid smokeless fuel according to the proposed technology, three options for the composition of the charge were used. The charge consisted of low-metamorphosed coal, bituminous coal (W^r = 9.0%; A^d = 8.5%; S_t^d = 1%; V^{daf} = 32.0%; Y = 18 mm), and coal wastes (W^r = 12.0%; A^d = 10.0%; S_t^d = 1.2%; V^{daf} = 31.0%; Y = 10 mm). An experimental sample of solid smokeless fuel was used in the experimental smelting to produce ferrosilicon manganese at furnace № 8 of the Nikopolskiy Ferroalloy Plant and furnaces № 1 and № 2 of the Stakhanovskiy Ferroalloy Plant (Table 9.2).

As shown by experimental smelting at the Nikopolskiy Ferroalloy Plant, the use of highly reactive smokeless solid fuel did not lead to a change in the gas and electric modes of ferrosilicon manganese smelting (Table 9.3). During the experiments, the fit of the electrodes into the charge was 150–200 mm lower. This fact illustrates the more excellent manufacturability of using solid smokeless fuel.

During the experiment, the weight of coke decreased from 360–370 to 320–330 kg. The quality of ferrosilicon manganese obtained during the experiment, the content of manganese (72–73%), silicon (17.6–17.8%), and phosphorus (0.25–0.28%) did not differ from the standard values. The specific consumption of raw materials manganese, coke, and quartzite decreased by 1.1, 16.6, and 4.4 kg/t, respectively, and electricity fell 41 by 41.9 kWh/t. The manganese content in the slag decreased by 1.2%, the manganese extraction increased by 0.7%, and the furnace productivity increased by 1.2% (Table 9.3).

During the experiment at the Stakhanovskiy Ferroalloys Plant, the specific consumption of quartzite and reducing agents (in terms of carbon) decreased by 0.4% and 1.06%, respectively. The electricity consumption decreased by 375 kWh/t (Table 9.4).

As a result of the industrial experiments at the Nikopolskiy Ferroalloys Plant and the Stakhanovskiy Ferroalloys Plant, the optimum quality indicators of solid smokeless fuels recommended for ferroalloys were established (Table 9.5). Depending on the brand of alloys and the type of new reducing agents, the proposed technology allows:

- To increase the efficiency of ferroalloy furnaces by 5–10%
- Proportional reduction in the specific energy consumption by 5–7%
- To increase the proportion of the main elements that pass into the ferroalloys

A study of the properties of solid smokeless fuels has shown that this fuel meets the requirements for reducing agents for ferroalloys' production plants.

9.4 Pilot Studies of Solid Smokeless Fuel

Based on the results of laboratory, experimental, and pilot studies of solid smokeless fuel, requirements for its quality indicators have been developed for use in gasification technologies and production of sugar, soda, and cement and during limestone burning/calcination.

Table 9.2 The quality indices of the charge and solid smokeless fuel

Consumers	Charge composition, %			Technical analysis of the charge, %			Technical analysis of the solid smokeless fuel, %		Screen composition of the solid smokeless fuel, %				CRI, %	CSR, %
	G	Coal Proc. waste	Zh	A^d	S^d	V^d	A^d	S^d	>40 mm	25–40 mm	10–25 mm	0–10 mm		
Nikopolskiy Ferroalloys Plant	80	10	10	8.2	0.80	40.3	11.7	1.2	4.0	8.0	77.0	6.0	65	18
Stakhanovskiy Ferroalloys Plant	80	10	10	8.0	0.75	39.4	11.5	1.1	2	12	78.0	8.0	61	21
Stakhanovskiy Ferroalloys Plant	100	0	0	7.2	0.50	41.5	10.5	0.7	2	10	78.0	10.0	76	9

Table 9.3 The smelting indicators of the MnC17 grade ferrosilicon manganese in furnace № 8 of the Nikopolskiy Ferroalloys Plant using highly reactive solid smokeless fuel

Indicators	Melting process options	
	Standard	Pilot
Unit consumption of raw materials, kg/t:		
Manganese raw materials (48% Mn)	1667.4	1666.3
Ordinary coke	440.9	–
Solid smokeless fuel	–	424.3
Quartzite	393.9	389.5
Iron ore pellets	37.4	29.9
Secondary manganese raw materials	247.7	244.5
Waste	367.4	188.3
Screenings	60	62.3
Unit consumption of electric power, kW·h/t	4331.4	4289.5
Manganese to alloy transition, %	86.6	87.3
The manganese content in slag, %	12.3	11.5

Table 9.4 The smelting indicators of the FS65 grade ferrosilicon manganese using highly reactive solid smokeless fuel

Indicators	Melting process options	
	Standard	Pilot
Working hours:		
Rated, days	52.5	26
Hot downtime, %	0.6	0.14
Actual, days	52.21	25.86
Production FS 65, t	3578.8	1881.14
The average silicon content, %	66.3	66.5
Active power, kW	20,960	21,146
Productivity, t/days	68.2	72.7
Unit consumption, t/t:		
Quartzite	1.517	1.511
Ordinary coke	0.620	–
Solid smokeless fuel	–	0.609
Gas carbon	0.148	0.116
Carbon reducer	0.5829	0.5567
Steel cuttings	0.340	0.343
Scrap metal of the ferrosilicon	0.058	0.048
Limestone	0.014	0.115
Electric power consumption, kW·h/t	7338	6963
Silicon extraction, %	92.1	92.5

Generating energy by combusting solid smokeless fuel, which is produced by the new technology, enables to reduce the anthropogenic impact on the environment. This effect is characterized by the changes in the level of streams of harmful chemicals in the environment. The main substances that pollute the environment

Table 9.5 Recommended quality indicators of the solid smokeless fuel for ferroalloys production

Quality indicators	Symbols	Value
Technical analysis		
Ash content, %	A^d	23.0
Total sulfur content, %	S_t^d	1.15
Light constituents yield, %	V^{daf}	1.9
Crushing strength of coke, %	M_{25}	81.3
Coke abrasion resistance, %	M_{10}	13.1
Apparent density, g/cm^3	d_k	0.935
Actual density, g/cm^3	d_r	1.9301
Porosity, %	p^d_r	51.6
Reactive capacity, cm^3/g c	K_m	1.67
Coke reactivity, %	CRI/CSR	69.8 /30.1
Size distribution: +80 mm – 32.7%; 60–80 mm – 30.0%; 40–60 mm – 23.6%; 25–40 mm – 6.7%; <25 mm – 7.0%		

and are formed during the combustion of solid smokeless fuel in boilers are nitrogen oxides, sulfur oxides, and dust. Their amount is limited by the law in the vast majority of countries.

According to the nature of the formation, there are three types of nitrogen oxides: thermal, fuel, and fast ones. The rapid growth (almost exponential) of the thermal nitrogen oxide volume with the rise of the combustion temperature is the critical feature. The formation of fuel nitrogen oxides occurs due to the presence of the nitrogen component in coal. Their volume depends less on the combustion temperature compared to thermal oxides of nitrogen. The number of fast nitrogen oxides in the total volume of emissions and their dependence on the combustion temperature is insignificant.

For temperatures of 800–900 °C (in comparison with 1200–1700 °C at pulverized coal combustion), fuel and fast nitrogen oxides make the main contribution to the formation of nitrogen oxides. Thermal nitrogen oxides are hardly ever formed. Usually, 5–20% of fuel nitrogen is converted into nitrogen oxides. They consist of about 90% of nitrogen monoxide and 10% of nitrogen dioxide. Thermal oxides contribute to the formation of nitrogen oxides during fuel combustion in boilers at high temperatures (1500–1700 °C). They are formed by the interaction of nitrogen and oxygen, which are part of the air. This situation is typical for all large thermal power plants in Ukraine. The contribution of fuel and fast nitrogen oxides, in this case, is insignificant.

Emissions of sulfur oxides that are formed during the combustion of coal at thermal power plants are determined by the presence of sulfur in the burned coal. The quantitative estimation of emissions is performed based on the stoichiometry of the combustion reaction. The thermal power plants in Ukraine are not equipped with sulfur-nitrogen purification systems. Designing a desulfurization system is relevant for thermal power plants in Ukraine. In particular, the fee for gross emissions of sulfur oxides is significant, and it constantly increases. The use of low-sulfur

Table 9.6 Technical and environmental indicators of the boiler on regular and solid smokeless fuels

Indicators	Value	
	Regular fuel	Solid smokeless fuel
Steam productivity of the boiler, tons/hour	5	6
Specific flue gas consumption, m^3/hour	10,400	12,400
Concentration:		
Sulfur dioxide, g/m^3	0.497	0.338
Nitric oxide, g/m^3	0.404	0.202
Carbon monoxide, g/m^3	0.55	0.154
Dust, g/m^3	0.44	0.300
Fine for environmental pollution by main types of emissions:		
Sulfur dioxide, $/ton	42	42
Nitric oxide, $/ton	42	42
Carbon monoxide, $/ton	1.6	1.6
Dust, $/ton	1.6	1.6

smokeless solid fuels will significantly affect the global problem of environmental protection from sulfur compounds in industrial emissions. The industrial experiments on combusting solid smokeless fuel in the ASB-2 (air supply barrel) boiler were done at "Pokrovsk" mine. During the experiments, a significant reduction in emissions of harmful substances into the atmosphere with flue gases was determined (Table 9.6).

As a result of industrial tests of solid smokeless fuel produced by the proposed technology, a reduction in the main emissions into the atmosphere was achieved:

- Sulfur dioxide – 32.5%
- Nitric oxide – 50%
- Carbon monoxide – 72%
- Dust – 32%

Reducing the concentration of harmful substances affects the technical and economic performance due to reducing the environmental charges for environmental pollution.

A significant reduction in industrial emissions allowed the company to reduce the environmental charge for environmental pollution by 41% for the main types of emissions (Tables 9.7 and 9.8).

Solid smokeless fuel has higher productivity compared to conventional pulverized coal fuel. The use of this fuel in boilers is environmentally and economically feasible.

Table 9.7 Calculation of payment for harmful emissions on the example of the boiler of "Pokrovsk" mine during the working on coal

Indicators	Value
Coal consumption, tons	22,250
Specific coal consumption, tons/hour	1.032
Boiler operation time, hours	21,560
Steam production, tons	107,800
Flue gas yield, thousand m^3	224,225
Composition of harmful emissions:	
Sulfur dioxide, tons	111
Nitric oxide, tons	91
Carbon monoxide, tons	123
Dust, tons	98
Environmental pollution charge for the main types of emissions:	8825
Sulfur dioxide, $	4675
Nitric oxide, $	3800
Carbon monoxide, $	195
Dust, $	155

Table 9.8 Calculation of payment for harmful emissions on the example of the boiler of "Pokrovsk" mine while using solid smokeless fuel

Indicators	Value
Solid smokeless fuel consumption, tons	13,241
Specific solid smokeless fuel consumption, tons/hour	0.737
Boiler operation time, hours	17,967
Steam production, tons	107,800
Flue gas yield, thousand m^3	222,787
Composition of harmful emissions:	
Sulfur dioxide, tons	75
Nitric oxide, tons	45
Carbon monoxide, tons	34
Dust, tons	67
Environmental pollution charge for the main types of emissions:	5207
Sulfur dioxide, $	3159
Nitric oxide, $	1888
Carbon monoxide, $	55
Dust, $	105

The economic evaluation of the efficiency of solid smokeless fuel is proposed to be carried out in two stages. At the first stage, the total cost of fuel use should be calculated. The indicators of fuel cost, railway tariff for fuel transportation, and pollution charge are used for calculation.

The total cost of burning 1 ton of fuel in the boiler is proposed to be calculated by the formula:

Table 9.9 Technical and economic indicators of "Pokrovsk" mine and current tariffs for fuel use

Indicators	Value
Coast of the low-metamorphic coal, $/ ton	71.3
Coast of the solid smokeless fuel, $/ ton	99
Low-metamorphic coal consumption per year, tons	22,250
Solid smokeless fuel consumption per year, tons	13,241
Environmental pollution charge for using the low-metamorphic coal per year, $	8825
Environmental pollution charge for using the solid smokeless fuel per year, $	5207
Cost of the railway tariff for the low-metamorphic coal per year, $	44,535
Cost of the railway tariff for the solid smokeless fuel per year, $	59,604
Coefficient of operating costs for using the low-metamorphic coal	1.12
Coefficient of operating costs for using the solid smokeless fuel	1.09

$$T = R \cdot C_{\mathrm{f}} + C_{\mathrm{p}} + T_{\mathrm{r}}; \tag{9.2}$$

T – the total cost of burning, $
C_{f} – the cost of 1 ton of fuel, $
C_{p} – pollution charge by using 1 ton of fuel, $
T_{r} – railway tariff for the transportation of 1 ton of fuel, $
R – coefficient of operating costs

The technology of fuel used in the boiler house involves transportation, grinding, drying, and loading of fuel in the boiler furnace, which is accompanied by the operating costs. These costs can significantly vary, depending on the type of fuel. This fact must be taken into account when calculating the total cost of fuel use based on the unique indicator, i.e., the coefficient of operating costs. This coefficient considers the peculiarities of the use of different types of fuel in the boiler. The use of solid smokeless fuel is more technological than coal. This technology requires less labor and material costs, including energy ones. The value of the coefficient of operating costs also depends on the type of boiler and the peculiarities of its operation, the mode of operation of the boiler, and some other conditions. Therefore, the coefficient value for each type of boiler is individual and is determined by the experimental-statistical method.

The calculation of the total cost of using solid smokeless fuel in comparison with the low-metamorphic coal was carried out for the ASB-2 (air supply barrel) boiler of the "Pokrovsk" mine (Table 9.9).

The total annual costs for the use of low-metamorphic coal for the "Pokrovsk" mine are as follows.

$$C_{\mathrm{lmc}} = 71.3 \cdot 22250 \cdot 1.12 + 8825 + 44,535 = 1,830,156\,.$$

The total annual costs for the use of solid smokeless fuel for the "Pokrovsk" mine are as follows.

$$C_{\mathrm{ssf}} = 99 \cdot 13241 \cdot 1.09 + 5207 + 59,604 = 1,493,647\,.$$

The total amount of savings on annual fuel costs is

$$S = 183,0156 - 1493,647 = 336,509 \quad (+18.4\%).$$

Thus, the costs of using solid smokeless fuel in the boiler are lower by 18.4% than the costs of using low-metamorphic coal due to it is higher efficiency. The calculated economic result shows the economic component of the industrial use of solid smokeless fuel in boilers and the reduction of pollution charges. At the same time, this indicator does not consider the result of environmental protection. The direct effect of the environmental component is the economic effect of non-environmental damage. Taking into account the indirect effects, such as productivity, health, and quality of life, is also considered to be an essential evaluation of the use of solid smokeless fuel as an energy product.

9.5 Conclusions

The technology of production of smokeless carbon fuel has been developed during research. The result of the processing is lumpy carbon fuel, which has a homogeneous solid structure, high combustion, and mechanical strength, high reactivity, high speed, and complete combustion of the fuel particles. Proposed technology offers the opportunity to reutilize carbon wastes and improve environmental indicators of air pollution. Smokeless lumpy carbon fuel has a significant advantage over other fuel types: combustion of fuel particles in high-temperature zones (more than 800 °C) proceeds combustion mode to a phase close to exterior diffusion. Based on the experimental/industrial results, the optimal quality indicators of solid smokeless fuels recommended for ferroalloy production have been established. Methodical bases of ecological and economic estimation of solid smokeless fuel at combustion in boiler units were carried out. The calculation of the financial result of solid smokeless fuel at the enterprise was fulfilled. The total amount of savings on annual fuel costs for the "Pokrovsk" mine was 336,509 $ (which represents 18.4% compared to the costs of using low-metamorphic coal). As the use of solid smokeless fuels in boilers expands, research needs to focus on the socioeconomic impacts on people and the environment.

References

Alvarez R, Barriocanal C, Casal M et al (1998) Weathering study of an industrial coal blend used in cokemaking. ISIJ Int 38(12):1332–1338

Gordienko O, Yemchenko A, Zbykovskyy Y et al (2007) The method of producing lumpy carbon fuel. UA Patent 25,469, 10 Aug 2007

Gordienko A, Ilyshov M, Zbykovskyy Y et al (2009) The method of obtaining smokeless lumpy carbon fuel. RU Patent 2,367,681, 20 Sept 2009

Ilyshov MA, Gordienko AI, Starovoit AG et al (2009) Directions for the use of coke-chemical capacities of coal processing. Coke Chem 6:28–32

Korchevoi IP, Maistrenko AI, Topal AI (2004) Eco-friendly coal energy technologies. Naukova Dumka, Kyiv, p 187

Zbykovskyy YI (2018) The ecologic and economic evaluation of using solid smokeless fuel during combustion in boiler equipment. J Coal Chem 3:11–18

Zbykovskyy Y (2019) Resource-saving technology of complex energy-chemical and technological coal processing in coke production. Donetsk National Technical University, Pokrovsk, p 251

Zbykovskyy Y, Saranchuk V, Gordienko A et al (2008) Power-generation fuel with coal preparation waste. 25 Annual International Pittsburgh Coal Conference, USA, p 58.

Zolotaryov I, Tamko V, Zbykovskyy Y et al (2011) Means of improvement of CSR and CRI indexes of blast-furnace coke. Karbo, Wydawnictwo Gornicze, Katowice. 10:95–99

Chapter 10
Improvement of Diesel Engine Parameters by Using of Alcohol Conversion

Sviatoslav Kryshtopa, Liudmyla Kryshtopa, Myroslav Panchuk, Volodymyr Korohodskyi, Igor Prunko, and Ivan Mykytii

Nomenclature

CO	Carbon monoxide
CO_2	Carbon dioxide
H_2	Hydrogen

10.1 Introduction

A significant part of vehicles and other types of transport use internal burning motors with diesel systems. This reason requires fuel-air mixture consumption of expensive diesel fuel-air mixture with significant parameters of gas exhaust toxicity; thus it is obvious the feasibility of switching to more environmentally friendly and inexpensive alternating fuel-air mixtures (Panchuk et al. 2020b). Diesel motors have a long resource. There is no doubt that after the total cessation of the production of new diesel motors, the already produced diesel motors will be in operation for a long time. Thus, the problem of converting diesel energy drives to existing gas fuel-air mixtures, which are more environmentally friendly and inexpensive alternatives to diesel fuel-air mixture, deserves special attention (Kovalov 2020). Converting diesel motors to gas motor fuel-air mixture is possible according to different schemes. In general, the diesel motor can be converted to gas-diesel or mono gas regimes (Panchuk et al. 2019a, b).

The technology of converting diesel motors to the gas-diesel regimes is already known and well-established at present when both gas mixture and diesel fuel-air mixture supply are used simultaneously. In this case, an incendiary dose of diesel

S. Kryshtopa (✉) · L. Kryshtopa · M. Panchuk · V. Korohodskyi · I. Prunko · I. Mykytii
Ivano-Frankivsk National Technical University of Oil and Gas, Ivano-Frankivsk, Ukraine

S. Boichenko et al. (eds.), *Chemmotological Aspects of Sustainable Development of Transport*, Sustainable Aviation, https://doi.org/10.1007/978-3-031-06577-4_10

fuel-air mixture is supplied to the gas fuel-air mixture. It is positioned by the manufacturers of diesel-gas systems at the range of 25–40% and at actual operating conditions reaches 45–50%. This technology, along with its advantages, has many disadvantages (Vershina and Bystrenkov 2019), and one of the main ones is the necessity to supply diesel fuel-air mixture, but it impairs the economic feasibility of converting. The second, less tested variant of converting diesel motors to gas motor fuel-air mixture is a purely gas or mono gas regime. The converted diesel motors are additionally equipped with ignition spark systems. The supply of pure gas or mono gas is more profitable alternating to the diesel-gas regime, given the absence of necessity to supply expensive diesel fuel-air mixture. Diesel motor, according to its construction, can be converted to work with gas equipment on both methane (liquefied or compressed) and propane-butane (Boretti 2019).

The economic advantages of converting diesel engine drives of motor transport to the mono gas regime are related to the following (Jurkovič et al. 2019): mono gas fuel-air mixture is significantly inexpensive than diesel fuel-air mixture, and diesel fuel-air mixture is not supplied at all; resource of cylinder-piston group engine of convertible motors, due to smoother increase at burning pressures, increases by 1.4–1.6 times; and frequency of replacement of oil filters and engine oil of gas motors is approximately doubled, due to the reduction of carbon making, the absence of the process of washing the oil and reducing vacuum of motor oil.

Thus, the converting of diesel engine drives of motor vehicles to mono gas in terms of reducing the cost of fuel-air mixture and lubricants and overhaul of motors is actual. A study of domestic and foreign developments experience shows that diesel motors, which are converted into gas, have high economic and traction-dynamic characteristics. According to environmental parameters, they are significantly superior to basic diesel motors (Mateichyk et al. 2020). Diesel motor converting to mono gas one requires significant changes at the design of the base diesel (Gorski and Smigins 2011) in contrast to the converting of gasoline motors. Ignition of the fuel-air mixture at diesel motor is carried out by warming from compressing. Thus standard diesel motor cannot work on the gaseous fuel-air mixture, as gas motor fuel-air mixture has a significantly higher ignition temperature compared to diesel fuel-air mixture (e.g., diesel fuel-air mixture, 300–320 °C; propane, 460–470 °C), but it cannot be achieved with compressing ratios used at diesel motors.

The second reason why a diesel motor will not work on the gas fuel-air mixture is the phenomenon of detonation – it is an explosive burning of the fuel-air mixture, which occurs at a high degree of compressing. For diesel motors, the degree of compressing of the fuel-air mixture ranges from 14.5 to 22.5 and for gas motors must have a degree of compressing from 12.5 to 13.5. It should be noted that reducing compressing degree at converted into gas diesel motors is performed, as a rule, by boring the burning chamber at the piston of the base diesel engine. However, experimental studies show that even minor changes in the shape of the burning chambers at the pistons lead to significant changes in the processes of warmth, mass, and gas exchange and burning (Kryshtopa et al. 2019). Thus, optimization of the

a) Freightliner 120 on propane-butane   b) Volvo-460 on methane

Fig. 10.1 Transport vehicles powered with gas fuels. (**a**) Freightliner 120 on propane-butane. (**b**) Volvo-460 on methane

burning chamber of a convertible motor requires serious experimental and calculation work to ensure high economic, energy, and environmental parameters.

The creation of mono gas motors by the world's leading manufacturers of vehicle motors and cars, which are already widely used in road transport, testifies to the significant prospects of converting diesel motors into gas with spark ignition (Smigins 2016). Moreover, diesel motors conversion of is carried out both on the liquefied propane-butane mixture (Fig. 10.1a) and on compressed or liquefied natural gas (Fig. 10.1b). Well-known world concerns such as Cummins, Freightliner, Volvo, MAN, Mercedes, Iveco, Scania, and others are working in this direction. They have developed spark ignition gas motors for commercial vehicles based on existing diesel motors.

In general, to convert diesel motors to gas fuel-air mixture, the following must be performed: to install gas equipment, to reduce the compressing ratio of the diesel motor, to assemble the ignition system, and to configure motor management systems. It should be noted it is needful to answer many questions of technical, scientific, and commercial nature because there is only prior experience in converting diesel motors into mono gas ones (Kryshtopa et al. 2017; Gorski and Smigins 2018). The conversion of chemical fuel-air mixture energy into work plays an important role in choosing the class of gaseous fuel-air mixture. Converting of chemical energy of any class of fuel-air mixtures into work at internal burning motors is carried out at two stages: at the first stage, it is converted into the warmth; at, the second stage, the warmth is realized at work.

During these transformations, the main energy losses of the fuel-air mixture occur, and it can significantly increase the effectiveness of the internal burning motor (Afanasev and Tret'yakov 2016). It is needful to reduce the waste of chemical energy of the fuel-air mixture at internal burning motors at both stages of the transformation to increase the effectiveness of its usage. At actual motor construction, mainly different processes are used to reduce energy waste of fuel-air mixture effectiveness at the second stage of energy converting (Panchuk et al. 2019a, b). In regimen motors' temperature grade of the work, the body is such that its further increase causes a significant problem of providing the required thermic strength

Table 10.1 Converting temperature and cost of the main alternating fuel-air mixtures that can be used at motors

Class of fuel-air mixture	Ethanol	Methane	Octane	Propane	Methanol
Cost, Euro/kg	0.2–0.3	0.05–0.1	0.55–0.8	0.4–0.75	0.15–0.25
Converting temperature, T_c, K	600	1000	1000	700	570

(Abbondanza et al. 2020). Thus, increasing fuel-air mixture effectiveness usage at a warmth motor by increasing the upper thermodynamic temperature of the working fluid at the cycle depends on the ability to further increase the warmth resistance and warmth resistance of structural materials of motor parts. According to these conditions, it is recommendable to develop a process to reduce irreversible losses at the first stage of converting of chemical energy of the fuel-air mixture into warmth. Moreover, the implementation of this process should not be associated with an increase in the temperature grade of the working fluid (Cherednichenko 2019).

In the case of internal burning motors, to ensure the preliminary endothermic stage of converting as a starting convertible production, it is recommendable to use such classes of alternating fuel-air mixtures that have a converting temperature below the average temperature of the exhaust gases. Such fuel-air mixtures may include lower alcohols, alkanes, and simple ethers (Tselischev et al. 2020). The choice of the alternating fuel-air mixture as a production for the converting process is a compromise that considers its energy value; the temperature conditions of the process, formed during the processing of the gas; and their cost (Bureika et al. 2020; Korpaniuk et al. 2019).

The most promising product for the organization of converting in the first place should include methyl alcohol (methanol), which in world practice has long been used as an inexpensive substitute for expensive traditional motor fuel-air mixtures (Table 10.1).

For comparison, the average cost of diesel fuel-air mixture in Europe (Bildirici and Gökmenoğlu 2016) ranges from 1.15 to 2.0 Euros/kg (1–1.65 Euros/liter). From an economic point of view, it is recommendable to use methanol alcohol as an alternating fuel-air mixture for a diesel motor (Bahman et al. 2018). Methanol alcohol is a renewable natural resource, i.e., there is an excellent raw material base to increase its production and much wider use as an energy source. Methanol alcohol is widely used in the chemical industry, and considerable volumes are used in the production of fuel-air mixtures for vehicle motor vehicles (Havrysh et al. 2019). Usage of this alcohol as an alternating biofuel-air mixture for vehicles is possible due to its production at affordable and inexpensive ways from food waste and agricultural, from the gaseous fuel-air mixture. Nevertheless, one of the most important reasons for using methanol alcohol is to reduce emissions of components from the toxic exhaust gases of vehicle transport (Plashikhin et al. 2019). Methyl alcohol, with its small molecule size and more straightforward structure, is one of the determining factors for a "cleaner" burning of fuel-air mixture (Zhang 2010).

In comparison to other alternating fuel-air mixtures, cost of methyl alcohol is relatively low. In addition, the usage of methanol alcohol as a fuel-air mixture for diesel motors can significantly reduce emissions of nitrogen oxides and soot particles. It is because the burning of methanol alcohol at the diesel engine cylinder does not form intermediates that promote the formation of aromatic hydrocarbons and acetylene, which lead to the formation of soot (Li 2013).

Methanol alcohol has a higher octane number and a low grade of fire hazard. It is poorly soluble in petroleum fuel-air mixtures. The temperature of exhaust gases of the motors working on a methanol alcohol mixture is 30–70 °C lower than the basic internal burning motors (Liu 2018). Usage of liquefied methanol alcohol as a fuel-air mixture is accompanied by increased wear of the cylinder group caused by the ingress of its droplets on the walls of the engine cylinder and the destruction of the oil film.

Therefore, due to economic, environmental, and technical points of view, it is desirable to use methanol as an alternative fuel-air mixture for a diesel engine.

10.2 Materials and Processes

Thermochemical essentiality of the energy increase of primary fuel-air mixture represents the introductory provisions of thermodynamics, in particular, Hess's principle. It will be shown on the analysis of the basis of thermic effects from the burning of methanol, which is based on two processes. According to the first and the second processes of methanol alcohol oxidation, the primary and final states of systems are the same: primary, 1 kmol CH_3OH; final, 1 kmol CO_2; and 2 kmol H_2O.

According to the first process, methanol alcohol is burned directly at the burning chamber of an internal burning motor:

$$3/2O_2 + CH_3OH \rightarrow CO_2 + 2H_2O + Q_M, \tag{10.1}$$

where Q_M = exothermic thermic effect from the burning of the methanol-air mixture and $Q_M = 62.944$ MJ (Vershina and Bystrenkov 2019).

As a result of reactions (10.1) by the first process, 3 kmol of burning productions are formed.

According to the second two-stage process of methanol alcohol converting, alcohol decomposed at first:

$$CH_3OH \rightarrow CO + 2H_2 - Q_c. \tag{10.2}$$

It produces 2 kmol H_2 and 1 kmol CO with endothermic warmth conversion Q_c. Then burn at 3 kmol oxygen obtained by the reactions (10.2) of converting productions of methanol:

$$O_2 + 2H_2 \rightarrow 2H_2O + 481.5\text{MJ}; \tag{10.3}$$

$$1/2O_2 + CO \rightarrow CO_2 + 282.6\text{MJ}. \tag{10.4}$$

Total thermic effect

$$Q_\Sigma = 282.6 + 481.5 = 764.1\ \text{MJ/kmol}.$$

Thermic effects of reactions (10.3) and (10.4) are given according to the data (He et al. 2019). Then, according to Hess's law, total thermic effects for different routes of methanol alcohol oxidation must coincide:

$$629.4\frac{\text{MJ}}{\text{kmol}} = -Q_c + 764.1\ \text{MJ/kmol}. \tag{10.5}$$

Hence the endothermic thermic effect of the converting (methanol alcohol dissociation reaction) will be

$$Q_c = 134.6\ \text{MJ/kmol}.$$

Thus, the calculation showed that the thermic effect from burning of the converted mixture of CO and H_2 exceeds effect from the burning of the same amount of nonconvertible methanol alcohol (source fuel-air mixture) by $Q_c = 134.66$ MJ/kmol (≈21.4%), and it corresponds to the decay of alcohol energy.

The effectiveness of this converting, which is estimated by the thermic effectiveness coefficient of the engine cycle, depends on the ratio of the average supply T_1 (higher temperature) and removal temperatures T_2 (exit temperature) of warmth at the cycle (Kryshtopa et al. 2021a, b):

$$\eta = \frac{T_1 - T_2}{T_1}. \tag{10.6}$$

The higher warmth transfer temperature in the considered grade of warmth removal temperature, the higher the thermic effectiveness coefficient of the engine cycle.

It can be proved that implementation of such process is a priori possible only in the presence of alternating fuel-air mixtures burning, of which at the same temperature would be accompanied by different grades of irreversible losses. Such fuel-air mixtures can be converted into new (artificial) fuel-air mixtures with higher energy potential by thermochemical transformations. Converting the chemical energy of such artificial fuel-air mixture into warmth is accompanied by more minor irreversible losses.

This process of transferring energy is developed on fundamental thermochemistry principles and can be applied to any class of energy plant. The proposed process provides for the necessity to organize the endothermic process of fuel-air mixture converting, which is based on the engine cycle of its thermochemical converting, and the operating cycle of the energy plant provides for another process. If the chemical

Table 10.2 Comparative energy characteristics of the burning warmth of methanol alcohol and diesel fuel-air mixture

Parameter	Diesel fuel-air mixture, *Hu. d*	Methanol, *Hu. м*	Converted methanol, *Hu. p. м*
Absolute burning heat, MJ/kg	42.50	19.70	23.87
Relative burning heat, %	100	46.35	56.17

energy of the fuel-air mixture converts at a conventional energy plant into heat at one stage, then the proposed process will be at two ones.

In the first stage, the primary fuel-air mixture is converted, and at the second one, the converted fuel-air mixture is burned at a higher energy grade. Two stages of energy converting at the process of which the warmth removed from the engine cycle are used increasing the energy effectiveness of the primary fuel-air mixture. In the common case, the effectiveness of thermochemical recovery depends on the class of energy plant, process, and conditions of working process organization, as well as, the class of fuel-air mixture, and endothermic energy effect of the reaction system of its converting.

It should be noted that converting productions of traditional petroleum fuel-air mixtures obtained in thermoreactor based on the mechanism of exothermic reactions of incomplete oxidation of hydrocarbons have lower heat of burning than the primary fuel-air mixture, and this fact eliminates the effect of recovery (Zhennan et al. 2018). Let us calculate the burning heat increase of converting productions for liquefied methanol. Combustion heat of gaseous productions of methanol alcohol converting $Hu.\,p.\,м = 23.87$ MJ/kg (Mäyrä and Leiviskä 2018). The heat of burning of methanol alcohol $Hu.\,м = 19.67$ MJ/kg (Table 10.2).

Thus, burning at the motor of 1 kg of methanol alcohol-converting productions obtained from the same mass of liquid methanol alcohol releases additional warmth energy accumulated during converting of alcohol fuel-air mixture:

$$Hu.c = Hu.p.м - Hu.м = 4.20\ \text{MJ/kg} \tag{10.7}$$

Thus, 21.4% of primary fuel-air mixture energy is returned to the operating cycle of motors for helpful working. This energy is obtained from the utilization of thermic energy of exhaust gases during endothermic converting at conditions of the actual operation of the internal burning motor. According to the external warmth balance of the internal burning motor, a significant part of the chemical energy is not realized in the form of working and is discharged into the surrounding space with engine exhaust gases too. At diesel motors component of energy losses at the warmth balance reaches 25–40% (Yakovlieva and Boichenko 2020); at gasoline and gas motors, the part of irreversible heat loss lost from the motor with exhaust gases is 30–40%. This corresponds to 12–22 MJ of warmth per 1 kg of the consumed fuel-air mixture.

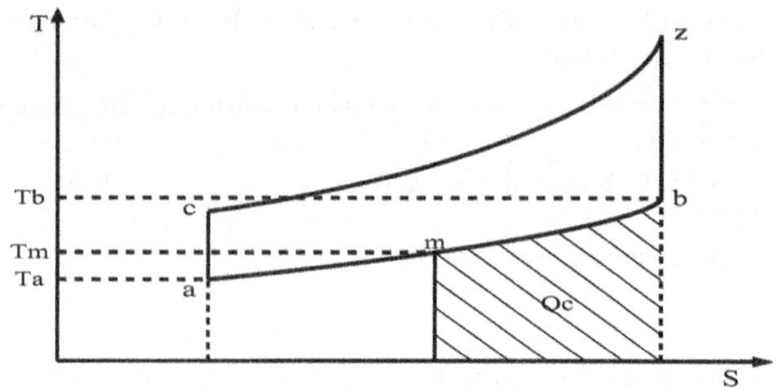

Fig. 10.2 Thermodynamic cycle for diesel motors converted to gaseous productions of methanol alcohol converting with warmth recovery

Warmth must be extracted from the exhaust manifold for ensuring the converting process, where the temperature of the motor parts reaches from 650 to 750 K at idling speed to 1000–1100 K at maximum loads (Wargula et al. 2019). Even considering the heat loss during transferring of energy from the motor to the working body, it is light to achieve the converting temperature of methanol alcohol to $T_{\mathrm{m}} =$ 568–572 K.

From the above, it follows that usage of pre-chemical converting systems of alternating fuel-air mixtures can increase the effectiveness of its use at internal burning motors by regenerating exhaust gases warmth. For example, the thermodynamic cycle (Otto cycle) for diesel motors converted to gas with spark forced ignition with warmth recovery can be represented by a T-S diagram (Fig. 10.2).

In the diagram, warmth suitable for recovery at the cycle is expressed as a portion of the waste heat Q_{w}, i.e., the degree of recovery is equal to (Boichenko et al. 2020)

$$\eta_{\mathrm{p}} = \frac{Q_{\mathrm{c}}}{Q_{\mathrm{w}}}, \tag{10.8}$$

where

Q_{w} is the waste heat removed per cycle.

Warmth quantity removed per cycle Q_{w} (Fig. 10.2) is determined from the formula

$$Q_{\mathrm{w}} = C_{\mu v} \cdot M_{pr.c.}(T_{\mathrm{b}} - T_{\mathrm{a}}), \tag{10.9}$$

where

$M_{pr.\ c.}$ is the number of burning productions at a constant volume;
$C_{\mu v}$ – average molar warmth capacity of burning productions at constant volume.

Endothermic warmth converting quantity Q_{c} is determined from (7) and from the formula

$$Q_{\mathrm{c}} = C_{\mu v} \cdot M_{pr.c.}(T_{\mathrm{b}} - T_{\mathrm{m}}). \tag{10.10}$$

In other words, the recovery degree depends on the converting process temperature T_m, and it increases with decreasing of temperature T_m, and it can be determined from the formula

$$\eta_p = \frac{T_b - T_m}{T_b - T_a}. \tag{10.11}$$

These requirements for the exhaust gases temperature grade of the motor may not be met at all ranges of operation regimes of internal burning motors. For example, for a motor not warmed up to operating temperature, the recovery of the converting process degree will be reduced (Kryshtopa et al. 2021a, b). However, it should be noted that the operating time of the motor at the warm-up regime is relatively short. In addition, it should be noted that regimen catalysts allow implementing of the converting of alternating fuel-air mixtures at lower operating temperatures. For example, for methanol alcohol, it is 300–400 °C (Alarifi et al. 2015), and this fact determines the possible minimum temperature regime for organization the converting process of exhaust gases at internal burning motors. Thus, we can talk about the almost constant converting process at internal burning motors due to warmth recovery of engine exhaust gases for today.

Let us compare possibilities of energy converting at two energetically similar internal burning motors to analyze the ability of effectiveness increasing of chemical energy usage of the fuel-air mixture. The first one works according to the ordinary engine scheme, the second – using thermochemical recovery at two stages. The required thermic regime of fuel-air mixture converting is provided by supplying warmth to the exhaust gases.

The first motor works as follows. The primary fuel-air mixture at the environment parameters enters to the motor, where it burns with warmth emitting $Q_n = \mathrm{Nu}$. This warmth is transferred to the working body. Let us approximate that warmth transfer occurs at an average temperature T_{av1}. The exhaust gas is discharged into the air atmosphere after the cycle. To simplify the analysis, we assume that warmth dissipation occurs for an average temperature T_{av2}.

Then the working performed by the motor can be determined as follows from the formula

$$L_m = \frac{T_{av1} - T_{av2}}{T_{av1}} Hu.m. \tag{10.12}$$

The effectiveness of chemical energy used for fuel-air mixture is equal to

$$\eta = \frac{T_{av1} - T_{av2}}{T_{av1}} = \frac{L_m}{Hu.m}, \tag{10.13}$$

Dependence (10.14) determines the effective coefficient of the thermic energy plant, which is equal to the ratio of the quantity of energy converted into working to the whole energy received into the thermic energy plant. The effective coefficient of this idealized plant coincides with the thermic effectiveness coefficient of the cycle. It indicates that in a motor without thermochemical recovery, the energy effectiveness of the fuel-air mixture depends entirely on converting warmth into the work.

The second motor works with thermochemical recovery of exhaust gas warmth. Unlike the first motor, the fuel-air mixture enters the burning chamber not immediately but pre-undergoes the stage of thermo processing at a thermochemical reactor. There is thermochemical converting of the primary fuel-air mixture into a convertible in the reactor with the help of warmth coming at the average lower temperature T_{av2}. It goes to the burning chamber of the motor. Assuming that the warmth exchange at this motor is carried out similarly to the first variant, the working body of the second motor with thermoreactor must receive warmth even at medium temperature T_{av1}. Then in this case, the warmth taken by the working body *Hu. p. м* exceeds the burning warmth *Hu. м* of the primary fuel-air mixture by the amount *Hu. c* that was absorbed during the thermochemical converting of the source fuel-air mixture at the thermoreactor and is accordingly equal to

$$Hu.p.м = Hu.c + Hu.м. \tag{10.14}$$

We will assume that the motor with other reactor converts the warmth perceived by the working body with the same effectiveness as the motor without a thermochemical reactor, i.e., the thermic effective coefficient of both units is the same. Then the work with the thermoreactor will be equal to

$$L_{p.м} = \frac{T_{av1} - T_{av2}}{T_{av1}}(Hu.м + Hu.c). \tag{10.15}$$

Then the effectiveness of the fuel-air mixture energy in the motor containing the unit of thermochemical warmth recovery will be determined from the formula

$$\eta_{n.м} = \left(\frac{T_{av1} - T_{av2}}{T_{av1}}\right)\left(\frac{Hu.м + Hu.к}{Hu.м}\right) = \left(\frac{T_{av1} - T_{av2}}{T_{av1}}\right)\frac{Hu.p.м}{Hu.м}. \tag{10.16}$$

Let us analyze the effectiveness of fuel-air mixture energy usage at a motor containing the thermochemical recovery unit and working on methanol alcohol in comparison to the ordinary motor. Let us take the average upper temperature of the cycle $T_{av1} = 2100$ K, and the lower average one – $T_{av2} = 1260$ K. Then, according to dependence (10.14), we have the thermic effectiveness coefficient of an ordinary motor $\eta_{м} = 40\%$. As shown above, the warmth of the burning methanol alcohol is *Hu. м* 19.67 MJ/kg. The warmth of burning of gaseous productions of methanol alcohol converting is *Hu. p. м* 23.87 MJ/kg. For a motor with thermochemical recovery, its thermic effective coefficient $\eta_{n.\ м}$ will exceed the thermic effective coefficient $\eta_{М}$ of the first motor at (23.87/19.67) = 1.214 times, i.e., for a motor with thermochemical recovery, its thermic effective coefficient will be equal to $\eta_{p.\ м} = 48.5\%$.

From the above analysis, it follows that the external irreversible losses of the fuel-air mixture chemical energy converting into the warmth in case of using the process of thermochemical warmth recovery of the exhaust gases is always less than corresponding losses at direct burning of the fuel-air mixture without prior thermo-chemical process. Moreover, the reduction of the irreversible waste is adequate to the energy that must be spent to compensate for the total thermic endothermic effect of converting reactions of the primary fuel-air mixture (Panchuk et al. 2020a).

The theoretical foundations of such energy supply systems for motors in regimen research practice remain poorly understood (Tang et al. 2020; Cherednichenko 2019; Tartakovsky et al. 2014; Korohodskyi et al. 2020). However, they are of significant interest in terms of assessing the potential applications of thermochemical recovery motors in energy supply systems and improving the energy effectiveness of the alternating fuel-air mixtures, at particular for existing diesel motors, that will be converted to gases. The implementation of a thermochemical process of the warmth utilization for the operating cycle conditions of internal burning motors with spark ignition is possible if the primary fuel-air mixture is hydrocarbon compounds with a relatively low temperature of converting reactions (alcohols, ethers, and similar compounds). Gaseous converting productions can be used as the basic fuel-air mixture to energy the gas motor. It is essential to implement this process at energy supply systems of diesel motors that are converted to the gaseous fuel-air mixtures.

Let us analyze the conditions for the engine achieving the maximum possible degree of recovery. The conditions for achieving the maximum possible degree of recovery are implemented when the endothermic effect of the converting reaction corresponds to supply of an equivalent quantity of the warmth to the reaction space from an external source – warming coolant and it acts exhaust gases warmth of the motor and can be the warmth of the cooling system. At actual conditions of the converting process at the exhaust system of the internal burning, motor coolant (the exhaust gas and coolant) must have a higher potentially grade of energy, which is used not only to compensate by the endothermic effect of the converting reaction but also to organize its auxiliary cycles. Energy is required to pre-warmth the source fuel-air mixture to the boiling point, its evaporation, increase the vapor temperature to the dissociation temperature, compensate for warmth loss to the environment through the walls of the thermochemical reactor, and supply pipelines.

For example, the total warmth consumption for a fully completed conversion of 1 kg of methanol alcohol reaches 7 MJ (Panchuk et al. 2020b). At the same time, to compensate for the engine endothermic effect of the alcohol converting is spent about 60% of total energy waste. A significant portion of them (about 22–27%) is spent on the energy-intensive processing of vaporization (the warmth of evaporation of methanol alcohol 1.1 MJ/kg). The maximum possible degree of recovery is achieved when the thermic energy of the coolant exceeds the total energy required for the converting process.

10.3 Results and Discussion

The aim of experimental studies is using methanol alcohol-converting productions at alternating fuel-air mixture mixtures at existing diesel motors to save petroleum motor fuel-air mixture and improve their environmental characteristic. Evaluation of the effectiveness of the thermochemical process of warmth utilization at the operating cycle of the diesel class D21A was performed at the laboratory on a diesel motor stand. A brief technical description of the diesel experimental motor D21A1 is shown in Table 10.3.

A scheme of the experimental diesel stand based on the motor D21A1 for research of characteristics of its working to productions of methanol alcohol converting is shown in Fig. 10.3. The experimental stand motor could work at three regimes: supply of diesel fuel-air mixture at the liquefied phase through the fuel injectors, supply of methanol alcohol-converting productions through a gas reducer, and supply of methanol alcohol at the liquefied phase through the injectors. Methanol alcohol was supplied from the fuel-air mixture tank by an electric fuel-air mixture pump through the fuel-air mixture pressure regulator.

The fuel-air mixture pressure regulator supplies of methanol alcohol at a pressure of 0.5 MPa to compensate relatively low warmth of methanol alcohol burning and increase its supply. A throttle pipe with an air throttle valve was mounted on the experimental motor to ensure regulation of the methanol-air mixture. Methanol alcohol is warmed to convert productions by exhaust gases or liquid coolant warmer during the cold motor start-up and warm-up. To ensure cold start of the motor at the regime of operation from productions of methanol alcohol converting at the front of the fuel, gas reducer was assembled liquid coolant warmer Webasto Thermo Top C (5 kW). The liquid coolant warmer was turned off when the exhaust gas reached the temperature 280–300 °C. The catalytic methanol alcohol-converting reactor was installed at the diesel exhaust system at the immediate vicinity of the exhaust manifold (Fig. 10.4).

Table 10.3 Brief characteristics of the experimental motor D21A1

The name of the motor parameters	Unit of measurement	Value
Class of diesel motor	–	Two-cylinder, four-stroke, air-cooled
Diesel motor displacement	Liter	2.08
Diesel motor weight	kg	280
The process of mixing	–	Direct injection of diesel fuel-air mixture, integral burning chamber
Rated motor energy	kW (hp)	18.4 (25)
Efficient specific fuel-air mixture consumption, (g/hph)	g/kWh	248 (182)
Motor crankshaft speed at rated energy	rpm	1800
Motor crankshaft speed in idling speed	rpm	775–825

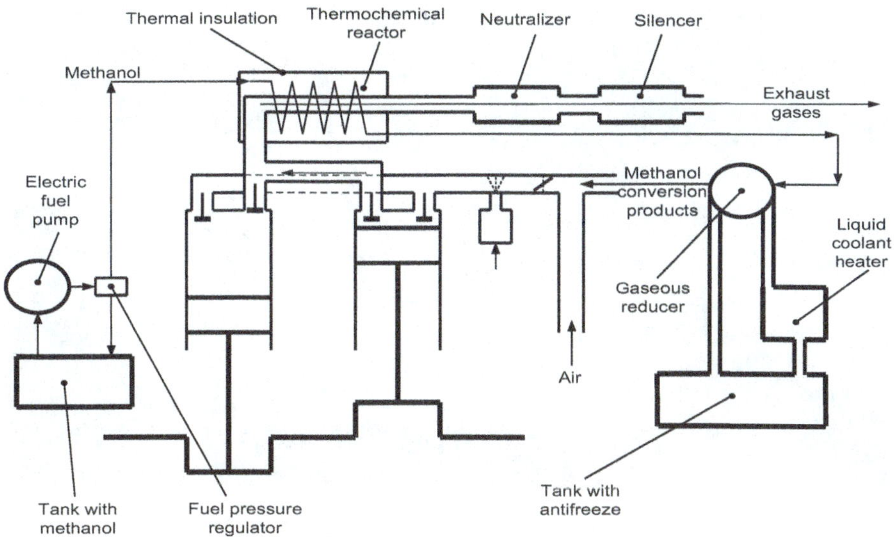

Fig. 10.3 Scheme of the experimental diesel stand based on the motor D21A1 for study parameters of work of methanol alcohol-converting productions

Fig. 10.4 The appearance of the experimental diesel stand based on the motor D21A1 for the study of characteristics on alternating fuel-air mixtures: (1) thermoreactor; (2) loading device of the motor; (3) D21A1 diesel motor; (4) gas reducer

The fuel-air mixture was ignited using an electronic ignition system of own design. The developed original ignition electronic system consists of a control unit 1 (Fig. 10.5), which changes the angle of ignition advance depending on the regime of operation of the motor, switch 2, ignition module 3, high-voltage wires 4, and spark plug 5.

Fig. 10.5 The ignition system of the converted diesel motor D21A1: (1) electronic control unit; (2) commutator; (3) spark ignition module; (4) starter; (5) high-voltage wire; (6) plug spark; (7) sensor of position crankshaft

Fig. 10.6 D21A1 diesel motor with dismantled unit heads

The D21A1 diesel motor was converted to work on an alternating gases mixture. For this reason, diesel fuel injectors were dismantled from the engine unit heads, other threads were cut in the spray injector channels, and spark plugs were installed. The motor D21A1 with the dismantled heads of the engine block is shown at Fig. 10.6.

The design of the motorhead is redesigned so that instead of spark plugs, the diesel fuel injector can easily screw back. The convertibility of the experimental motor allows switching to diesel or methanol alcohol fuel-air mixture during 10–15 min by replacement of diesel fuel injectors to spark plugs. Converted from diesel fuel-air mixture to gas heads of the motor block, D21A1 are shown in Fig. 10.7.

a b

Fig. 10.7 Converted from diesel to gas fuel-air mixture heads of the unit. View from the timing valves head (**a**) and the burning chamber (**b**): (1) the head of the block engine before re-equipment with established diesel fuel injectors; (2) the head of the engine block after re-equipment with dismantled injectors and established spark plugs; (3) plugs spark; (4) diesel fuel injectors

Fig. 10.8 Experimental dependences of the effective energy *N* on the motor crankshaft speed *n* for different motor fuel-air mixtures ------ operation of the motor on diesel fuel-air mixture; ········ motor operation on methanol alcohol-converting productions

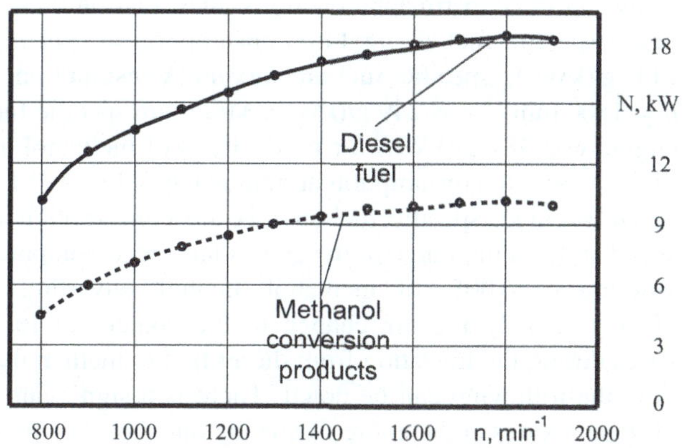

The purpose of experimental research was to compare the primary energy, fuel-air mixture-economic, and environmental characteristics of the converted diesel motor during its operation on diesel fuel-air mixture and methanol alcohol-converting productions. Experimental studies were performed on the motor diesel stand of the D21A1 motor to evaluating the energy-saving effect. Experimental dependences of external speed characteristics of the diesel motor D21A1 converted for diesel fuel-air mixture and methanol alcohol-converting productions are shown in Fig. 10.8. It was found, analyzing experimental energy values, that at nominal speed engine (n = 1800 min^{-1}), the effective energy N on diesel fuel-air mixture was 18.1 kW, but on the productions of methanol alcohol converting, it was equal to 10.1 kW. On average, the value of the effective energy *N* of the motor at the entire frequency range of the crankshaft compared to the diesel fuel-air mixture working on methanol alcohol-converting productions decreased by 45%.

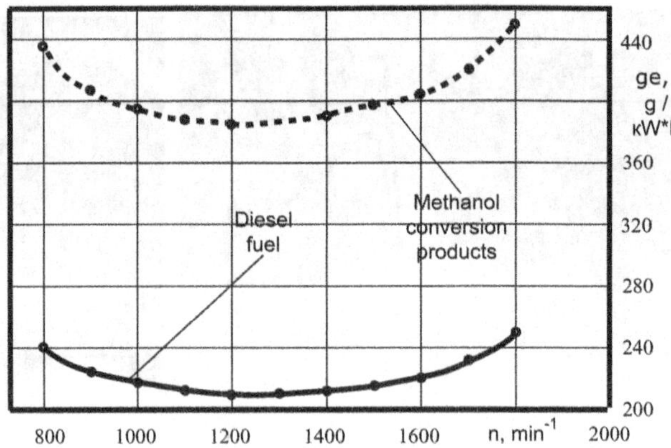

Fig. 10.9 Experimental dependences of specific fuel-air mixture consumption on the motor crankshaft speed n for different motor fuel-air mixtures: ------ operation of the motor on diesel fuel-air mixture; ········· motor operation on methanol alcohol-converting productions

The experimental dependences of specific fuel-air mixture consumption on the crankshaft speed n of the diesel motor D21A1 worked on diesel fuel-air mixture and methanol alcohol-converting productions are shown in Fig. 10.9. It was found, analyzing the experimental energy values, that on diesel fuel-air mixture, the minimum specific fuel-air mixture consumption at rpm n = 1220–1230 min^{-1} was 211 g/(kWh); specific fuel-air mixture consumption at nominal speed engine n = 1800 min^{-1} was 249 g/(kWh). Minimum specific fuel-air mixture consumption engine was 387 g/(kWh) for productions of methanol alcohol converting; specific fuel-air mixture consumption at nominal speed was 451 g/(kWh).

On average, specific fuel-air mixture consumption from the motor crankshaft speed at the entire range of the crankshaft speed compared to diesel fuel-air mixture working on 100% of methanol alcohol-converting productions increased by 80–84%. Analyzing the change in the content of toxic components at exhaust gases during the transition from diesel fuel to methanol alcohol-converting production, the following can be noted. There is a significant decrease in the content of nitrogen oxides at the entire frequency range of the engine crankshaft (Fig. 10.10).

The content of nitrogen oxide engine in n = 800 min^{-1} decreased from 1136 ppm when the motor is running on diesel fuel-air mixture to 460 ppm when the motor is running on the productions of methanol alcohol converting. The decrease in the content of nitrogen oxide engine was 59.5%. At n = 1800 min^{-1} the content of nitrogen oxide engine decreased from 826 ppm when the motor is running on diesel fuel-air mixture to 305 ppm when the motor is running on the productions of methanol alcohol converting. The decrease in the content of nitrogen oxides was 53.2%. The decrease in the content of nitrogen oxide engine during the operation of the motor on diesel fuel-air mixture compared with the operation of the motor on the productions of methanol alcohol converting is explained by lower warmth dissipation rates and more minor increase in burning pressure at the motor.

There is a significant decrease in the content of the carbon monoxide at the entire frequency range of the crankshaft (Fig. 10.11). Thus, the content of the carbon monoxide engine at n = 800 min^{-1} decreased from 0.226% when the motor is running on a diesel fuel-air mixture to 0.150% running on the productions of

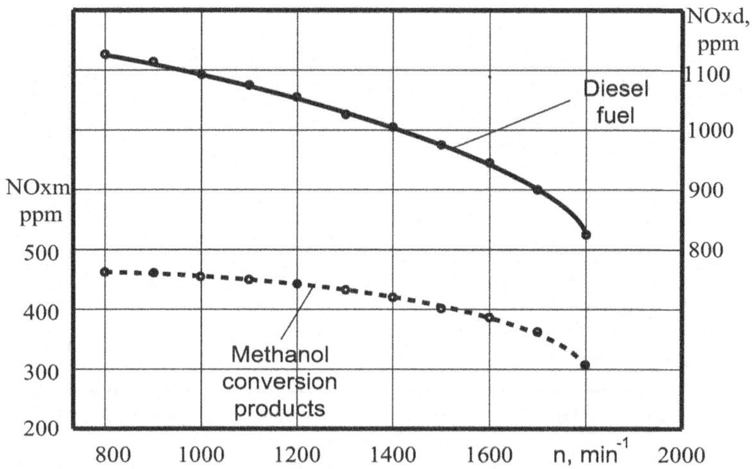

Fig. 10.10 Experimental dependences of the content of nitrogen oxides at exhaust gases on the crankshaft speed of the motor *n* for different motor fuel-air mixtures: ------ operation of the motor on diesel fuel-air mixture; ········ motor operation on methanol alcohol-converting productions

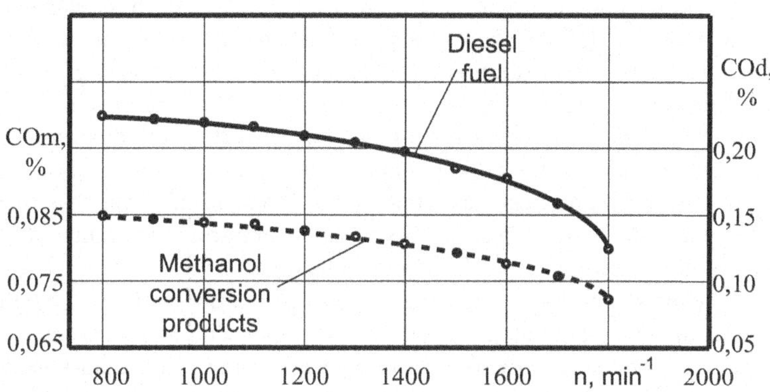

Fig. 10.11 Experimental dependences of the content of carbon monoxide at the exhaust gases engine on the motor crankshaft speed *n* for different fuel-air mixtures: ------ operation of the motor on diesel fuel-air mixture; ········ motor operation on methanol alcohol-converting productions

methanol alcohol converting. At $n = 1800$ min^{-1}, the carbon monoxide engine content decreased from 0.086% when the motor is running on diesel fuel-air mixture to 0.072% when the motor is running on methanol alcohol-converting productions. That is, the decrease in the carbon monoxide content occurs in the range of 62.2–52.0%.

The hydrocarbon content increases slightly at the entire range of crankshaft speed (Fig. 10.12). Thus, at $n = 800$ min^{-1}, the hydrocarbon engine content varies from 0.043% when running on diesel fuel-air mixture to 0.031% when the motor is running on methanol alcohol-converting productions. That is, the decrease in the

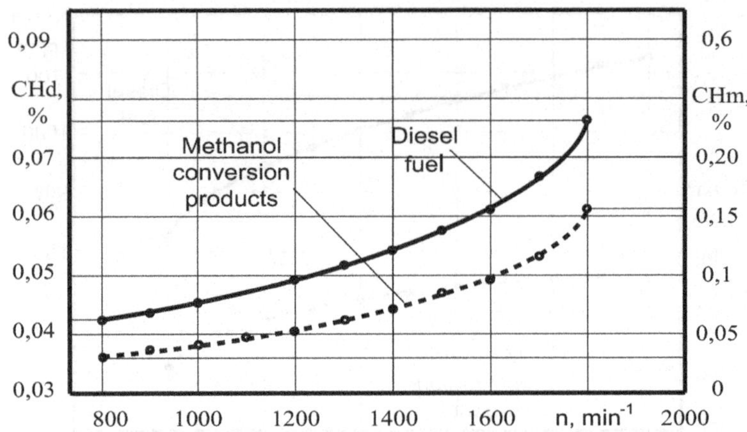

Fig. 10.12 Experimental dependences of the hydrocarbon content at the exhaust gases on the crankshaft speed of the motor n for different motor fuel-air mixtures ------ operation of the motor on diesel fuel-air mixture; ········ motor operation on methanol alcohol-converting productions

hydrocarbon engine content was 26.1%. At $n = 1800\ \text{min}^{-1}$, the hydrocarbon engine content increases from 0.076% when the motor runs on the diesel fuel-air mixture to 0.155% when the motor is running on methanol alcohol-converting productions. That is, increasing in the hydrocarbons was 2.01 times. Comparing motor work on liquefied methanol alcohol and methanol alcohol-converting productions were also performed. On average, during experimental studies, the motor running on methanol alcohol-converting productions was compared to motor running on liquefied methanol. There was a decrease of fuel-air mixture consumption at the range of 10–14% depending on the crankshaft speed and motor loading.

To identify the share of each factor at improving the effectiveness of the motor, it was tested using a methanol alcohol-converting reactor with an autonomous electric warmer, i.e., without usage (recovery) of thermic energy of exhaust gases. It is established that energy saving for this diesel due to the thermochemical process of warmth utilization was 6–8% and due to the improvement of the working process, was about 4–6%.

There was the additional experimental study devoted to identifying characteristics of improving fuel-air mixture economy of the experimental motor. At the first stage of tests using a balloon gas supply system, the synthesis gas was fed into engine cylinders, which has a component composition (capacity fractions) similar to the composition of methanol alcohol-converting productions: 33% CO and 67% H_2. It is evident, that in this case, the effect of utilization of thermic energy of exhaust gases is not shown. At the second stage of testing, the motor worked with the thermoreactor. The utilized energy of engine exhaust gases converted into a new class of the gaseous fuel-air mixture (methanol alcohol-converting productions) with higher energy parameters was used for additional helpful work. The motor worked on the scheme of the regenerative cycle engine.

Because in both cases, the fuel-air mixture had the same component composition, its burning parameters, and, hence, the nature of the impact of kinetic characteristics of the engine operating cycle were identical. Thus, the increase of motor effectiveness, which was observed in experimental studies, clearly corresponded to the effect of term recovery of the engine exhaust energy.

10.4 Conclusions

The experimental engine studies have shown that converting diesel motors to work using methanol alcohol-converting productions is quite profitable. It is established that the effectiveness of the motor at idling regime depends on its engine speed. On average, the specific fuel-air mixture consumption from the motor crankshaft speed at the entire range of the engine crankshaft speed compared to diesel fuel-air mixture working on 100% of methanol alcohol-converting productions increased by 81–83%. As the price of methanol alcohol is, on average, 15–20% of the cost of fuel diesel engines, the converting of diesel motors for work using methanol alcohol-converting productions is very profitable.

Economic characteristics of operating engine cycle at all loading regimes of the motor with thermoreactor were higher (on average by 10–15%) than it worked on methanol alcohol without a thermoreactor. Operating engine cycle parameters almost corresponded to the basic parameters of the motor during its work on liquefied methanol alcohol at low engine crankshaft speeds (from 800 to 900 min^{-1}), characterized by energy grades and low temperature of engine exhaust gases. The most significant increase in effectiveness (by 15% compared to work on liquefied methanol) was observed at the range of shaft speed 1300–1600 min^{-1} at an exhaust gas temperature of 420–460 °C. It is because at this range of speeds of the thermoreactor, consumed thermic energy and the energy of the coolant are almost the same.

The improvement of engine economic parameters of the studied diesel motor with thermochemical process of warmth utilization running on the mixed hydrogen-containing fuel-air mixture is due to the influence of two factors: the effect of partial utilization of engine exhaust gas warmth and improvement of kinetic burning at the presence of hydrogen.

Reduction of fuel-air mixture consumption was accompanied by an improvement at the environmental parameters of the diesel motor, which works in conjunction with a thermochemical methanol alcohol-converting reactor. In particular, depending on the engine crankshaft speed and the load on the motor formation of nitrogen oxides at the exhaust gases decreased by 52–60%, carbon monoxide occurred at the range of 53–63%.

Studies have shown that at loading regimes, when the temperature of the exhaust gases at the inlet to the thermochemical reactor exceeds 400–420 °C, the thermochemical reactor productivity including the target component reaches its highest grade significantly increasing the effectiveness of the operating engine cycle.

Application of the considered process of energy utilization of exhaust gases at transport motors is quite promising. Its implementation does not require great financial investments and radical re-equipment of existing motor production due to technical simplicity. Any serial regime of diesel motors can be used as the base motor. This applies to both in-service motors and new motors. The main element of the converting system is the thermoreactor which has the simplest design of the warmth exchanger; mass and dimensions of which at the capacity of a conventional silencer provide the convenience of its installation at the exhaust system of the motor.

An important stimulus for the further development of such systems is determining the ability of cumulative improvement of motor car characteristics on a set of parameters. Their implementation on automobiles allows, in particular, to utilize waste thermic energy, improve burning processes, and improve the environmental property of vehicles, providing the opportunity to replace traditional petroleum fuel-air mixture with alternating energy from including biological renewable sources, thus helping to solve global problems of resource conservation.

References

Abbondanza M, Cavina N, Corti E et al (2020) Development of a combustion delay model in the control of innovative combustions. E3S Web of conferences, vol. 197, Article 6013

Afanas'ev A, Tret'yakov A (2016) Simulation of diesel engine energy conversion processes. J Min Inst 222:839–852

Alarifi A, Alsobhi S, Elkamel A et al (2015) Multiobjective optimization of methanol synthesis loop from synthesis gas via a multibed adiabatic reactor with additional interstage CO_2 quenching. Energy Fuel 29(2):530–537

Bahman N, Sina F, Shahaboddin S et al (2018) Application of ANNs, ANFIS and RSM to estimating and optimizing the parameters that affect the yield and cost of biodiesel production. Eng Appl Comput Fluid Mech 12(1):611–624

Bildirici M, Gökmenoğlu S (2016) Environmental pollution, hydropower energy consumption and economic growth: evidence from G7 countries. Renew Sust Energ Rev 75:68–68

Boichenko S, Zubenko S, Konovalov S et al (2020) Synthesis of camelina oil ethyl esters as components of jet fuels. East Eur J Enterp Technol 1(6):42–49

Boretti A (2019) Advantages and disadvantages of diesel single and dual-fuel engines. Front. Mech. Eng. 5:64

Bureika G, Matijošius J, Rimkus A (2020) Alternative carbonless fuels for internal combustion engines of vehicles. Lect Notes Netw Syst 124:1–49

Cherednichenko O (2019) Efficiency analysis of methanol usage for marine turbine power plant operation based on waste heat chemical regeneration. Probl Reg Energ 39(1):102–111

Gorski K, Smigins R (2011) Impact of ether/ethanol and biodiesel blends on combustion process of compression ignition engine. In: 10th international scientific conference on engineering for rural development, Jelgava, pp 260–265

Gorski K, Smigins R (2018) Selected physicochemical properties of diethyl ether/rapeseed oil blends and their impact on diesel engine smoke opacity. Energy and Fuel 32(2):1796–1803

Havrysh V, Kalinichenko A, Minkova O et al (2019) Agricultural feedstock for solid and liquid biofuel production in Ukraine: cluster analysis. Procedia Environ Sci Eng Manag 6(4):649–658

He L, Fu Y, Lidstrom M (2019) Quantifying methane and methanol metabolism of "Methylotuvimicrobium buryatense" 5GB1C under substrate limitation. Msystems 4(6): 748–719

Jurkovič M, Kalina T, Jancosek L et al (2019) Proposal of conversion the tugboat engines to diesel – LNG operation. Adv. Sci. Technol. Res. J. 13(4):129–142

Korohodskyi V, Kryshtopa S, Migal V et al (2020) Determining the characteristics for the rational adjusting of an fuel-air mixture composition in a two-stroke engine with internal mixture formation. East Eur J Enterp Technol 2(5-104):39–52

Korpaniuk T, Ishchenko Y, Koval N (2019) Backgrounds for improving resource management of agricultural enterprises based on economic diagnostics of biofuel consumption. J Soc Sci Res 5(2):367–380

Kovalov S (2020) Designing the shape of the combustion chambers for gas engines converted on the basis of the diesel engines. East Eur J Enterp Technol 2:1 (104):23–31

Kryshtopa S, Kryshtopa L, Melnyk V et al (2017) Experimental research on diesel engine working on a mixture of diesel fuel and fusel oils. Trans Probl 12(2):53–63

Kryshtopa S, Melnyk V, Dolishnii B et al (2019) Improvement of the model of forecasting heavy metals of exhaust gases of engine vehicles in the soil. East Eur J Enterp Technol 4(10-100):1–8

Kryshtopa S, Górski K, Longwic R et al (2021a) Increasing parameters of diesel engines by their transformation for methanol conversion products. Energies 14(6):1710

Kryshtopa S, Kryshtopa L, Panchuk M et al (2021b) Composition and energy value research of pyrolise gases. IOP Conf Ser Earth Environ Sci 628(1):012008

Li Y (2013) Numerical study on the combustion and emission characteristics of a methanol/diesel reactivity controlled compression ignition (RCCI) engine. Appl Energy 106(2):184–197

Liu Z (2018) Economic analysis of methanol production from coal/biomass upgrading. Energy Sources B: Econ Plan Policy 13(1):66–71

Mateichyk V, Saga M, Smieszek M et al (2020) Information and analytical system to monitor operating processes and environmental performance of vehicle propulsion systems. IOP Conf Ser Mater Sci Eng 776(1):012064

Mäyrä O, Leiviskä K (2018) Modeling in methanol synthesis. Elsevier, Methanol, pp 475–492

Panchuk M, Kryshtopa S, Panchuk A et al (2019a) Perspectives for developing and using the torrefaction technology in Ukraine. Int J Energy Clean Environ 20(2):113–134

Panchuk M, Kryshtopa S, Sladkowski A et al (2019b) Efficiency of production of engine biofuels for water and land transport. Nase More 66(3):6–12

Panchuk M, Kryshtopa S, Panchuk A (2020a) Innovative technologies for the creation of a new sustainable, environmentally neutral energy production in Ukraine. In 2020 international conference on decision aid sciences and application, 9317165, pp 732–737

Panchuk M, Kryshtopa S, Sładkowski A et al (2020b) Environmental aspects of the production and use of biofuels in transport. Lect Notes Netw Syst 124:115–168

Plashikhin S, Zelenin E, Semenyuk N et al (2019) Purification of a flue gas from solid particles and acidic impurities. Int J Energy Clean Environ 20(3):247–259

Smigins R (2016) Perspectives of low level ethanol and biodiesel/diesel fuel blends on diesel engine emission reduction. In Transport means. Proceedings of the international conference, pp 337-341

Tang G, Wang S, Zhang L et al (2020) Diagnosis and improvement of combustion characteristics of methanol miniature reciprocating piston internal combustion engine. Micromachines 11(1):96

Tartakovsky L, Baibikov V, Veinblat M (2014) Modeling methanol steam reforming for internal combustion engine. Energy Power 4(1A):50–56

Tselischev O, Kudryavtsev S, Loriya M et al (2020) Modification of engine gasoline with bioethanol in the cavitation field. Voprosy Khimii i Khimicheskoi Tekhnologii 6:171–178

Vershina G, Bystrenkov O (2019) Influence of diesel fuel ignition portion value on working process parameters of gas-diesel engine. Sci Tech 18(5):395–400

Wargula Ł, Waluś Konrad J, Krawiec P (2019) The problems of measuring the temperature of the small engines (SI) on the example of a drive for non-road mobile machines. In Matec Web of conferences, vol. 254, Article 04004

Yakovlieva A, Boichenko S (2020) Energy efficient renewable feedstock for alternative engine fuels production: solutions for Ukraine. Stud Syst Decis Control 298:247–259

Zhang Z (2010) Experimental investigation on regulated and unregulated emissions of a diesel/methanol compound combustion engine with and without diesel oxidation catalyst. Sci Total Environ 408(4):865–872

Zhennan H, Sulong G, Xi Z et al (2018) Reaction decoupling in thermochemical fuel conversion and technical progress based on decoupling using fluidized bed. Carbon Resour Conver 1(2): 109–125

Chapter 11
Influence of Microbiological Pollution on Properties of Motor Fuels

Olena Shevchenko and Daryna Popytailenko

Nomenclature

FAME	Fatty acid methyl esters
CO_2	Carbon dioxide
H_2S	Hydrogen sulfide
NH_3	Ammonia
N_2O_5	Dinitrogen pentoxide

11.1 Introduction

Increasing environmental pollution, environmental degradation, and declining oil inventories around the world have led scientists to search for alternative fuels derived from renewable sources. Currently, fuel based on fatty acid methyl esters (FAME) is intensively used. FAME can be added in any concentration to traditional fuel, or completely replace it, as they have similar performance properties.

There is more and more controversy surrounding the production of fuel based on FAME. One of its main advantages is environmental friendliness, which is to reduce the concentration of harmful substances emitted from exhaust gases and its better biodegradability compared to mineral diesel fuel.

Biodegradability of FAME is an advantage in the context of bioremediation but is a disadvantage for fuel quality control. Mineralization of FAME is two to four times higher than the mineralization of petroleum diesel fuel. The chain length of fatty acids, the number and position of C=C double bonds, and the presence of

O. Shevchenko · D. Popytailenko (✉)
State Higher Educational Institution, Ukrainian State University of Chemical Technology, Dnipro, Ukraine

S. Boichenko et al. (eds.), *Chemmotological Aspects of Sustainable Development of Transport*, Sustainable Aviation, https://doi.org/10.1007/978-3-031-06577-4_11

antioxidant compounds contribute to FAME's oxidative capacity and biological stability. The higher the concentration of saturated fatty acids, the greater the oxidative stability.

When FAME is added to petroleum diesel, the rate and completeness of decomposition of petroleum diesel is increased several times, because microorganisms use fatty acids as a source of energy (Krivushina 2012).

There are a number of problems with the use of FAME that need to be addressed before FAME becomes a viable alternative to fossil fuels.

FAME and mineral diesel have similar calorific value, viscosity, and density but are more prone to microbiological contamination. FAME have a higher rate of degradation caused by microorganisms and a higher rate of microbiological corrosion of the fuel system components.

The problem of microbiological damage to fuels and materials of fuel systems, operated in different climatic zones, especially in areas with humid climates, is relevant. The share of bacteria and fungi account for more than half of all biological lesions. Fungi are found in soil, water, air, and on various surfaces.

Statistics shows that a significant part of all corrosion damage to materials and technical equipment is caused by processes involving microorganisms that secrete metabolic products (CO_2, H_2S, NH_3, SO_3, N_2O_5, etc.) and accelerate corrosion (Matveeva et al. 2011).

Nowadays, it is difficult to find a group of materials that would not be destroyed by microorganisms; in addition, the uncontrolled development of microorganisms on the materials is harmful to human health. Bacteria and fungi can cause allergies, skin diseases, and breathing problems.

Microbiological damage is divided into two types. The first type causes a reversible change in properties. After removal of microbiological deposits, all detected changes in the material are restored to their original values.

The second type includes microbiological damage that leads to irreversible changes in properties (reduction of rubber strength, corrosion of paints, defects of metals and alloys, changes in physical parameters of fuels and lubricants and oils) (Semenov et al. 2008).

As a result of the development of microorganisms, the following problems arise:

- Accumulation of sludge in the bottom of the fuel tank.
- Deterioration of fuel condition and formation of stable water emulsions.
- Increase in acidity, change of physicochemical properties of fuel (increase of kinematic viscosity, refractive index, pH, actual resins content, etc.), and the appearance of sediment, turbidity, and pungent odor (Matveeva et al. 2011).
- Sedimentation on the inner walls of fuel systems.
- Destruction of protective coatings under the cluster of colonies of microorganisms.
- Damage to paint and varnish covers (swelling, destruction, decrease of adhesion) and elastomeric products.

- Colonies of microscopic fungi on rubber and sealants intensify the condensation of water vapor and impair the mechanical and dielectric properties of these materials (Onuorah et al. 2015).
- Occurrence and development of corrosion processes on parts of oil or hydraulic systems.
- Increase in friction and wear in joints and bearings.
- Corrosion of power elements of a design (power beams).
- Increased engine wear.

Deposits that occur during the life of microorganisms, contamination of materials, and changes in their properties necessitate an increase in maintenance costs of the device (cleaning, repair, replacement of parts, etc.).

The costs of car maintenance and operation can be significantly reduced by controlling microbiological contamination before problems arise.

11.2 Main Material

Nowadays, 200 species of microorganisms are known, including 30 families that can use hydrocarbon fuels as the sole source of carbon and energy. The growth and development of microorganisms is influenced by a number of factors, among which the most important are temperature and humidity.

The optimum temperature for the development of the most fungi and bacteria is 25–35 °C. In addition, the higher the moisture content in the environment, the more actively the microorganisms grow (Shkilniuk and Boichenko 2020).

Species composition and properties of strains (aggressiveness to the material) differ depending on the material, design and technological features of the fuel system, and the climatic area of operation of the product. Fungi of the genera *Aspergillus*, *Penicillium*, *Trichoderma*, *Fusarium*, and *Alternaria* and bacteria of the genera *Bacillus* are able to destroy a big variety of materials. At the same time, many microorganisms have a destructive capacity only in relation to certain materials. The main microorganisms that cause biodegradation of fuels are bacteria of the genera *Pseudomonas*, *Micrasossis*, and *Mycobacterium* as well as fungi of the genera *Cladosporium*, *Aspergillus*, *Penicillium*, *Alternaria*, and others. At the same time, the bacterium *Aeruginosa* and the fungus *Cladosporium resinae* ("kerosene fungi") are found more often than others in oil products.

More than half of all cases of microbiological destruction of oils and lubricants occur as a result of fungi of the genus *Aspergillus*, *Penicillium*, *Fusarium*, and *Scopulariopsis*. Microbiological damage of polymeric and paint materials is a consequence of the activity of fungi of the genus *Penicillium*, *Stemphylium*, *Chaetomium*, and *Trichoderma* (Matveeva et al. 2011).

"Kerosene fungi" reproduce by spores, which can be in a passive state for a long time, waiting for favorable conditions for growth. Because the size of the spores is smaller than the pore size of the fuel filters, it is impossible to clean the fuel by filtration. Kerosene fungi grows in the form of fibers that can reach considerable lengths and form a tortuous layer – mycelium. Compared to other fungi, *Cladosporium resinae* produces much more biomass and is thus much more likely to cause technical malfunctions in the fuel system.

Since this fungus grows on the border of the "fuel-water" phases, only a drop of moisture is enough for it to grow. The mycelium then begins to cover the drop, holding it in place, and continues to grow, producing even more water due to its metabolism. While growing mycelium can firmly attach to the fuel tank (Semenov et al. 2008).

All microorganisms can be divided into three groups: active destructors (microorganisms that actively grow in fuel), potential destructors (strains with medium growth activity), and partially adapted to the environment or random micromycetes (strains that do not grow in fuel because they do not use hydrocarbons as a source of energy).

It is proved that biological contamination of fuel is due to microbial enzymatic oxidation of hydrocarbons to form organic acids that have surface-active properties. The rate and depth of microbial oxidation of fuel depends on its hydrocarbon composition. Hydrocarbons with a linear structure of molecules break down faster than their branched isomers. Aliphatic hydrocarbons are less biostable than aromatic (Shkilniuk and Boichenko 2020).

Bacterial and fungal spores are carried through the air, attaching to dust particles or water droplets. The degree of air pollution depends on the season. The total number of fungi in summer is greater than in winter. The air is a constant and universal source of microbiological pollution. Fuel becomes contaminated during transportation and manual cleaning of fuel systems, so spores can go unnoticed for a significant period of time. During operation, the fuel is mixed simultaneously with micromycetes and bacteria, but the presence of micromycetes, which are the dominant destructors, is masked by bacteria, which develop more actively.

FAME is more prone to microbiological contamination due to several factors, the main one being hygroscopicity. The presence of water in residual quantities is due to the stability of the emulsion formed after the transesterification process. Another factor is the higher bioavailability of FAME than traditional diesel fuel. Biofuels are easily hydrolyzed to fatty acids by both chemical and microbial reactions.

During studies on the rate of moisture saturation of FAME, it was found that while the first 96 h of the experiment, there is an intense saturation of FAME with moisture (more than 2 mg of moisture/g FAME) (Krivushina 2012). The main sources of moisture in the fuel:

- Atmospheric moisture from the air
- Rain or snow that may get into the fuel tank through holes for sampling, ventilation valves, or loose cover
- Water that accumulates at the bottom of tanks

The reaction of microorganisms that develop in the fuel to sudden changes in temperature can be varied and depends on their physiological properties. “Kerosene fungi” remains viable at a temperature of minus 30 °C. Border positive temperature is 50 °C; at this temperature, it stops the growth of colonies, and at 60 °C, it kills spores.

Micromycetes are able to restore vegetative growth after mechanical destruction of the mycelium into individual fragments, which leads to their spread through the fuel system. A number of special operational measures are being developed to prevent microbiological contamination:

1. Maintaining proper fuel storage conditions
2. Timely and complete drainage of sludge from fuel tanks, determination of the presence, and nature of pollution
3. Prevention of contact with water and atmospheric moisture
4. Timely drainage of fuel and removal of water from the bottom of tanks and fuel tanks
5. Use of biocidal additives
6. Use in warehouses of such order of a filtration which provides necessary degree of purity of fuel, according to regulatory documents
7. Regular inspection of the condition of filters to determine the presence and nature of contaminants on the filter elements and their quality cleaning (flushing)
8. Timely cleaning of technological equipment during fuel storage
9. Disinfection of fuel tanks with the help of chemicals that have disinfectant properties (sodium hypochlorite, chlorhexidine, solutions of organic compounds)
10. Heat treatment – purging the fuel system with hot air (in the absence of fuel), heated to temperature 70 °C
11. Introduction of antiseptics into the materials of fuel systems in order to ensure resistance to microorganisms (phenols, quaternary ammonium compounds, oxyphenolate derivatives)
12. Ultraviolet treatment with special lamps.

Biocidal additives are chemical compounds able to kill microorganisms or inhibit their reproduction, preventing fuel degradation. Currently, active work is underway to develop biocidal fuel additives. The main difficulties in the use of biocides are toxicity and danger to human health.

As a result of the processes of adaptation to new substrates, new types of microorganisms can appear in the fuel. Micromycetes are able not only to withstand the effects of many biocides but also to begin to produce more biomass. If the fungi survived in the environment with the biocide and began to grow, then further development occurs regardless of the concentration of the biocide. This is explained not only by adaptation but also by the fact that the shell around the mycelium is able to selectively prevent the penetration of certain substances into the vegetative cells and/or adsorb biocides and neutralize their effects. So, it’s important to explore micromyceta that affect fuel and fuel materials.

Selection of biocide additive is a complex task. The additive must be heat-resistant, nontoxic, and as universal as possible in relation to various microorganisms. The additive should not affect the physicochemical properties of the fuel.

One of the objectives of this study was the selection of a biocidal additive to protect hydrocarbon fuels from microbiological damage, which provides high efficiency of inhibiting the growth of microorganisms and preserves the physicochemical and operational properties of the fuel during long-term storage. Regular sludge drainage is used as the main prevention measure for all fuel tanks. If microbiological contamination is suspected, it should be tested.

The method of determining the microbiological damage to motor fuel requires the detection of active cultures of microorganisms and research conditions, as well as a reliable and rapid method of detection of microorganisms.

Methods for determining microbiological contamination include:

- The method of direct counting of the number of cells under a microscope. This method is rarely used due to the presence of emulsion and uneven distribution of cells in the fuel. It is impossible to isolate viable cells among the dead.
- Method of counting microorganisms on fixed colored Gram smears. The disadvantages of this method are the uneven distribution of the emulsion on the slide and the difficulty of preparing a drug of known area.
- Method of determining the presence of microorganisms by changing certain fuel parameters (pH of the aqueous phase, acidity, sulfide content, component composition, metal content, changes in flash point). This method gives inaccurate information due to indirect determination of the presence of microorganisms.
- Indicator method. Direct fuel analysis using indicators: ninhydrin (color change from blue to pink) or triphenyltetrazolium chloride (color change from red to brown). The Lowry method is singled out: the solution turns blue as a result of the interaction of proteins produced by microorganisms with Folin's reagent. This method provides only qualitative information on biological contamination.
- Gravimetric method. The mycelium of fungi is separated on filters under vacuum. The resulting biomass is washed, dried, and adjusted to constant weight. The method gives an accurate result only in the presence of microbiological contamination, which can be seen with the naked eye.
- Centrifuge method. Based on centrifugation contaminated fuel samples that were previously homogenized. The length of certain layers in a test tube is read from the calibration graph.
- Method of determining the presence of products of metabolism of microorganisms: protein compounds (determined by IR), carboxylic acids (chromatography), vitamin B_6 (spectrofluorescent method), and enzymes (biochemical method).
- Luminescent method. When the irradiation of UV light microorganisms previously stained fluorochrome, the greenglow can be seen.

Express methods are currently being developed that significantly reduce the time required for this test.

The study develops an improved method for quantitative counting of colonies of microorganisms in fuel, based on a combination of microbiological and gravimetric methods. Using paper filters can improve the accuracy of microscopic examination, avoiding errors due to uneven distribution in fuel cells, and, in turn, using seeding samples fuel on Petri dishes can detect colonies of microorganisms present on the filter, in concentrations that cannot be determined by gravimetric method.

The following reagents were used for microbiological studies: alcohol, enzymatic peptone, glucose 5%, microbiological agar, sodium citrate, yeast, hydrochloric acid 1 mol./l, and sodium hydroxide 10%. The following equipment was used in this study:

- Glass measuring cylinders with nominal capacity of 100, 400, and 1000 ml. and glass lids of appropriate size
- Glass bottles with a nominal capacity of 1000 ml
- Sterile glass pipettes with a nominal capacity of 1 and 2 ml with a graduation of 0.1 ml
- Sterile plastic tips
- Paper sterile filters "blue tape" with a pore size of 3–5 mcm.
- Forceps with blunt tips
- Thermostat capable of maintaining the temperature 25 ± 2 °C
- Alcohol burner
- Drying cabinet capable of maintaining the temperature 180 ± 5 °C
- Laboratory microscope "Biolam" C-11
- Glass spatulas
- Aluminum foil
- Sterile cotton wool
- Paper stickers and marker
- Parchment

The experimental part of the study included the following steps:

- Sterilization
- Basic preparation
- Determination of the degree of biological pollution
- The effect of biocidal additive on the degree of contamination

11.2.1 Sterilization

The glassware was wrapped in parchment paper; the pipette holes were closed with cotton plugs; and the paper filters were wrapped in parchment paper envelopes one by one and then placed in an oven and kept for 2 h at a temperature of 180 ± 5 °C.

Metal tools were placed in a glass with alcohol so that the working ends are completely covered.

The nutrient medium was sterilized by the method of stepwise sterilization. The nutrient medium was brought to a temperature of 100 °C and kept at this temperature for 10 min. During this time, all vegetative cells die, and only spores remain viable. The medium was then cooled to a temperature optimal for spore germination (25 ± 2 °C) and reheated to 100 °C after a few hours. This cycle was repeated three times.

11.2.2 Basic Preparation

The fuel was left at rest for 1 h, after which it was visually inspected for visible mechanical impurities or free water. If there is foreign contamination in the sample, the fuel was purified and then carefully moved for uniform distribution of microorganisms.

11.2.3 Determination of the Degree of Biological Pollution

To prepare the medium, 60 g of glucose-peptone agar in 1 l of distilled water was suspended and heated to complete dissolution. The medium was poured into a sterile Petri dish and after complete solidification was checked with a pH meter, so that the pH was within 5,4 ± 0,2. This range is optimal for the good growth of most bacteria and fungi. If the value is out of range, the pH was adjusted using hydrochloric acid or sodium hydroxide. After preparation, the medium was poured into sterile glass bottles and sterilized.

In sterile Petri dishes poured melted in a boiling water bath agar medium, 20 ml each. To prevent the formation of condensate, the cups were left on a horizontal surface until completely solidified and then kept for 3 days at 25 ± 2 °C with the lids down to dry the surface of the medium and check its sterility. Paper labels with the date and number of seeding were pasted on the cups.

The degree of microbiological contamination was established for FAME samples from sunflower and rapeseed, as well as diesel fuel. For each type of fuel were carried out three crops with volumes of 0.15, 0.20, and 0.24 ml. These volumes were selected by a series of crops to find the optimal count of colonies in the cup (maximum 300 colonies).

A known volume of fuel premixed with a sterile pipette was placed on the surface of the agar medium in a Petri dish and distributed over its surface with a sterile spatula. The mixed fuel is distributed on the surface of the medium in a circular motion, simultaneously with the rotation of the cup. Glass and metal tools were immersed in alcohol and held over an alcohol burner before and after use.

Petri dishes were kept for 7 days at a temperature of 25 ± 2°C without access to direct sunlight. To prevent reseeding of the surface by fungal spores, the samples were stored with the lids facing up.

During and after incubation, the number of colonies of each class of microorganisms was counted, and the number of viable bacteria and fungi present in this volume of the initial fuel sample was calculated from them. Measurement of bacteriological contamination was performed after an incubation period of 48 h and mycological after 7 days.

To count the number of colonies of forming elements (CFE), the Petri dish was placed on a light background, and the total number of all visible colonies was counted, marking them at the bottom of the cup with a marker. Once a colony has been counted, it is not counted again, even if it has increased in size. To calculate the number of colonies of cells of microorganisms in 1 l of the initial suspension summed, the results of parallel seeding and determined the average number of colonies by the formula (Debarati 2017):

$$M = \frac{a \cdot 1000}{V}$$

where

a – the average number of colonies at seeding
V – the volume of suspension used for seeding, ml

To increase accuracy, parallel crops were performed on Petri dishes.

11.2.4 Identification of Microbiological Contamination

Bacterial isolates are characterized morphologically. Evaluation of morphological and cultural characteristics of microorganisms was performed on a dense medium. Smears were prepared, and a small part of the bacterial material was removed from the largest colony with a spatula, smeared evenly on the glass, fixed in the flame of an alcohol burner, and examined using an immersion microscope "Biolam" C-11 according to conventional methods.

Dishes with crops were initially examined with the naked eye or through a magnifying glass and then put upside down and examined with a microscope with a slight increase and narrowed aperture. The colonies were described using a color scale indicating the pigmentation of the colonies and the surrounding agar, as well as morphological features (shape and size) (Levanova and Zakharova 2017).

Fungi were characterized and identified based on the morphology of their colonies and microscopic features. To characterize and identify fungi and yeasts, methods of cultivation and microscopic examination with a description of the size, shape, structure of colonies, and their color were used.

The relief of the colonies and their contours were determined with a magnifying glass. Colonies that differed in at least one feature were considered as different types. A mycological atlas was used to confirm the identity of each fungus (Vorobev 2008).

After 48 h of incubation at 25 °C, there was the growth of fungi with the formation of colonies with a diameter of 1–5 mm. Each type of microorganism has a certain type of colony, so the number of types of colonies was judged on the diversity of microbiological damage to the studied fuel samples

11.2.5 Influence of Biocidal Additive on the Degree of Biological Pollution

The study of the effectiveness of the proposed biocidal additive to the fuel was performed by the method of cultivation on dense nutrient media. The filter paper was cut into equal strips and sterilized. The degree of microbiological contamination was established for FAME samples from sunflower and rapeseed, as well as mineral diesel fuel. Mixtures with different additive concentrations were prepared for each type of fuel: 1%, 3%, 5%, 7%, and 10%.

A drop of pre-mixed fuel with a sterile spatula was placed on the surface of the paper strip and evenly distributed over its surface. Petri dishes with striped samples were kept for 7 days at a temperature of 25 °C ± 2 °C without access to direct sunlight. Three repetitions of the test were performed to confirm reproducibility.

During and after incubation, the growth rate of colonies and the degree of microorganism damage were assessed. Conclusions on the degree of microbiological contamination and the effectiveness of the additive were made after 7 days of incubation. Resistance to microbiological damage of fuels with biocidal additive was also determined by the following criteria: transparency and the presence of sediment after long-term storage for three months.

11.3 Results and Discussion

The results of determining of the degree of microbiological contamination photos of crops of 0.20 ml of each type of fuel, which reflect the degree of microbiological damage at different points in time, are shown in Figs. 11.1, 11.2, 11.3.

The results of calculations of microbiological contamination of different fuels are given in Table 11.1.

Fig. 11.1 The degree of microbiological contamination of diesel fuel

Analyzing the results, we can say that the samples of diesel fuel and FAME from sunflower oil have a medium contamination with fungal spores, and the sample FAME from rapeseed oil, heavy pollution, while all fuel samples are heavily contaminated with bacteria.

11.3.1 The Results of Identification of Microbiological Contamination

Photos of the appearance of all types of bacterial isolates present in different fuels are shown in Fig. 11.4; a description of their morphology is given in Table 11.2.

Fig. 11.2 The degree of microbiological contamination of FAME from sunflower oil

Photos of the appearance of isolates of all species of fungi present in different fuels are shown in Figs. 11.4 and 11.5; a description of their morphology is given in Table 11.3. There were a variety of staining of colonies – a number of colors were clearly species-specific (Table 11.3). Colonies began to acquire color on day 3–4 of the experiment, and the final intense color was acquired after 7 days.

Some representatives of fungi were found once in the studied fuels; others, on the contrary, were isolated from two samples at once.

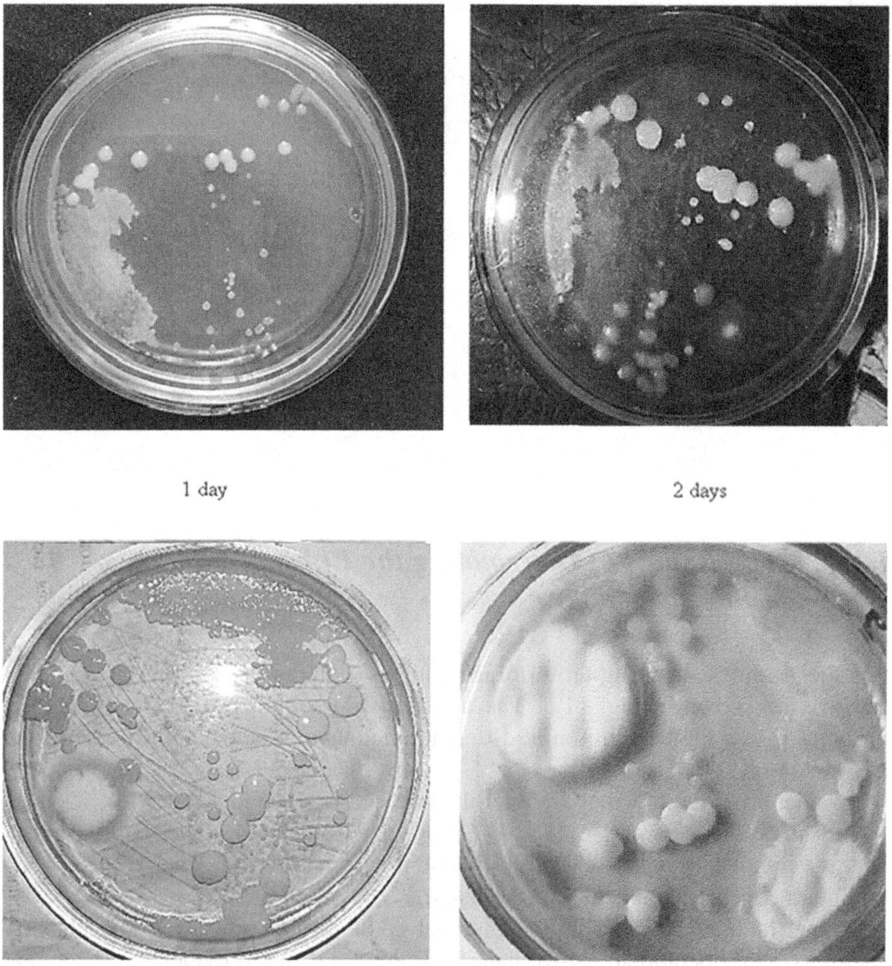

Fig. 11.3 The degree of microbiological contamination FAME from rapeseed oil

The figure shows that all samples infected with bacteria of the genera *Pseudomonas*, which do not significantly affect the performance of the fuel. In diesel fuel, there are fungi *Cladosporium resinae*, while in FAME from rapeseed oil, *Cladosporium resinae* and *Aspergillus*. Both genera of fungi have a destructive effect on fuel and structural materials of the engine and fuel system; in addition, fungi of the genera *Aspergillus* are pathogenic.

Table 11.1 Microbiological contamination of different fuels

Sample	Sample volume, ml	Bacteria Number of colonies	Contamination cfe/l	Fungi Number of colonies	Contamination cfe/l
Diesel fuel	0.15	42	280,000	1	6667
	0.20	130	650,000	1	5000
	0.24	150	625,000	1	4167
Average value		–	518,333	–	5278
FAME from sunflower oil	0.15	17	113,333	0	0
	0.20	33	165,000	0	0
	0.24	58	241,667	1	4167
Average value		–	173,333	–	1389
FAME from rapeseed oil	0.15	34	226,667	3	20,000
	0.20	50	250,000	4	20,000
	0.24	76	316,667	6	25,000
Average value		–	264,444	–	21,667

11.3.2 The Results of Determining the Effect of Adding a Biocidal Additive

The fuel samples after long-term storage with and without the addition of a biocidal additive are shown in Fig. 11.6. The degree of microbiological contamination of different fuels with the addition of different amounts of biocidal additives is shown in Figs. 11.6, 11.7, 11.8.

Evaluation of the biocidal effect of the additive is given in Table 11.4.

As can be seen from Figs. 11.7, 11.8, 11.9 and Table 11.4 in the most cases, the additive completely suppresses the growth of micromycetes of fungi and bacteria at a minimum concentration of 5% of the mass.

It was found that the usage of biocidal additive does not affect the physico-chemical and operational performance of the fuel and provides high biocidal efficiency.

Thus, based on the analysis of the literature on the subject of the study, currently the question of the degree of microbiological damage to traditional and alternative fuels is beginning to develop, but there is no method for quantitative assessment of microorganisms. Currently, indicator methods for determining microbiological contamination are used, but these methods give only a qualitative assessment, not a quantitative one, which does not fully determine the degree of microbiological contamination of fuel (Shkilniuk and Boichenko 2020). We have developed an improved method for detecting the presence of microbiological

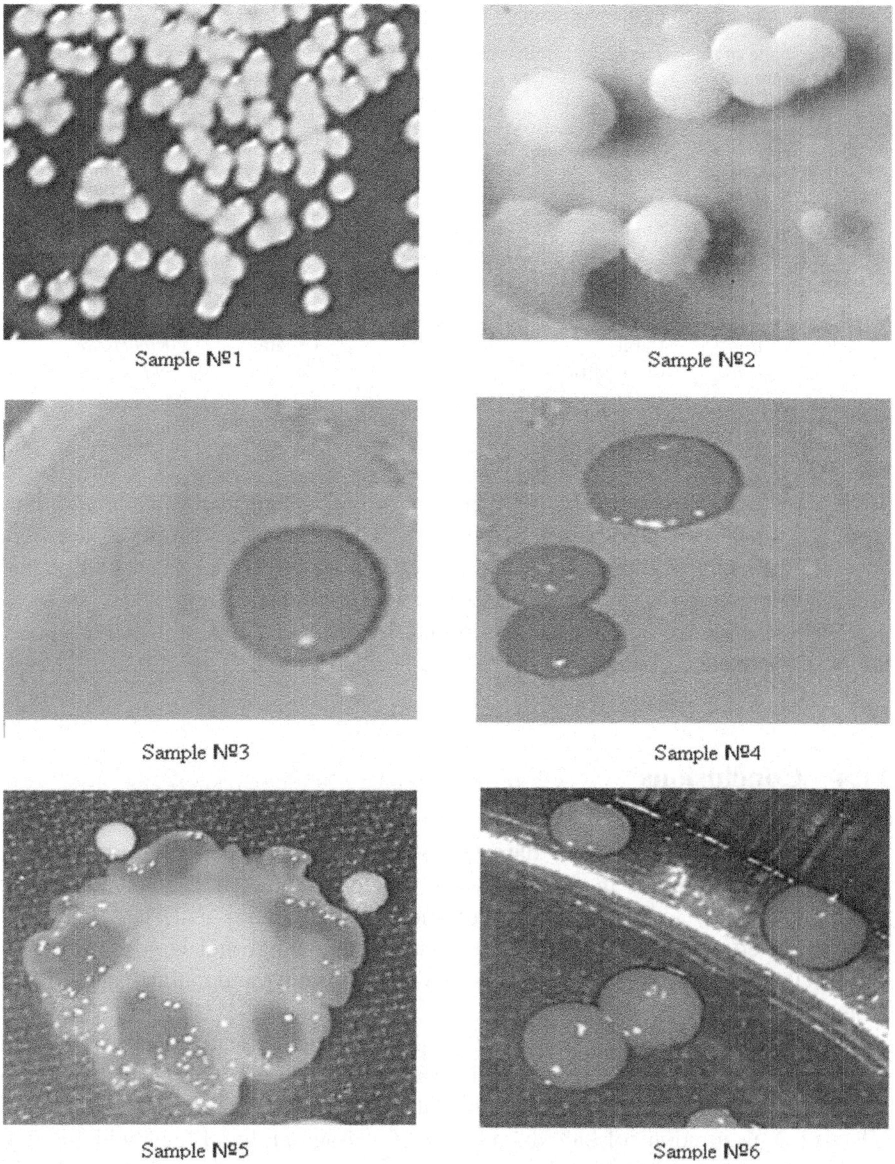

Fig. 11.4 Appearance of bacterial isolates in fuel: Sample No 1, bacteria present in diesel fuel; samples No 2–4, types of bacteria present in FAME from sunflower oil; samples No 5–6, types of bacteria present in FAME from rapeseed oil

Table 11.2 Description of the morphology of bacterial isolates

№	The shape of the colony	The size of the colony	Relief of the colony	Color	Genus and representative
1	Rounded	Small (1–2 mm)	Flat	Transparent, cloudy	*Alcaligenaceae*
2	Complex shape	Large	Hilly with a convex center	Lemony	*Pseudomonas putida*
3	Rounded	Small	With a convex center	Lemony	*Micrococcus luteus*
4	Rounded	Small	Hilly	Light pink	*Pseudomonas chlororaphis*
5	Rounded	Large (4–6 mm)	Domed	Cream	*Klebsiella aerogenes*
6	Rounded	Medium (2–4 mm)	Convex	Red and pink	*Serratia marcescens*

damage to fuel and quantifying its degree. Currently, there the biocidal additives based on quaternary ammonium compounds. Their main disadvantages are high cost and difficulty in use, as such additives when added to fuel cause sludge and affect performance (Matveeva et al. 2011). In the course of this study, an inexpensive and effective additive was proposed that does not have a negative impact on the performance of fuels.

11.4 Conclusions

The degree of microbiological damage to various motor fuels was determined. Microorganisms present in the fuel were isolated, characterized, and identified. Among them are active destructors of fuels and structural materials. The fuel samples tested were found to be contaminated with fungi and large numbers of bacteria that could cause fuel problems. Severe monitoring and maintenance of fuel systems is recommended.

Of scientific and practical interest is the proposed biocidal additive, which contributes to a significant reduction in the degree of microbiological damage to motor fuels. The effectiveness of the biocidal additive has been tested. The minimum effective concentrations of biocidal additives for usage in diesel fuels with alternative components are determined.

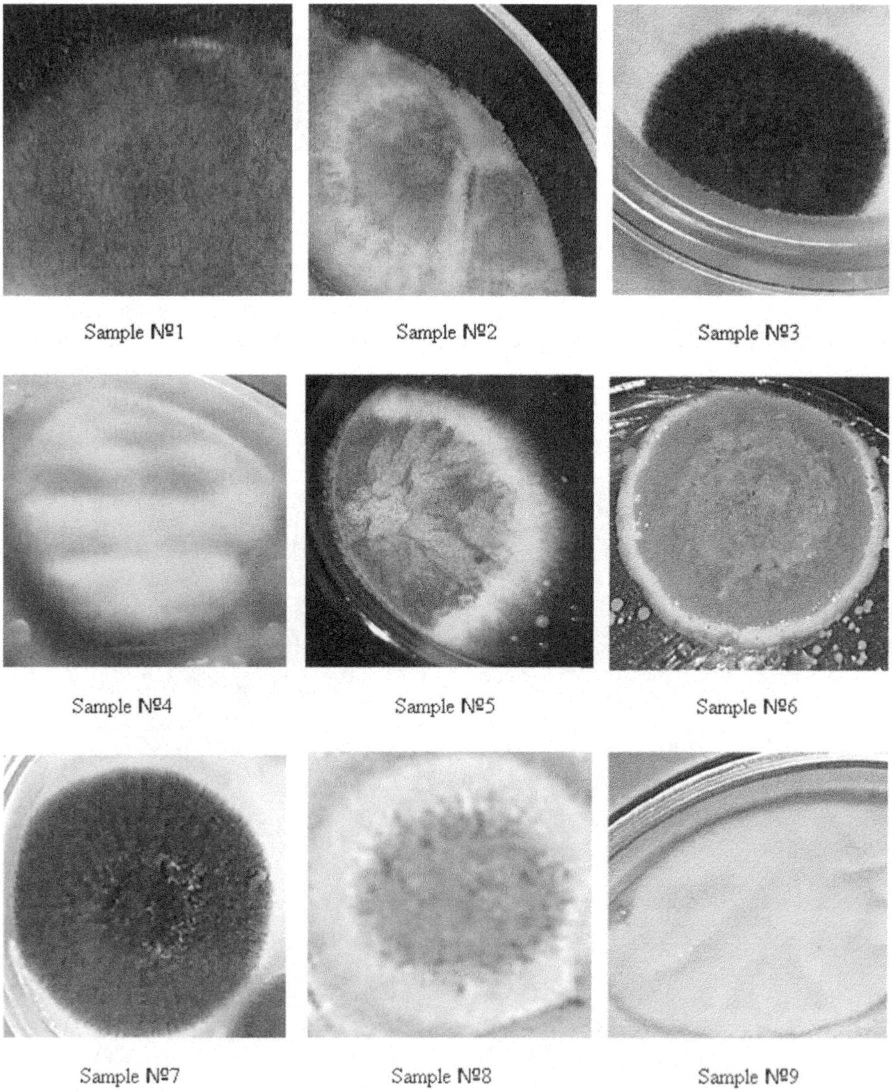

Fig. 11.5 Appearance of fungal isolates in fuel: Samples No 1–3, species of fungi present in diesel fuel; samples No 4–5, species of fungi present in FAME from sunflower oil; samples No 6–9, species of fungi present in FAME from rapeseed oil

Table 11.3 Description of the morphology of fungal isolates

№	Relief of the colony	Color	The contours of the edge	Genus and representative
1	Difficult to determine	Translucent, dark gray	Wavy	*Rhizopus* sp.
2, 5	With a convex center, wavy	Light yellow with brown spots and white spray, white outline	Wavy	*Aspergillus baeticus*
3, 7	With a clear center with waves from the center	Black with a translucent white outline	Straight	*Aspergillus niger*
4, 9	Mesh, hilly	White	Ragged	*Aspergillus candidus*
6	With a clear center	Dark green with a light outline	Wavy	*Cladosporium cladosporioides*
8	Rough	Gray-blue with a white outline	Straight	Penicillium commúne

Fig. 11.6 FAME with the addition of biocidal additive (right) and without it (left)

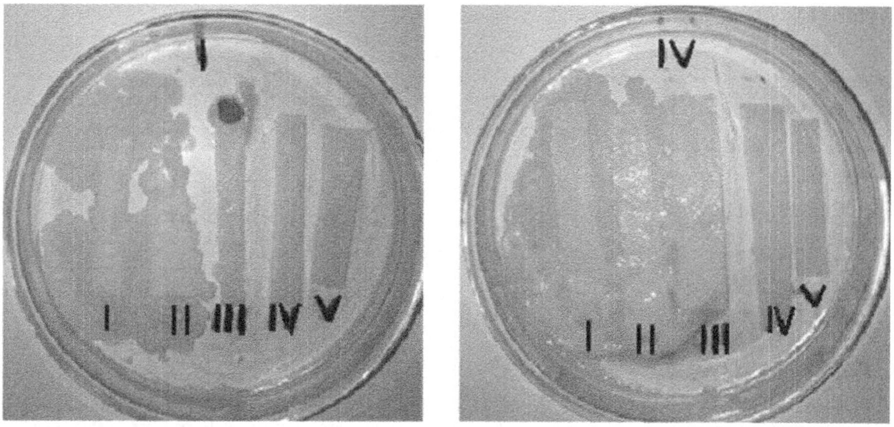

Fig. 11.7 The degree of microbiological contamination of FAME from sunflower oil with a biocidal additive: strips No I, II, III, IV, and V – fuel with the addition of 1, 3, 5, 7, and 10% vol. additives, respectively

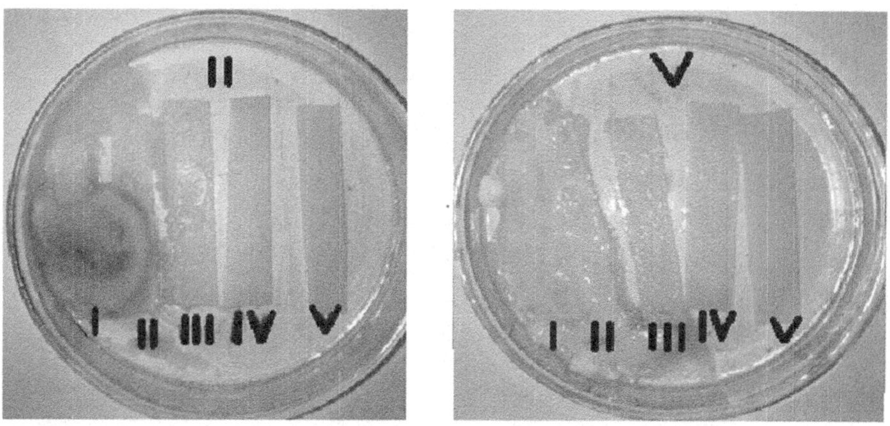

Fig. 11.8 The degree of microbiological contamination of FAME from rapeseed oil with biocidal additive: strips No I, II, III, IV, and V – fuel with the addition of 1, 3, 5, 7, and 10% vol. additives, respectively

Table 11.4 Processing of test results of the additive

Composition	Biocide concentration, %	Microbiological damage
FAME from sunflower oil	–	+
FAME from rapeseed oil	–	+
Diesel fuel	–	+
FAME from sunflower oil + biocidal additive	1	+
	3	+
	5	Partial growth
	7	–
	10	–
FAME from rapeseed oil + biocidal additive	1	+
	3	+
	5	Partial growth
	7	–
	10	–
Diesel fuel + biocidal additive	1	+
	3	+
	5	–
	7	–
	10	–

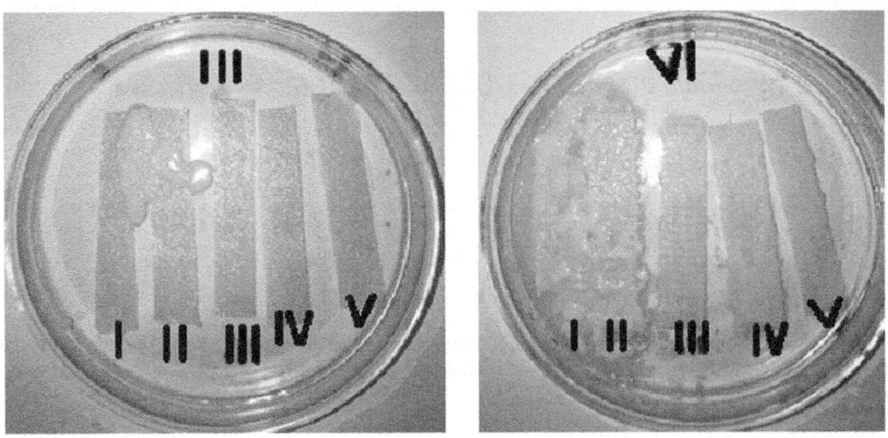

Fig. 11.9 The degree of microbiological contamination of diesel fuel with a biocidal additive: strips No I, II, III, IV and V – fuel with the addition of 1, 3, 5, 7 and 10% vol. additives, respectively

References

Debarati D (2017) Essential practical handbook of cell biology & genetics, biometry & microbiology: a laboratory manual. Academic Publishers, Kolkata, p 176

Krivushina AA (2012) Mikromicety v aviacionnom toplive. Dissertation, Moskovskij gosudarstvennyj universitet imeni M.V. Lomonosova

Levanova L, Zakharova Y (2017) Systematics, taxonomy and classification of bacteria. Fundam Clin Med 2(1):91–101

Matveeva EL, Vasilchenko OA, Demyanko DA (2011) Mikrobiologicheskoe porazhenie aviacionnyh topliv. Systemy ozbroiennia i viiskova tekhnika 2:152–156

Onuorah S, Obika I, Orj M et al (2015) Microbial contaminants in the commercial aviation fuel obtained from Benin City airport. Niger Universal J Microbiol Res 3:31–35

Semenov SA, Gumargalieva KZ, Zaikov GE (2008) Harakteristika processov I osobennosti povrezhdeniya materialov tekhniki mikroorganizmami v usloviyah ekspluatacii. Vestnik MITHT 2:23

Shkilniuk I, Boichenko S (2020) Biological Risk of Aviation Fuel Supply. In: Babak V, Isaienko V, Zaporozhets A (eds) Systems, Decision and Control in Energy I. Studies in Systems, Decision and Control, vol 298. Springer, Cham

Vorobev A (2008) Atlas-rukovodstvo po bakteriologii, mikologii, protozoologii i virusologiis immunologiej i allergologiej. Pervyj MGMU im. Sechenova, Moskva

Chapter 12
System for Monitoring Microbiological Contamination of Jet Fuels and Fuel Systems

Iryna Shkilniuk, Sergii Boichenko, Tetyana Kondratiuk, and Kazimierz Lejda

Nomenclature

AF	Aviation fuel
NAU	National Aviation University

12.1 Introduction

The reliability of a fuel system depends on the quality of the fuel and its physical and chemical stability during all phases of the life cycle. Practical experience and numerous studies show that hydrocarbons are not resistant to certain microorganisms. Destructive microorganisms cause degradation (microbiological damage) to operating materials, which leads to a change in the primary performance characteristics.

An analysis of the scientific literature shows that fuels and lubricants are among the available products for certain species of microorganisms. All types of fuel and lubricant are subject to microbiological destruction: motor fuel, oil, lubricant, lube-coolant, and bitumen.

I. Shkilniuk (✉) · S. Boichenko
National Technical University of Ukraine "Igor Sikorsky Kyiv Polytechnic Institute", Kyiv, Ukraine

T. Kondratiuk
Taras Shevchenko National University of Kyiv, Kyiv, Ukraine

K. Lejda
Rzeszow University of Technology, Rzeszow, Poland
e-mail: klejda@prz.edu.pl

S. Boichenko et al. (eds.), *Chemmotological Aspects of Sustainable Development of Transport*, Sustainable Aviation, https://doi.org/10.1007/978-3-031-06577-4_12

Scientists in many countries are actively engaged in studying the issue of the biological stability of petroleum products, in particular aviation fuels (AF). In the last decade, there have been many reports on the results of studies by scientists from South Korea, Vietnam, and the People's Republic of China.

The widespread interest of scientists and researchers in many countries shows the relevance of the problem of microbiological contamination of petroleum products, particularly aviation products. Thus, improving the chemical additives and developing technological solutions have not solved this problem.

12.2 Microbiological Contamination of Fuels and Biofilm

Microorganisms are highly active organisms. Microorganisms live by general biological law: the smaller the size of an organism, the more intensive its metabolism.

More than 150 species of microorganisms capable of degrading hydrocarbons have been discovered and identified. This property of microorganisms is a significant problem in oil production, refining, and petrochemistry, especially in the use of petroleum products (Shkilniuk and Boichenko 2014).

Fungi, bacteria, and yeasts are microbial contamination of AF. This has been confirmed by numerous studies by scientists from around the world (Shkilniuk 2016).

Microorganisms are a specific form of existence of living matter. They are characterized by their abundance, diversity of forms, distribution, wide sphere of interaction with the environment, and the scale of their impact on it (Shkilniuk 2016; Harold 2003; Vasilyevaa et al. 2012; Gaylarde et al. 1999; Krivushina 2012; Robbins and Levy 2004; Behbahany-Pour and Radice 2017). The prominent representatives of microorganisms-destructors of hydrocarbons are:

1. Bacteria – *Achromobacter, Alcaligenes, Arthrobacter, Bacillus, Bacterium, Brevibacterium, Citrobacter, Clostridium, Corynebacterium, Desulfovibrio, Enterobacter, Escherichia, Flavobacterium, Methanobacterium, Micrococcus, Micromonospora, Mycobacterium, Nicrococcus, Pseudomonas, Sarcina, Serratina, Spirillum, Vibrio, and Thiobacillus*
2. Fungi – *Alternaria, Aspergillus niger, Aspergillus fumigatus, Hormoconis resinae, Monacus floridanus, Phialophora sp., Cephalosporium,* and *Penicillium*
3. Yeast – *Candida, Debaryomyces, Endomycopsis, Hansenula, Rhodotorula, Saccharomyces, Torula, Torulopsis, Trichoderma, and Trichosporon*

The results of numerous studies of aviation fuels indicate a greater prevalence of mycelial fungi in aviation fuels and fuel systems and the danger of these microorganisms to the normal functioning of fuel systems (Kondratyuk et al. 2007; Pirog 2004; Passman 2013; Kozlova et al. 2008).

Hormoconis resinae is an active fungus, also called the "kerosene" fungus (Fig. 12.1). This fungus lives in subtropical and tropical soils in natural conditions.

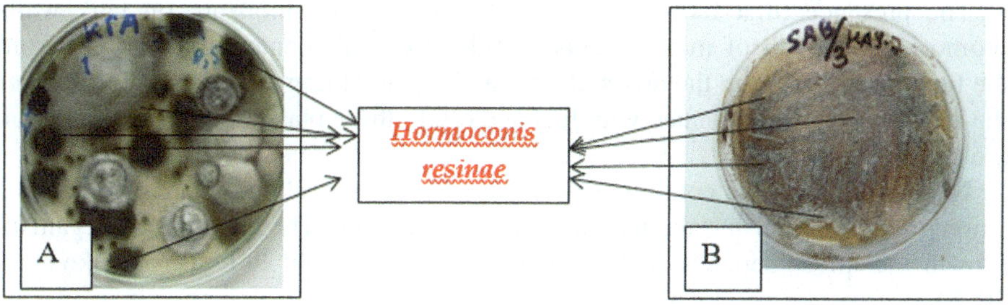

Fig. 12.1 *Hormoconis resinae* in an enrichment culture (**a**, in samples with aviation fuel TC-1; **b**, diesel fuel)

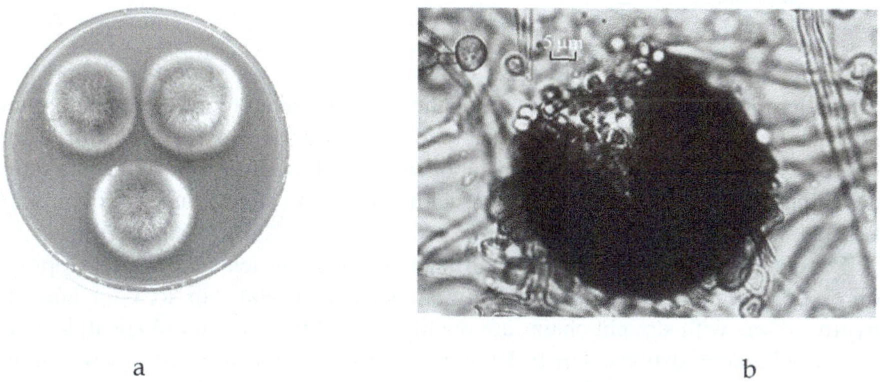

Fig. 12.2 *Monascus floridanus*: (**a**), on agar; (**b**), under the microscope

The "kerosene fungus" is an active hydrocarbon destructor of aviation fuels (Kondratyuk et al. 2007). The presence of these microorganisms was found in jet fuel samples from Australia, Brazil, USA, Grat Britain, Denmark, India, Syria, Nigeria, Japan, and New Zealand (Prince et al. 2007; McVea and Solly 1991). The development of aviation and the expansion of the geography of flights contribute to the spread of the fungus in other geographical areas. Other names of this microorganism are *Hormodendrum resinae*, *Cladosporium resinae*, and *Amorphotheca resinae*.

Scientists have isolated another active destructor from the aircraft's filtration system and identified it (Fig. 12.1). The fungus was identified as *Monascus floridanus* (Fig. 12.2), which can also be called a "kerosene" fungus because of its aggressive effect on aviation fuels (Vasilyevaa et al. 2012; Krivushina 2012).

Microorganisms are highly active organisms. Microorganisms live by general biological law: the smaller the size of an organism, the more intensive its metabolism. Microbial enzymes can act on a large volume of nutrient substrate per unit of time. Microorganisms can produce adaptive enzymes in the process of adapting to a new food source if they do not have the appropriate enzyme systems to oxidize the unknown product (Pirog 2004).

The presence of a fuel-air interface and condensed moisture on tank walls promote the growth of microorganisms, which can also enter from the air, from the fuel line system, and the hoses. Spores and mycelial fragments of fungi can enter the air from soil contaminated with fuel and oil. In these places, pockets of infection are created.

Natural and climatic conditions determine the numerical and qualitative composition of microorganisms and their activity in the habitat. The rate of microorganism growth and spread depends on the following factors (Harold 2003; Boychenko et al. 2008):

- Oxygen
- Humidity
- Temperature
- Nutrients
- pH

The microorganism needs water (moisture), an organic food source, inorganic nutrients, an appropriate temperature, and an appropriate pH to develop. Some microorganisms need oxygen for metabolism. All of these factors contribute to the development of organisms in aviation fuel handling and storage areas (Robbins and Levy 2004).

Petroleum hydrocarbons contain sufficient organic nutrients for the development of oil-degrading microorganisms. Low molecular weight (up to C_{18}) aliphatic hydrocarbons with straight chains are readily amenable to microbiological degradation. The biodegradability of petroleum products tends to increase inversely with the aromatic content and the distillation temperature. That is, mid-distillate fuels and gasoline are particularly biodegradable.

Water is the dominant factor for the growth and metabolic activity of microorganisms in aviation fuels. The life activity of microorganisms depends on the water since water constitutes 75–90% of the vegetative cell mass (Pirog 2004). The nutrients are metabolized by the water molecules in the microorganism's cell. Water is one of the metabolic products of the biodegradation of hydrocarbons. According to scientists' studies, for example, *Hormoconis resinae* can excrete 0.94 g of water per 1 l of fuel during 4 weeks of growth (Hill and Thomas 1975).

A certain kind of microorganism has a range of minimum, optimal, and maximum temperatures for growth. The metabolism of microorganisms accelerates with an increase in temperature within this range (Pirog 2004). Once the maximum temperature is exceeded, the microorganism ceases to function, and the microorganism dies. The optimum temperature for most oil-degrading microorganisms is 25–30 °C. The growth of microorganisms at 2 and 55 °C has been recorded. Microorganisms are divided to temperature into:

- Psychrophiles – microorganisms that grow at temperatures up to 10 °C
- Mesophiles – microorganisms that grow at moderate temperatures 25–40 °C
- Thermophiles – microorganisms that grow actively at temperatures above 50 °C

a b

Fig. 12.3 Samples of aviation fuel with microbiological contamination (**a**, fuel RT from aircraft tank; **b**, fuel TC-1 from tank bottom)

Most oil-destroying microorganisms belong to the mesophilic group. Their growth is active at a temperature of 20–50 °C. The metabolic potential of microorganisms increases with rising temperature in an optimum range. Temperatures below the optimal range slow down the development of microorganisms. Microbial cells may die or retard their growth for some time if the temperature exceeds the optimal range.

The moisture in the air and the amount of water in the environment depend on geographic location, climate, and precipitation. The scientist F. Passman (2013) developed the criteria for assessing the risk of microbiological damage to petroleum products based on the average annual precipitation. The average annual rainfall is 64 cm at low risk, between 64 and 190 cm at medium risk, and over 190 cm at high risk. The number of days it occurs is also taken into account. Low risk is less than 100 days per year, medium risk is 100–200 days per year, and high risk is more than 200 days per year.

Microorganisms and their waste products accumulate in tanks at the water-fuel interface (Fig. 12.3). A jellylike, amorphous, gray-brown mucous mass is formed as a result of microbiological contamination of the fuel. The dark brown slime-like contamination is a cluster of filament-like, bacilliform, and oval shapes (Fig. 12.4).

Microbial cells are not in the space of the nutrient medium in individual growth but the form of specifically organized groups (biofilms). Microorganisms in the biofilm form 5–35%; the remainder is an intercellular matrix (Kozlova et al. 2008). A biofilm is a complex and dense form of coexistence of microorganisms of different species, which facilitates cell metabolism and protects cells from external negative influences.

Fuel cleanliness is the permissible level of impurities of various origins in the composition of fuels, at which the operation of oil-regulating and fuel-dispensing equipment is uninterrupted. The purity of fuels is a variable quality characteristic of fuels and can vary at different stages of the fuel life cycle. Thus, the microbiological

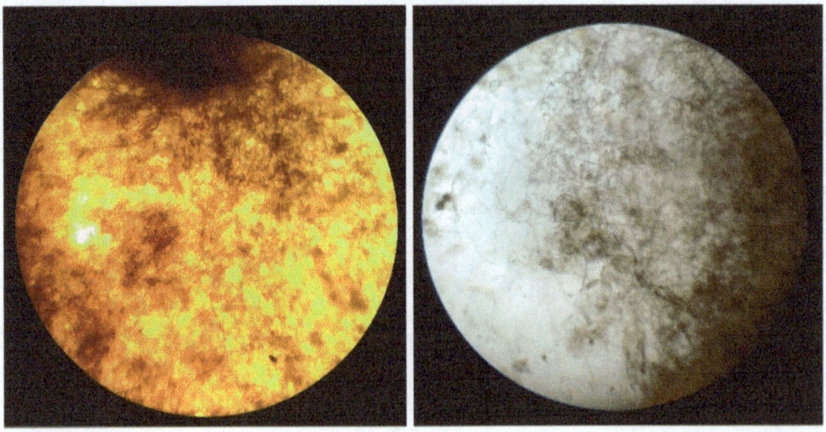

Fig. 12.4 Microbiological contamination of aviation fuel by a microscope (100× zoom)

phase should be regarded as a type of contamination of fuels. The purity of aviation fuels is determined by the presence of mechanical impurities, moisture, and the microbiological phase.

The impact of microorganisms on fuel purity can be explained by several factors. Firstly, because individual microbial cells are small enough not to pose a filtration risk, groupings of microorganisms (colonies, biofilms) within fuels can be of human-visible size. Secondly, surfactants are produced during microorganisms' metabolism and help to coagulate the mechanical contaminants in the fuel, especially in the presence of moisture. Thirdly, the sticky properties of microbial colonies or biofilms.

It can be concluded that the occurrence of microbiological contamination leads to changes in the chemical stability of aviation fuels and physical and chemical parameters and poses a particular risk for aviation fuels of long-term storage.

12.3 Impact of Microbiological Contamination on Fuel Quality and Fuel System Operation

The reliability of fuel systems and fuel management facilities is determined by the ability of aviation fuel to maintain quality in the regulatory parameters.

The biodegradability of hydrocarbons depends on the physiological properties of the particular microorganism. Thus, hydrocarbons can be classified according to their biodegradability (Table 12.1). Aviation fuels consist predominantly of highly sensitive and sensitive hydrocarbons. Engine fuels can be ranked in order of microbiological spreading rate: (1) motor gasoline, (2) jet fuel, (3) diesel fuel, and (4) aviation gasoline.

The biodegradation of hydrocarbons occurs intracellularly by specific oxidative enzymes of the oxygenase class. Oxygenation catalyzes the utilization of one oxygen

Table 12.1 Classification of the biological destruction capacity of hydrocarbons

Group	Group name	Degree of bio-destruction, %	Hydrocarbons
I	Highly sensitive	80–100	n-Alkanes, isoalkanes
II	Sensitive	60–80	Cyclones with 6, 1, 5, 2 pins, S-aromatics, monoaromatics,
III	Moderately sensitive	45–60	Three aromatics
IV	Resistant	30–45	Tetra-aromatics, triterpenes, naphthenic-aromatic compounds
V	Highly resistant	0–30	Penta-aromatic, asphaltene, resins

atom from its molecular form in the terminal methyl groups of hydrocarbons. Bonds with weak breaking energy (C—C, C—H) are replaced by bonds with intense breaking energy (C—B, H—O). Such oxidation mode is characteristic of aliphatic, acyclic, and aromatic hydrocarbons.

Problems caused by microbiological contamination can be classified (Shkilniuk et al. 2019):

- Problems caused by the physical presence of microbial colonies
- Problems caused by the metabolism of oil-destroying microorganisms
- Problems caused by metabolites of oil-destroying microorganisms
- Problems caused by the corrosive nature of microorganisms

Signs of microbiological contamination in fuels:

- The presence of clumps of slimy biomass, fibrous formations in the water-sediment
- The presence of lumps of sticky slime on the inner walls of tanks
- Swelling of the sealant and corrosion of the surface of the fuel tank
- Clogging with a sticky mass of filters and pump screens installed in tanks
- Malfunction of fuel metering equipment
- Unpleasant smell

The consequences of microbiological contamination are clogging of filters, pump nets, and malfunctioning of fuel gauges by the biomass. It can cause the failure of the fuel system, damage to the anticorrosion coating, and corrosion damage to fuel tanks, which, in turn, leads to detachment of the sealant in the aircraft tank and corrosion of structural power elements (Shkilniuk 2016).

It has been detected that the pressure drop across the filter increases faster in the case of microbiological contamination of fuel compared to typical mechanical impurities (at the exact content of contaminants in the fuel (up to 0.005% wt.%) (Skribachilin et al. 1993). It has been determined that the speed of increase in pressure drop across the filter is more significant in the case of contamination by bacteria than by fungi.

Thus, contamination of biological nature is more dangerous than ordinary contamination for fuel pumping and the service life of the filter itself. It is due to the physiological properties of microorganisms to produce biosurfactants and adhere to the surfaces of filter material and other aircraft functional materials. So, increased pressure difference and fuel flow disturbance can be considered an easily overlooked symptoms of microbiologically intensive fuel contamination.

Microorganisms can colonize drain valves on tanks where condensation provides the necessary water activity, using fuel hydrocarbon vapors as a carbon source. Biofilms usually look like slimy stalactites on tank valves.

Microbiological effects are one of the most critical manifestations of corrosive and aggressive environmental influences. Microorganisms cause more than 50% of all corrosion processes in functional materials. The analysis of cases of microbiological damage of functional materials shows that its occurrence, character, and intensity depend on the properties, condition and use of the material, the aggressiveness of the oil-destroying microorganism, the duration and conditions of the interaction between the material-microorganism pair, and a series of factors that contribute to this occurrence. Microbial corrosion is a complex process of interaction between microorganisms and metal, which is realized in a biofilm. That is, the biofilm is the main factor in microbial corrosion. Biofilm and its characteristics influence the formation of corrosion processes through metabolic activity and electrochemical reactions. The microorganisms directly "corrode" the metal and affect chemical, electrochemical, and mechanical factors, enhancing or weakening any type of damage. A typical scheme of aluminum biocorrosion is presented in Fig. 12.5.

Therefore, biochemical corrosion is hardly detectable, although it is widespread. In summary, microbiological corrosion of metals is one component of the complex problem of microbiological contamination of fuels. Biochemical corrosion is caused by the by-products of microorganisms present in the damaged fuel, the adhesion of the microorganism to the metal, and the implementation of electrochemical reactions (Skribachilin et al. 1993).

A change in fuel chemistry due to a decrease in alkanes impacts density, fractional composition, volatility, and flash points. Fuel combustion rate and completeness depend on paraffinic hydrocarbons, as they have a lower flash point. The change in the chemical composition of fuels is influenced by microbiological contamination. The paraffinic composition of aviation fuels affects the energy properties of the fuel since the highest combustion heat in alkanes. The combustion heat affects the efficiency of the jet engine (Shkilniuk and Boichenko 2014). These indicators affect the flight and technical and technical-economic characteristics of aircraft. Density is an important physical characteristic and operational indicator for fuel calculation and accounting. Flash point is a very significant property because it is directly related to fire safety when handling fuel. Vaporability is a fuel property that determines the rate at which a combustible fuel-air mixture is created, which, in turn, affects the completeness of fuel combustion and the ease of jet engine starting. Fractional composition is an indicator related to vaporability, characterizing the fuel evaporation ability and starting properties, and the safe operation of the aircraft fuel

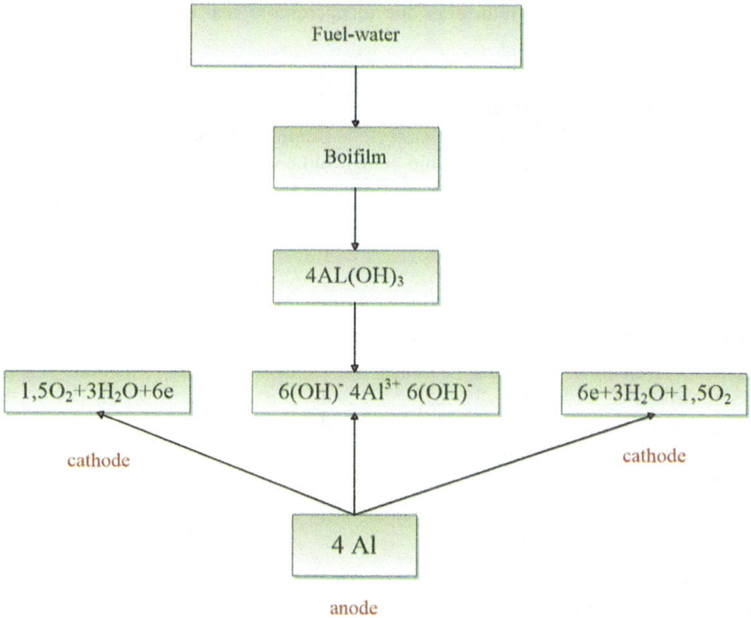

Fig. 12.5 Typical scheme of aluminum biocorrosion

system. The test results show the negative influence of microbiological contamination on the quality of modern fuels (Table 12.2).

The test results show a substantial increase in acidity, actual resin concentration, flash point, thermo-oxidative stability, an increase in crystallization temperature, and a decrease in the flash point. Testing on the copper plate showed an increased corrosion ability of the damaged samples.

The value of the acidity index in jet fuels is restricted. Acidity is an indicator that characterizes the content of organic acids in fuels and is determined by the amount of alkali consumed for their neutralization. On the one hand, organic acids improve the anti-wear and protective properties of fuels; on the other hand, they affect the corrosive properties of fuels and compatibility with operating materials.

An increase in the concentration of actual tar (2.6–4.2 times higher) poses particular risks for regular engine operation due to resin formation on fuel components, on valves, in the combustion chamber, and in the engine crankcase. The appearance of actual resin characterizes the ability of the fuel to form low-temperature resin sludge and oxidative deposits in the combustion chamber, reducing the capacity of fuel conduits.

Thus, damage from bio-damage manifests itself in deterioration of fuel quality, corrosion of tanks, fuel systems, pumps, clogging of fuel filters, and fuel control equipment.

Table 12.2 Changes in quality indicators due to microbiological contamination in fuel for jet engine

№	Name quality indicator	TC-1		RT		Jet A-1	
		Before bio cont.	After bio cont.	Before bio cont.	After bio cont.	Before bio cont.	After bio cont.
1	Acidity, mg KOH on 100 sm^3	0.2	6.8	0.2	6.5	0.1	6.6
2	The concentration of actual resins, mg/100 sm^3	2.5	8.7	1.8	7.6	3.8	9.8
3	Testing on copper plate	1	2a	1	2a	1	3a
4	The temperature of crystallization, °C	minus 61	minus 58	minus 59	minus 50	minus 51	minus 44
5	Density 20 C, kg/m^3	793	791	781	781	779	777
6	Cinematic viscosity, 20 °C, mm^2/s	1.35	1.4	1.38	1.41	1.36	1.42
7	Lower combustion heat, kJ/kg	43,313	43,004	43,254	42,822	42,911	42,411
8	Flash point, °C	34	30	39	33	41	35
9	Thermal oxidation stability precipitation rate, mg\100 SM^3	9	15	4	12	3	10

12.4 Monitoring of Microbiological Contamination of Aviation Fuel in a Fuel Storage Facility

Aviation fuel operations are complex. Aviation actors face some risks that affect not only the economic activities of the organizations but also the lives and health of the people using their services. Increased demand for air transport, the complexity of designs, new composite materials and technologies, and aircraft production lead to costlier incidents with aircraft stoppages.

Modern fuel storage facilities and bases must comply with all strict fire safety requirements and be technically equipped to efficiently receive, store, and refuel aircraft using conditioned aviation fuel. The volume of production fuel stocks should provide for an uninterrupted demand for aircraft refueling and for timely flights.

The storage equipment consists of a tank farm, pumps, filters, fuel control equipment, fuel lines, meters, and sensors. The analysis of the location of the occurrence and development of microbiological pollution of fuels (Fig. 12.6) suggests that tanks and stagnating areas of fuel lines are the most common and the most easily accessible elements of the equipment of fuel depots for the growth of microbiological destructors. Filtration plants are the main technical obstacle for the emergence and spread of microbiological contamination in fuel systems and fuels. Pathways to these elements should therefore be subject to monitoring.

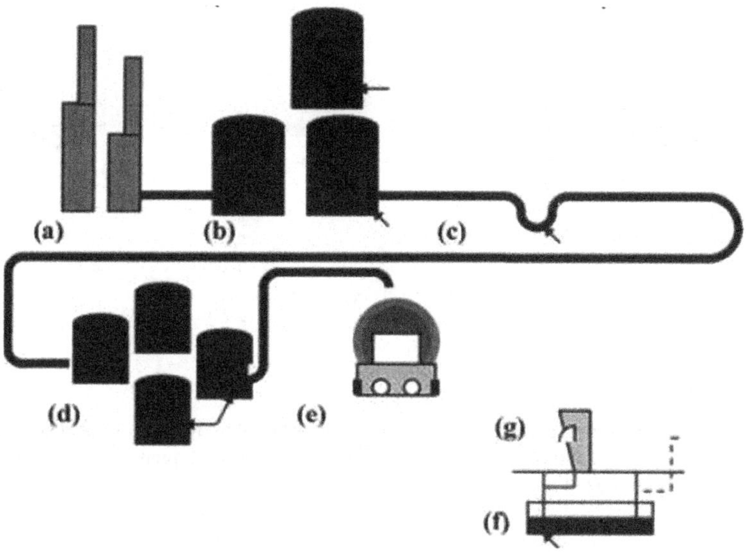

Fig. 12.6 A schematic of microbiological contamination sites in the fuel supply chain (arrows indicate where water and microbiological contamination tend to accumulate)

It is essential to identify the critical points for monitoring and the actions to be taken if biofuel contamination is detected.

Considering the specificity of microbiological contamination due to its ability to multiply, spread, and create solid biofilms and their adhesion to the surfaces of fuel system operating materials (ASTM 2017), as well as the negative impact on fuel quality, to prevent biological contamination of fuels and fuel systems and ensure flight safety, we developed a scheme for monitoring the microbiological contamination of aviation fuel at the fuel terminal (Fig. 12.7).

Thus, we have designed a four-stage model for monitoring the microbiological contamination of aviation fuel in fuel systems of fuel depots. The model provides:

- Control of the presence of microbiological contamination at points 1–4 (Fig. 12.7) in front of easily accessible locations for oil-degrading microorganisms
- Mandatory control of the presence of microbiological contamination in the fuels before the filters (points 1–3, Fig. 12.7) and of the presence of microbiological contamination in the fuels before the filters (points 1–3, Fig. 12.7)
- Mandatory fuel control before refueling into the aircraft tank (point 4, Fig. 12.7)

Detection of microbial contamination of AF (at critical point 1 of Fig. 12.7) enables to stop the shipment of microbiologically contaminated fuel to the fuel terminal in time. Repeated testing must be done in all cases of detection of microorganisms in fuel composition. The use of the microbiological contamination detection methodology (Boichenko et al. 2014) makes it possible to quickly determine the content of this type of contamination in the operational conditions of the airport and ensure

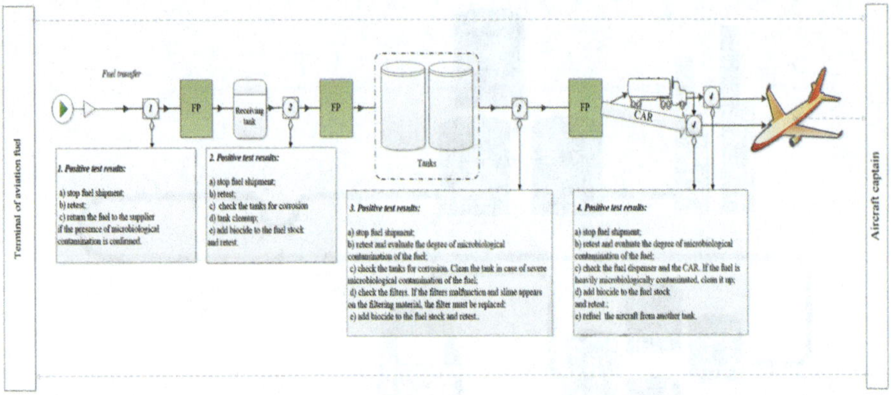

Fig. 12.7 A phenomenological four-step model for monitoring the microbiological contamination of aviation fuels, place of sampling for a rapid test for the presence of microbial contamination; *FP* filter point, *CAR* centralized aircraft refueling

timely flight. In case of repeated detection of microbiological contamination, the fuel must be returned to the supplier. In if in points 2–3 microbiological contamination of fuel is detected, it is necessary to check the fuel for conformity with normative requirements and add biocidal additives to the fuel. The final step is to continue further use of the fuel if it meets the requirements.

The purpose of developing this model is to ensure microbiological stability of aviation fuel, the prevention of bio-damage of the fuel storage system, and the safety of aircraft flights due to the use of conditioned aviation fuel.

12.5 Results and Discussion

Microbiological pollution is a specific type of pollution represented by fungi, bacteria, and yeasts. Specificity lies in the ability to multiply and spread, as well as in the complex impact on the environment (fuel), fuel storage facilities, and transport. The fungus *Hormoconis resinae* is the most common microbiological contaminant of aviation fuel and aircraft fuel systems.

The distribution of microorganisms in fuels is due to their biological characteristics:

- Speed of reproduction: cells of many microorganisms divide every 20–30 min when the conditions are favorable.
- Relatively high resistance to the action of various physical and chemical factors: high and low temperatures, the action of various kinds of radiation, desiccation, high osmotic pressure, lack of moisture, etc.
- Effortless adaptability to environmental conditions.

- An extensive variety of physiological properties, enabling them to use almost all-natural compounds for food or energy to live and multiply where other creatures cannot live

Microbiological contamination of aviation fuels is a process that consists of three successive interrelated steps: (1) interaction (adhesion) of microorganisms with fuels, (2) growth in the fuel hydrocarbon environment, and (3) change of fuel properties.

Water is a medium for the accumulation of nutrients and a source of microbiological fuel contamination. Water is a participant in biochemical reactions within the microbial cell, making water a determining factor for microbial growth.

The grouping of microorganisms in the biofilms is a significant factor in the microbiological corrosion of fuel system operating materials and fuel storage equipment. ICAO Doc 9977 AN489 "Manual on Civil Aviation Jet Fuel Supply" is a comprehensive document defining recommendations for aviation fuel supply in civil aviation. The document is recommendatory and provides for use in the light of national regulations. The directive states that microbiological contamination of fuels is a natural and costly problem that can affect the safety of aircraft flights. Therefore, the control of microbiological contamination in fuels and fuel systems should occur daily. The global trend toward globalization and integration of countries into a standard trading area makes it essential to unify and optimize aviation fuel quality requirements and internationally standardized procedures for fuel quality control. Ukraine's current regulatory system for the supply and quality control of aviation fuels is undergoing a reform process to harmonize national standards with international and European requirements, implement ICAO and IATA requirements, and require the development of national standards and guidance documents on the control of microbiological contamination of fuels to maintain a high technological level of aviation fuel supply system.

The use of biofuels is a trend in modern fuel supply. Aviation biofuels or aviation alternative fuels are a mixture of conventional petroleum fuels for jet engines and a biocomponent, which uses ethyl/methyl esters of fatty acids (EEFA/MEFA) of vegetable oils. Research results show that the biocomponent is a substance vulnerable to microorganisms. The biocomponent content in fuel (10%, 20%, 30%) accelerates the development of the microbiological phase. The research was conducted in the accredited NAU Test Laboratory "Aviatest." Therefore, the use of biocomponents in aviation fuels requires further research and the mandatory use of biocide additives.

12.6 Conclusion

The interconnectedness of microbial degraders of aviation fuel and the consequences for fuel quality and operation facilities, transport, and storage indicates the complex impact of microbiological contamination and the importance of ensuring the microbiological stability of fuel at every stage of its life cycle.

The implementation and execution of the developed monitoring system will make it possible to stop in time the transfer of aviation fuel with microbiological contamination, prevent the occurrence of biocorrosion of operating materials and the violation of the serviceability of filters, as well as exclude the refueling of aircraft with fuel with signs of biopollution.

Microbiological controls on fuel and fuel systems to detect viable microbial cells and to ascertain their quantitative content are highly relevant. Similar studies are advisable in disinfection and cleaning of the respective tanks (aircraft fuel tanks, fuel tankers, fuel trunk lines).

References

ASTM (2017) ASTM D 6469 standard guide for microbial contamination in fuels and fuel systems. ASTM

Behbahany-Pour MJ, Radice G (2017) Fuel contamination on the large transport airplanes. J Aeronaut Aerosp Eng 6(4):1–11. https://doi.org/10.4172/2168-9792.1000200

Boichenko S, Shkilniuk I, Novak A (2014) Methodology for the determination of microbiological contamination of aviation fuels. The Patent of Ukraine, 94190

Boychenko S, Shkilnuk I, Turchak V (2008) The problems of biopollution with jet fuels and the way of achieving solution. Transport 23(3):253. https://doi.org/10.3846/1648-4142.2008.23.253-257

Gaylarde CC, Bento FM, Kelley J (1999) Microbial contamination of stored hydrocarbon fuels and their control. Rev de Microbiol 30:1–10

Harold W (2003) Graef. An analysis of microbial contamination in military aviation fuel systems Technical report AFIT/GEE/ENV/03-10. Department of the air force Air university Air force institute of technology, Air Force Base W. p 221

Hill EC, Thomas AR (1975) Microbiological aspects of supersonic aircraft fuel. In: Proceedings of the 3rd international biodegradation symposium. pp 157–174

Kondratyuk TA, Kharkevich ES, Zaharchenko VA, Nakonechnaya LT, Roy AA, Zhdanova NN, Pashkevich RE (2007) Biodeterioration of aviation fuel TS-1 by microscopic fungi. Mycol Phytopathol 41(5):442–448

Kozlova I, Radchenko O, Stepura L, Kondratiuk T (2008) Geochemical activities by microorganisms and their applications: handbook – Science Opinion, p 528

Krivushina A (2012) Micromycetes in aviation fuels/Author's abstract. Izdatel'stvo Nauka, St. Petersburg

McVea GG, Solly RK (1991) Control of fuel microorganisms with magnetic devices: laboratory investigation with Hormoconis resinae. Aircr Mater Tech Memo 408:1–11

Passman FJ (2013) Microbial contamination and its control in fuels and fuel system. Int Biodeterior Biodegradation 1:88–104

Pirog T (2004) General microbiology: handbook. The National University of Food Technologies, Kyiv, p 471

Prince RC, Parkerson TF, Lee C (2007) The primary aerobic biodegradation of gasoline hydrocarbons. Environ Sci Technol 41(9):3316–3321

Robbins JA, Levy R (2004) A review of the microbiological degradation of fuel. In: Paulus W (ed) Directory of microbicides for the protection of materials. Kluwer Academic Publishers, Amsterdam, pp 177–202

Shkilniuk I (2016) Investigation of the microbiological stability of traditional and alternative aviation fuels. Int. J. Sustain Aviat 2(2):111–118

Shkilniuk I, Boichenko S (2014) Methodically organizational principles of biological stability providing of aviation fuel. Trans Inst Aviat Warsaw 4(237):76–83

Shkilniuk I, Boichenko S, Kondratiuk T, Shevchuk N (2019) Identification and Assessment of Biological Risk of Aviation Fuel Supply / Selected aspects of providing the chemotological reliability of the engineering: Monograph – Boichenko S, Aksionov O, Topilnitskyi P, Pushak A, Lejda K. – Kyiv (Ukraine), Center of Educational Literature: 197–214

Skribachilin V, Mikhailova L, Kachanov Y, Buriakovskaya T (1993) Effect of microbiological contamination of fuels on the functioning of fuel filters. Chem Technol Fuels Soils 6:12–13

Vasilyevaa AA, Chekunovaa LN, Bilanenkoa EN, Kachalkina AV, Polyakova AV (2012) Characterization of the strain monascus floridanus P. F. Cannon & E. L. Barnard, isolated from aviation fuel. Mikrobiologiya 81(2):266–272 [illegible] (20[illegible]) The potency [illegible] biodegradation of [illegible] hydrocarbons. [illegible] 3[illegible]

[illegible] (20[illegible]) A review of the [illegible] of [illegible] [illegible] (ed.) [illegible] Academic Publishers, Amsterdam, pp. 2[illegible]

[illegible] (20[illegible]) [illegible] of the [illegible] [illegible] [illegible] pp. [illegible]–13[illegible]

[illegible] (20[illegible]) [illegible]

[illegible]

[illegible]

Chapter 13
Kinetics of Dissolution of Oil Deposits

Olena Tertyshna, Kostiantyn Zamikula, Oleh Tertyshnyi, and Viacheslav Polishchuk

Nomenclature

ARPD	Asphalt-resin-paraffin deposits
HBTM	Hexane-benzene-toluene model
HBM	Hexane-benzene model

13.1 Introduction

In the process of oil transportation and storage of oil products in tanks, especially large volumes, there is an accumulation of asphalt-resin-paraffin deposits (ARPD), the amount of which sometimes in a few months forms a layer on the bottom of the tank of up to 10% of its volume (Teryshna et al. 2015).

The presence of deposits leads to the decrease of the oil tanks' volume, occurrence of corrosion-hazardous areas under the deposits, and complication of the reservoir inspection. In addition to reducing the useful volume of tanks, the accumulation of deposits complicates the process of their operation and the quantitative and qualitative accounting of oil and reduces the technical and economic performance of oil tanks and the transport system as a whole. It is necessary to preserve their useful volume to increase the efficiency of using the oil tanks' capacity.

O. Tertyshna (✉) · K. Zamikula · O. Tertyshnyi
State Higher Educational Institution, Dnipro, Ukraine

V. Polishchuk
Agrinol Scientific and Production Enterprise Limited Liability Company, Berdians'k, Ukraine

S. Boichenko et al. (eds.), *Chemmotological Aspects of Sustainable Development of Transport*, Sustainable Aviation, https://doi.org/10.1007/978-3-031-06577-4_13

ARPD have different components and are complex systems that include petroleum products, water, inorganic compounds, and mechanical impurities, the ratio of which varies widely.

The high-molecular hydrocarbons of the paraffin series, which are part of the ARPD, have high pour point, normally forming highly viscous precipitates, which sometimes turn into a solid state. The composition of resins and asphaltenes includes polycyclic aromatic structures that contain sulfur, oxygen, nitrogen, and various trace elements.

Sometimes refineries neglect the methods of disposal of ARPD, using the simplest methods of landfilling or incineration in landfills. This leads to harmful impacts on the environment. Therefore, the urgent task for the oil industry is to attract deposits as raw materials, which will increase the depth of oil refining and reduce the negative impact of industrial waste on the environment.

The most time-consuming is the process of preparation for the removal of deposits. Pumping requires compliance with increased requirements for safe operation. A number of methods are used to perform these works (Kononov and Mastobaev 2010):

- Use of a high-pressure water jet for erosion of deposits.
- Erosion of the deposit by oil.
- Dilution of the precipitate with solvents.
- Heating of bottom deposits by heat carriers.

13.2 Main Part

Two types of deposits were used as objects of study, which have differences in composition and structure: the first group of samples, with a high content of paraffins (ARPDp), and the second, with a high content of asphaltenes (ARPDa).

The characteristics of the deposits are given in Table 13.1.

For secondary use and utilization of accumulated and mechanically collected deposits in oil refining processes, it is necessary to address the issue of isolating their organic part in a technically pure form.

Table 13.1 Characteristics of ARPD

Properties	ARPDp	ARPDa
Part of organic compounds, % weight	93.12	94.90
Part of inorganic compounds, % weight	0.98	1.83
Density, g/ml	0.87	0.94
Melting point, °C	56	46
Part of resins, % weight	6.88	14.12
Part of asphalts, % weight	7.12	18.45
Part of paraffin, % weight	56.22	24,92
The content of mechanical impurities, % weight	5.90	3.27

Table 13.2 Physicochemical characteristics and group composition of distilled crude oil fractions

Hydrocarbon fraction:	Density at 20 °C, kg/m^3	Group composition, % weight			Average molecular weight
		Paraffin	Naphthenes	Arenes	
Petroleum 50–85 °C	671	90.12	9.88	2.12	93.47
Petroleum 85–110 °C	682	81.09	18.91	2.86	100.80
Petroleum 110–150 °C	701	80.41	19.59	3.39	108.34
Petroleum 150–200 °C	741	68.18	18.8	13.02	114.70
Kerosene 140–243 °C	778	61.2	22.11	16.69	177.26
Diesel 182–351 °C	823	–	–	28.21	231.82

Dissolution and dispersion of ARPD is a rather complex process, which depends on many factors, the most important of which are temperature, interaction time, composition and physicochemical properties of the solvent, as well as the composition and structure of ARPD.

The selection of reagents was mainly aimed at enhancing the solubility of the components of the ARPD, which are a complex dispersed system of refractory substances in both composition and structure with intramolecular bonds. From these positions, the greatest dissolution effect is achieved if the reagents contain components similar in composition and structure to the components of the deposits (on the principle of "similar dissolves in similar").

Crude distilled petroleum fractions (gasoline 50–85 °C; 85–110 °C; 110–150 °C, 150–200 °C, kerosene 140–243 °C, diesel 182–351 °C) were investigated as the most available and relatively cheap solvent. Fractions were obtained at the crude oil distillation unit of PJSC "Ukrtatnafta" (Table 13.2). The group of aromatic solvents is represented by benzene, toluene, n-hexane, and model compositions based on them in different ratios.

To increase the efficiency of the above solvents, compositions were investigated based on them with additives-inhibitors of paraffin deposits: a product of processing of vegetable oils (PPVO), Pachem P-505, Pachem P-501 (Tertyshnay et al. 2018).

The effectiveness of solvents was determined in the laboratory by methods (Ahmetov et al. 2008).

The research was conducted using the method of "baskets" using cartridges made of filter paper. The experiments were performed in static and dynamic modes.

In the static method, the volume of the solvent, its composition, temperature, and the duration of the experiment did not change. The ARPD sample cartridge was immersed in a 40-ml solvent for 1.5–2 h. At the end of a certain time, the sample was removed, dried in the open air, and weighed.

The effectiveness of the solvent was evaluated by the degree of dissolution using the formula:

$$\alpha = \frac{m_1 - m_2}{m_1} \tag{13.1}$$

where

α – dissolution degree, unit fraction,
m_1 – the mass of the ARPD sample before dissolution, g,
m_2 – the mass of the ARPD sample after dissolution, g,

Dissolution was performed at fixed temperatures of 10 °C, 25 °C, and 35 °C, which were stabilized by a thermostat. The maximum value of temperature and duration of research are caused by the necessity of development of technology of utilization of ARPD in industrial conditions (Tertyshna et al. 2018a, b).

In the dynamic mode method, the efficiency of solvents was evaluated depending on the duration of exposure.

For each experiment, six samples of ARPD weighing up to 1 g were prepared. The solvent or solvent composition was poured into six beakers. A cartridge with a sample of ARPD was immersed in each beaker with the solution, and the time was recorded. Every 20–30 min, one cartridge was removed in turn. After complete drying, the residual weight of the sample was determined, and the degree of dissolution was determined.

The process of ARPD dissolving corresponds to the first order of the reaction and is described by the equation Erofeev-Kolmogorov (Delmon 1969):

$$\alpha = 1 - e^{-k\tau} \tag{13.2}$$

where k is a constant that determines the rate constant of the dissolution process and

τ – the duration of the process.

To characterize the rate of first-order reactions, along with the rate constant, a quantity called the half-life was used. This value does not depend on the initial concentration of the starting material and was calculated for the first-order reaction by the Sakovich formula (Kolpakova et al. 2008):

$$\tau_{1/2} = \frac{\ln 2}{k} \tag{13.3}$$

The formula makes it possible to determine the time during which half of the initial amount of ARPD dissolves.

The preexponential multiplier (frequency factor) k_0 and the apparent activation energy of the ARPD dissolution were determined using the functional dependence of the logarithm of the dissolution rate constant on the corresponding inverse absolute temperature of the reaction medium:

$$\ln k = f\left(\frac{1}{T}\right) \quad (13.4)$$

To enhance the ability to dissolve and break the ARPD into small fragments, to create around the already dispersed particles of solvate shells that prevent their integration into the original conglomerates, and to reduce the interfacial tension at the boundaries of the ARPD solvent, additives were introduced into the basic solvent in an amount of 5% of the mass.

13.3 Results and Their Discussion

At the first stage, the efficiency of light hydrocarbon gasoline fractions in the processes of dissolution of deposits by static and dynamic methods of ARPD samples (Table 13.2) as a function of temperature was investigated.

The results of the experiments are summarized in Table 13.3 (static mode) and in Fig. 13.1 (dynamic mode). At temperatures of 10 °C and 25 °C, the soluble effect of all distilled crude oil fractions for samples of ARPDp and ARPDa is negligible (Table 13.4). The best result of the degree of dissolution does not exceed 0.25 for the sample ARPDp. The fraction of hydrocarbons 50–85 °C showed a greater ability to

Table 13.3 The efficiency of light hydrocarbon gasoline fractions in the dissolution of ARPD in static mode

Hydrocarbon fraction	Solvent temperature, °C	Dissolution degree, α	
		ARPDp	ARPDa
50–85 °C	10	0.06	0.02
	25	0.19	0.041
	35	0.42	0.162
85–110 °C	10	0.11	0.02
	25	0.16	0.035
	35	0.49	0.22
110–150 °C	10	0.12	0.04
	25	0.25	0.09
	35	0.55	0.37
150–200 °C	10	0.10	0.03
	25	0.21	0.07
	35	0.55	0.42
140–243 °C	10	0.09	0.026
	25	0.18	0.05
	35	0.48	0.45
182–351 °C	10	0.07	0.012
	25	0.16	0.04
	35	0.33	0.26

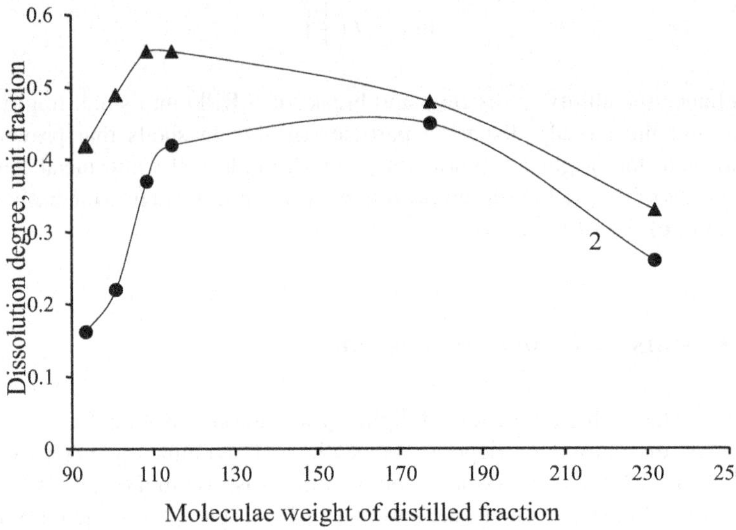

Fig. 13.1 Dependence of the efficiency of ARPD removal (τ = 90 min and t = 35 °C) on the molecular weight of hydrocarbon solvents: 1, ARPDp; 2, ARPDa

Table 13.4 The dissolution degree of ARPD at 10 °C under static conditions

Type of ARPD	Solvent	Degree of dissolution
ARPDp	Toluene	0.088
	Benzene	0.019
	n-Hexane	0.078
ARPDa	Toluene	0.227
	Benzene	0.201
	n-Hexane	0.183

dissolve the sample of paraffin-type ARPD. However, the relatively low degree of dissolution (0.041) and the slow process for dissolving the organic part of the ARPDa indicate the futility of using the gasoline fraction and hydrocarbon solvents of a similar nature for this type of deposit.

The solubility was affected by temperature. With increasing temperature, the solubility of the deposits improved, but the degree of dissolution for the sample ARPD was only 0.45 for the fraction 140–243 °C. When dissolving paraffin-type deposits, the degree of transformation reached 0.55 for fractions 110–150 °C and 150–200 °C.

It should be noted that in industrial conditions, it is necessary to maintain the increased temperature by using heaters, which entails energy costs. At the same time, at the subsequent suspended temperature, volatility of solvents, accordingly, and explosive fire danger as a whole increases. Therefore, we were limited to heating the solvent to 35 °C.

The dissolution degree of ARPDa sample in the gasoline fraction 110–150 °C is lower by almost 1.6 times relative to ARPDp. This fact is explained by the

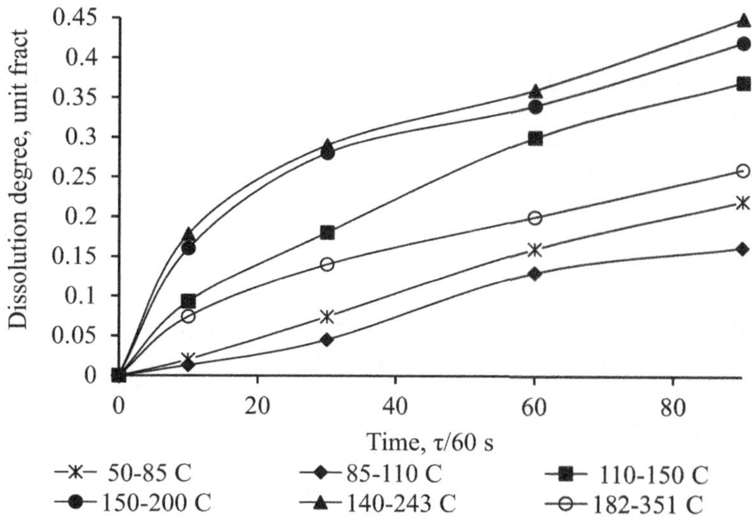

Fig. 13.2 Kinetic curves of ARPDa dissolution in distilled crude oil fractions depending on the process duration at the temperature of 35 °C

insufficient content in this fraction of arenes (Table 13.3), which are characteristic solvents for asphaltenes. Analysis of the results of the action of solvents with a boiling point of 200 °C and above revealed greater efficiency in the destruction of asphaltene-type deposits (the degree of dissolution increased to 0.45).

Comparison of the results showed that with increasing process temperature, the degree of dissolution of samples of asphaltene-type deposits is not significant and heating the solvent in this case is not justified. On the contrary, raising the temperature when dissolving paraffin-type deposits is more appropriate.

From a practical point of view, the connection between the efficiency of action of crude distillated oil fractions of a similar group composition depending on their average molecular weight when dissolving different types of ARPD is of interest. Analysis of the results of the study of the dependence of the degree of dissolution on the average molecular weight of solvents under static conditions is presented in Fig. 13.2.

The experiments (Fig. 13.1) have shown that the dependence of the dissolution degree on the average molecular weight of the ARPD is extreme. For ARPD of different group composition, these dependencies are slightly different from each other. With an increase in the composition of asphaltene substances, the maximum efficiency shifts toward solvents with a higher molecular weight. For ARPD with a high content of paraffinic hydrocarbons, the maximum efficiency is achieved when using crude distilled hydrocarbon fractions with an average molecular weight in the range of 90–120. With increasing molecular weight of solvents, their effectiveness decreases, even with increasing aromatic hydrocarbons (Table 13.4). For ARPDa, the maximum is the solvent with an average molecular weight in the range of 170–190.

Thus, the maximum solubility in relation to ARPD, especially paraffinic type, has a crude distilled gasoline fraction of 110–150 °C.

The investigation of dissolution in static conditions of clear patterns of intensification (temperature and contact time) of the process of oil deposit removal, taking into account the mutual influence of the composition and structure of distilled crude oil fractions and ARPD, is not detected. Only some dissolution trends for hard paraffins and asphaltenes were noted. ARPD are a complex system with intramolecular bonds. In addition, there are restrictions on temperature rise when removing ARPD using hydrocarbon fractions.

The processing temperature of the solvents should not exceed 30–40 °C. This is especially important when using crude distilled gasoline fractions. The system of such solvents and ARPD does not reach saturation. In addition, the basic properties of distillates differ from each other, even if they are obtained at different times in the same oil treatment plant. This affects the efficiency of dissolution and removal of ARPD.

The analysis of the process of dissolution of ARPD with high content of asphaltenes in the dynamic mode revealed the relationship between the efficiency of solvents and the composition of ARPD. In real conditions, the composition and structure of ARPD are variable, so it is necessary to study the patterns of ARPD removal using hydrocarbon compounds, taking into account extensive (composition and structure of solvent and ARPD) and intensive factors (temperature, contact time).

The process of dissolving ARPD in a dynamic mode at a temperature of 35 °C in distilled crude oil fractions (Table 13.2) was accordingly investigated to the method (Ahmetov et al. 2008). The results are summarized and presented in Fig. 13.2.

Dissolution of ARPD in the dynamic mode at 35 °C is more intense in the gasoline fraction of 150–200 °C and kerosene 140–243 °C and includes two stages. At the first stage of ARPD swells, absorbing solvent. This is followed by dissolution and dispersion. The maximum dissolution rates were 0.42 and 0.45, respectively. The dynamics of change in the dissolution degree in both cases is exponential. The maximum dissolution rate was observed in the first minutes of the process. After 30 min, the dissolution rate gradually decreases.

The hydrocarbon composition of gasoline fractions is characterized only by the ability to dissolve. At the interface between the solvent/ARPDa phases, equilibrium is quickly established, but complete destruction of the deposits is not achieved, because under the conditions of the experiment there is no dispersion.

The main components of ARPDa – resins and asphaltenes – are characterized by the complexity of structure, low stability, high reactivity, polarity, and surface activity, which is generally a characteristic of heterocyclic compounds. Asphaltenes are lyophilic (Kolpakova et al. 2008).

It can be assumed that the increase in the dissolution efficiency in kerosene and gasoline with a boiling point of 200 °C (Fig. 13.2) is associated with the presence of aromatic hydrocarbons in these fractions. The gradual inhibition of the dissolution of the samples is probably due to the fact that with the increase of the average molecular weight and boiling point of the solvent fractions, their content increases the content of high-molecular-weight condensed polycyclic structures. These structures are characterized by greater sorption activity (Traube rule) and form an adsorption layer on the surface of the ARPD, which prevents the penetration to the surface of deposits of low-molecular-weight hydrocarbons, which have a greater ability to dissolve.

In addition to temperature, the duration of contact of the solvent with the deposits has a significant effect on the process of removing the organic part of the ARPD. The investigation of the dissolution kinetics allowed to determine not only the rate of destruction of deposits but also the optimal contact time. At a fixed temperature, the kinetic characteristics of the solubility of a number of reagents for ARPD of different group composition were determined.

At the next stage, the effectiveness of individual substances (benzene, toluene, and n-hexane) was evaluated. The study was performed under harsh conditions: at a temperature of 10 °C for 90 min in static mode. The results are presented in Table 13.4.

The results of the experiments show that at a relatively low temperature, a slight level of dissolution is achieved for both types of deposits, depending on the selected solvent. The best of these for ARPDp was toluene (dissolution degree, 0.088) and n-hexane (dissolution degree, 0.0878). When dissolving ARPDa, a fairly high efficiency was confirmed by all three solvents.

Under dynamic conditions, the ability to dissolve ARPD samples in composite model mixtures based on n-hexane, benzene, and toluene (HBTM and HBM) at a temperature of 10 °C was investigated. Experimental kinetic curves of deposit dissolution are summarized in Fig. 13.3. The analysis showed that the best model mixture for ARPDp was HBTM (2:1:1), and for ARPDa, HBM (1:1). The composition of hexane and benzene HBM (1:1) proved to be a universal effective mixture of solvents for both types of deposits.

Comparison of the dissolution degree of ARPDa separately with benzene and separately with n-hexane (Table 13.4) with the action of their mixture (Fig. 13.3) showed that the efficiency of the model composite solvent is higher. This is due to the presence of a synergistic effect, which is manifested in the strengthening of the joint interaction of the components of the model mixture.

Studies of the effect of temperature on the dissolution kinetics were performed on samples of ARPDa using a model mixture of HBM (1:1). The experiments' results are presented in Fig. 13.4.

For a temperature of 35 °C, the initial stage of dissolution of deposits is characterized by a fairly high rate and, accordingly, a noticeable process efficiency. After 40 min, the degree of dissolution exceeds 82% and after 120 min, – 95%. The shape of the obtained curves indicates that the experimental model mixture in the first period, which lasts about 40 min, provided the maximum initial rate of dissolution of ARPDa. This is due to the high chemical activity of the solvent and the fact that the process takes place on the surface of the solid material.

Over time, the reaction rate gradually decreases. Inhibition of dissolution in the second stage indicates an increase in the influence of diffusion processes, which are associated with the penetration of the solvent into the middle of the solid material

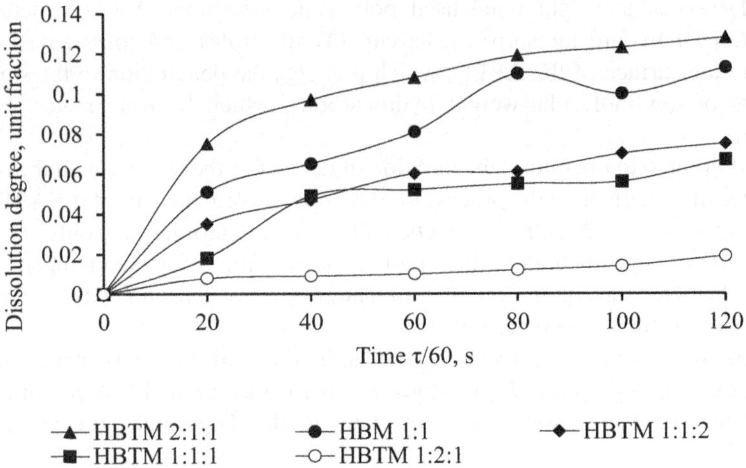

a)

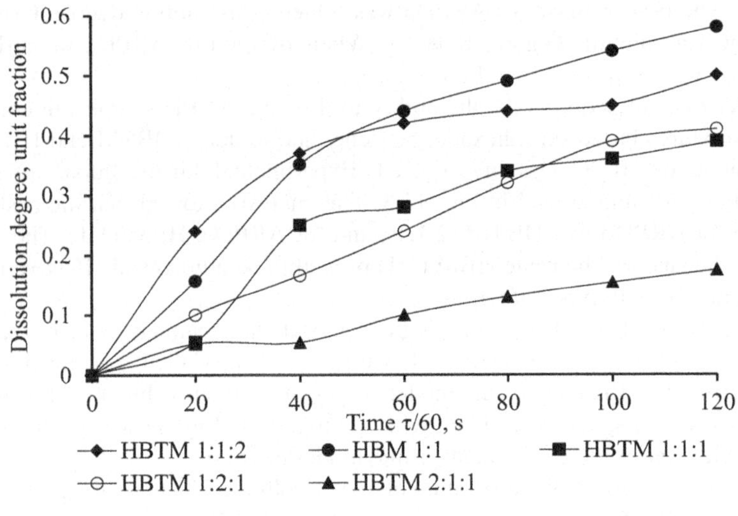

b)

Fig. 13.3 Experimental kinetic curves of dissolution of ARPDp (**a**) and ARPDa (**b**) deposits under dynamic conditions at a temperature of 10 °C

and the reverse movement of the solution formed to the surface and directly into the liquid phase. Excess of one of the components (solvent) reduces the order of reaction per unit. This suggests that the dissolution of ARPDa corresponds to the first order of the chemical reaction. With this in mind, the kinetic model of the process is adopted.

Analysis of the shape of the curves showed that the dissolution of ARPD belongs to the class of reactions with the maximum initial rate. However, with increasing dissolution degree, the speed of the process gradually decreases.

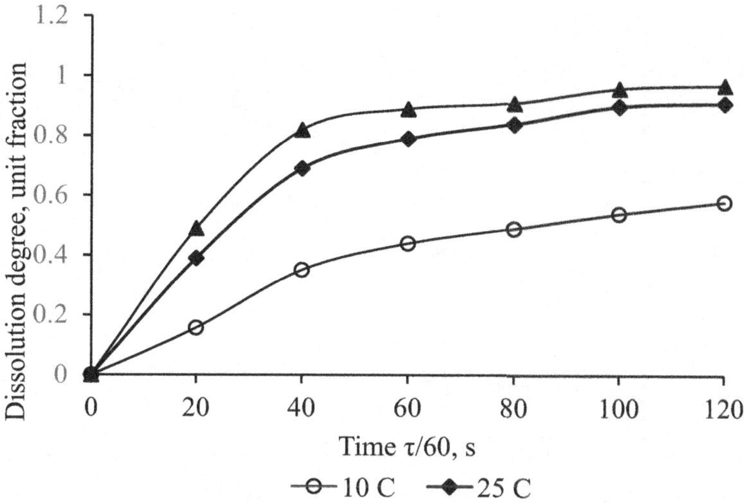

Fig. 13.4 Experimental kinetic curves of dissolution of ARPDa depending on the process duration in the model mixture of HBM (1: 1) at temperatures, °C: 10, 25, 35

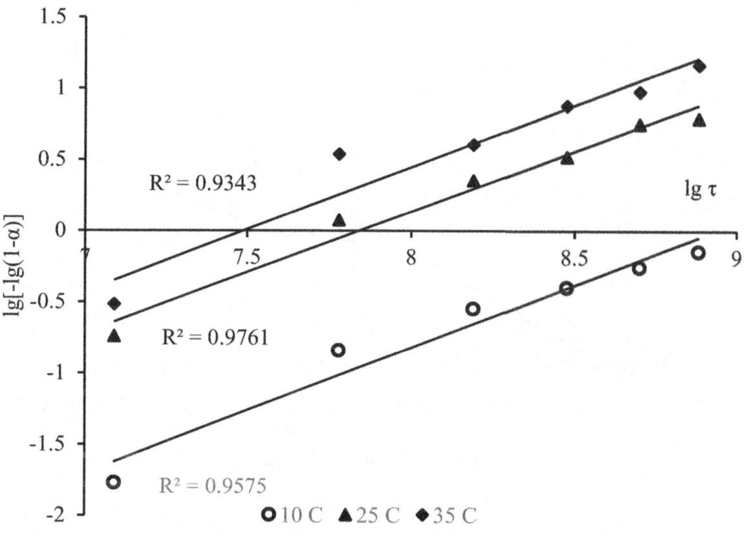

Fig. 13.5 Kinetic curves of ARPDa dissolution in HBM (1:1) at different temperatures

The data presented in Fig. 13.5 and the obtained values of the approximation reliability values demonstrate the validity of the decision on the choice of equation (13.1), because the experimental curves linearizes in the coordinates lg $[-\lg(1-\alpha)] - \lg\tau$ in a wide time range for each experiment temperature. (Fig. 13.5). After logarithmization of equation (13.1), the following expression was obtained:

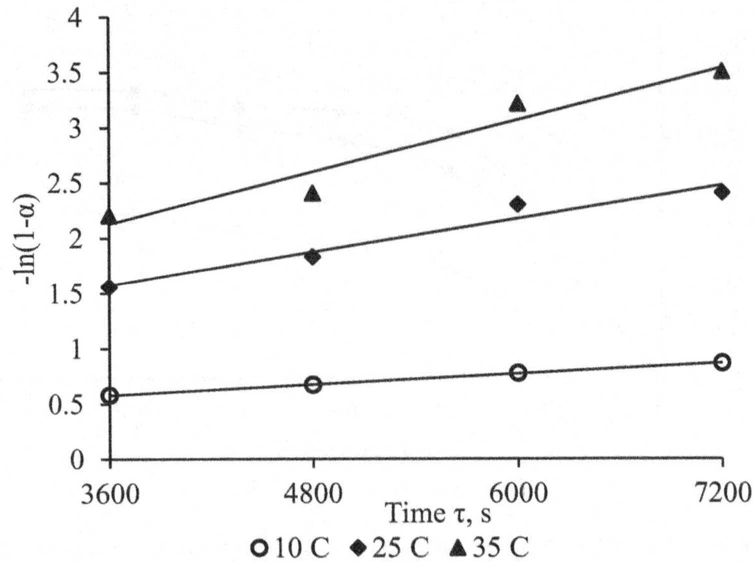

a

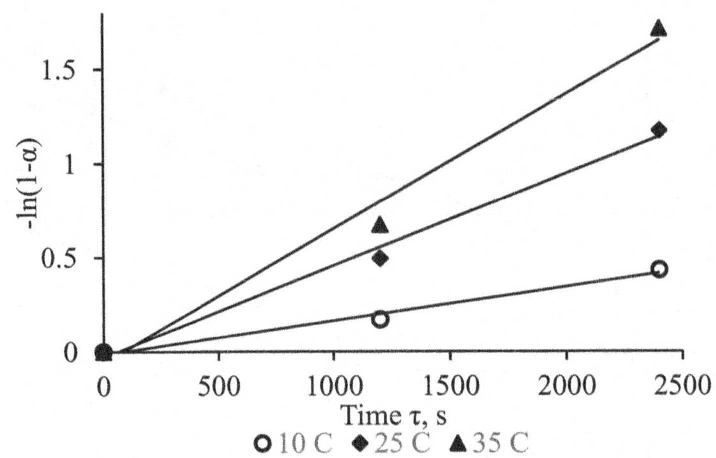

b

Fig. 13.6 Graphical determination of the constant of the dissolution rate of ARPDa in HBM (1:1) depending on the process temperature for the first (**a**) and second (**b**) stages of the process

$$-\ln(1-\alpha) = k \cdot \tau \tag{13.5}$$

The results of experimental kinetic studies of the dissolution of ARPDa deposit samples in the model mixture of HBM (1:1) depending on the duration of contact and temperature were processed by formula (13.2) and were presented graphically (Fig. 13.6a, b) to calculate the dissolution rate constant.

Table 13.5 Determination of the rate constant of dissolution of ARPDa depending on the process temperature

Solvent temperature, °C	Dissolution stage	Dissolution rate constant k, s^{-1}	$\tau_{1/2}$, min
10	1st	$1.8 \cdot 10^{-4}$	412
	2nd	$7.99 \cdot 10^{-5}$	
25	1st	$4.88 \cdot 10^{-4}$	24
	2nd	$2.35 \cdot 10^{-4}$	
35	1st	$7.14 \cdot 10^{-4}$	16
	2nd	$3.61 \cdot 10^{-4}$	

The graphical dependence on $-\ln(1-\alpha)$ for equation (13.5) is a straight line (Fig. 13.6), which is the main criterion that confirms the assumption that the dissolution of ARPDa corresponds to the first order. The calculated values of the dissolution rate constant of ARPDa are presented in Table 13.5.

Dissolution of ARPD in the temperature range of 25–35 °C is intense, which corresponds to the values of the rate constants and the half-dissolving time. At the temperature of 10 °C, the half-dissolving time was almost 7 h, which dictates the need to heat the solution. The possibility of effective use of aliphatic-aromatic solvents for the extraction of ARPD is confirmed by the values of $\tau_{1/2}$ at temperatures of 25 °C and 35 °C (24 and 16 min, respectively).

Decreasing the ARPDa dissolution rate constant for the second stage at all temperatures indicates a slowing down of the process. In the first period of time, resins and low-melting paraffins are mainly dissolved. Then the rate of destruction and dissolution of the ARPD decreases. Probably, the solvent-ARPD system approaches the state of saturation in a thin solvate layer of solvent, which is formed around the conglomerates of ARPDa with a high content of asphaltenes and high-molecular-weight hydrocarbons of hybrid structure. In addition, re-aggregation of the dispersed ARPD particles is possible (primarily as a result of swelling of asphaltene particles), which prevents the penetration of the solvent into the middle of the conglomerates of ARPD (Gurvich and Sherstnev 1994; Gorskov et al. 2012).

At a temperature of 10 °C, the change in the dissolution rate is weak, which confirms the dependence of the dissolution rate on the temperature of the solvent, especially in the initial period.

The results confirmed that the process of dissolving ARPDa in a composite mixture of HBM takes place as a first-order reaction. The rate of physicochemical interaction of the components of ARPDa with hexane-benzene mixture is equal to the rate of their diffusion from the surface of the condensed material into solution. This is observed at all research temperatures, which provides the maximum effect of using the solvent of the selected type.

The graphical dependence of the logarithm of the reaction rate constants on the inversed temperature is presented in Fig. 13.7.

The linearization of the experimental data allowed to determine the imaginary activation energy of the dissolution of ARPDa by the value of the angular coefficient

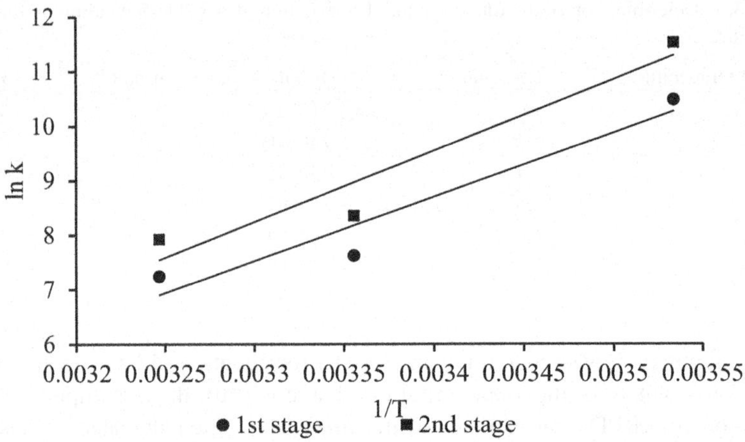

Fig. 13.7 Graphic dependence of the logarithm of the dissolution rate constants of deposits with high asphaltene content on the inversed temperature for the first and second stages of the process

Table 13.6 The main kinetic characteristics of the dissolution of ARPDa in HBM

Dissolution stage	k_0, c^{-1}	E, kJ/mol
1st	$3.84 \cdot 10^{13}$	97.76
2nd	$1.43 \cdot 10^{15}$	108.69

and the value of the preexponential factor by the value of the free term from the equation of trend lines for each stage separately.

The results of determining the kinetic parameters of the Arrhenius equation when dissolving ARPDa in the model hexane-benzene mixture are summarized in Table 13.6.

The obtained values of the apparent activation energy indicate the influence of diffusion processes on the dissolution rate of ARPD in a binary experimental solvent.

A detailed study of the dissolution kinetics of deposits with high paraffin content was performed under dynamic conditions using a distilled crude oil fraction of 110–150 °C, which during the experiment, conducted in static mode, provided the highest degree of dissolution (Table 13.3). The results of research at temperatures of 10 °C, 25 °C, and 35 °C are presented in Fig. 13.8.

The highest rate and the greatest efficiency of dissolution of ARPDp are observed at a temperature of 35 °C (0.57 particle), which confirmed the feasibility of the process at this temperature. The dissolution process was conditionally divided into two successive stages, the transition of which is observed approximately 50 min after the start of the experiment.

At a fixed contact time for deposits of asphaltene and paraffin types, similar dependences are observed: with increasing duration of contact, the dissolution rate gradually decreases. At the same time, solvents with a lower content of aromatic

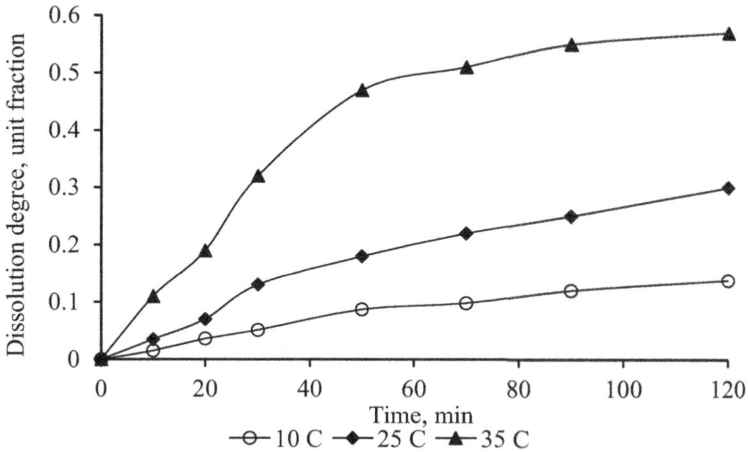

Fig. 13.8 Experimental kinetic curves of dissolution of ARPDp depending on the duration of the process in the distilled crude oil fraction of 110–150 °C at temperatures, °C: 10, 25, 35

hydrocarbons – distilled crude oil fractions of 110–150 °C and 150–200 °C – have the maximum efficiency on ARPD with a high content of hard paraffins.

The increase in the composition of the solvent of aromatic hydrocarbons and the increase in molecular weight lead to an increase in the ability to dissolve deposits with a predominant content of asphalt-resinous substances (Fig. 13.9).

Differences in the composition of deposits mainly determine the features of their dissolution in reagents of different chemical structure. The higher content of the asphaltene component, the less strong the deposits are, because compounds of this type have an increased ability to retain the liquid phase. Methane hydrocarbons – especially high-molecular-weight paraffins – on the contrary, form strong, dense structures and create serious complications, the elimination of which takes a lot of resources and time.

The results of experimental kinetic studies of the ARPDp sample dissolution in the distilled crude oil fraction of 110–150 °C depending on the duration of contact and temperature were processed similarly to the previously selected algorithm. Graphical interpretation of the results for calculating the dissolution rate constant is presented in Fig. 13.9. The kinetic characteristics of the process are summarized in Table 13.7.

In the analysis of the kinetic characteristics of dissolution by the distilled crude oil fraction of 150–200 °C ARPD with the maximum content of paraffins, a slightly different dependence is observed in comparison with the behavior of asphaltene-type deposits. For ARPD paraffin base, the decrease in the dissolution rate constant with increasing temperature is observed to a greater extent. This is due to the fact that paraffins create weaker bonds between the components of the deposits, the rupture of which requires less energy, as evidenced by the greater half-life of the samples at 25 °C and 35 °C (Tables 13.5, 13.7).

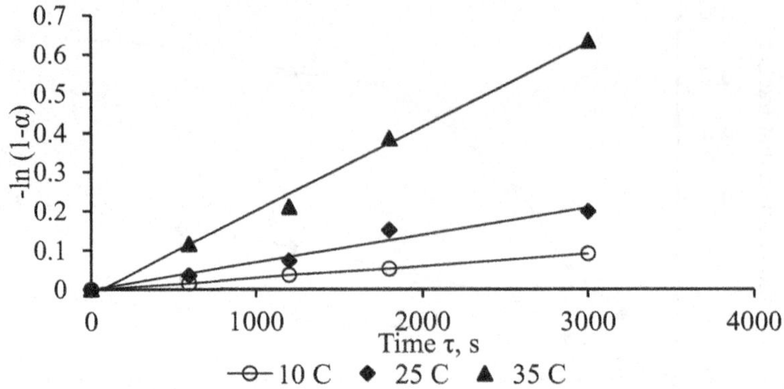

a)

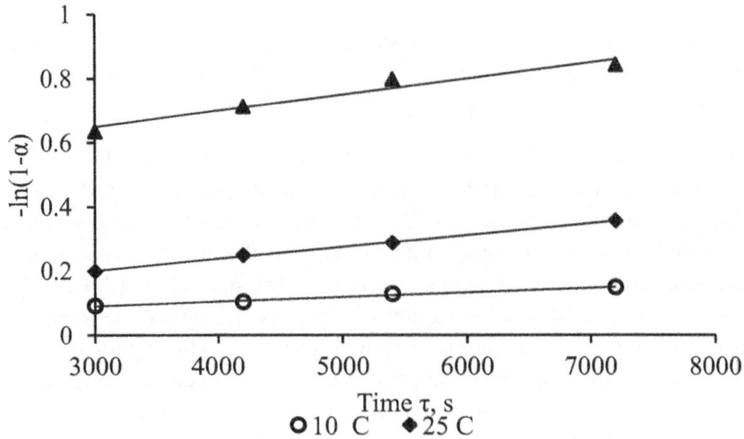

b)

Fig. 13.9 Graphical determination of the dissolution rate constant of ARPDp in the distilled crude oil fraction of 110–150 °C depending on the duration of the process and temperature for the first (**a**) and second (**b**) stages of the process

Table 13.7 The value of the dissolution rate constant of ARPDp depending on the process temperature

Solvent temperature, °C	Dissolution stage	Dissolution rate constant k, s^{-1}	$\tau_{1/2}$, min
10	1st	$3.43 \cdot 10^{-5}$	337
	2nd	$1.21 \cdot 10^{-5}$	
25	1st	$7.18 \cdot 10^{-5}$	161
	2nd	$4.22 \cdot 10^{-5}$	
35	1st	$2.05 \cdot 10^{-4}$	56
	2nd	$5.31 \cdot 10^{-5}$	

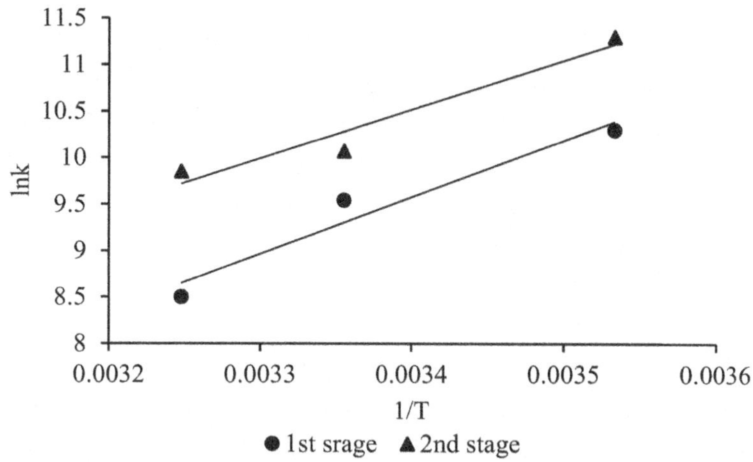

Fig. 13.10 Graphic dependence of the logarithm of the dissolution rate constants of deposits with high paraffin content on the inverse temperature for the first and second stages of the process

Table 13.8 The main kinetic characteristics of the dissolution of ARPDp distilled crude oil fraction of 110–150 °C

Dissolution stage	k_0, c^{-1}	E, kJ/mol
1st	$5.63 \cdot 10^4$	50.14
2nd	$2.04 \cdot 10^3$	44.38

The preexponential factor k_0 and the apparent activation energy of the ARPDp dissolution were determined by processing the graphical dependence (Fig. 13.10) of the logarithm of the reaction rate constants from the inverse temperature.

The results of determining the kinetic parameters of the Arrhenius equation when dissolving ARPDp in a distilled crude oil fraction of 110–150 °C are summarized in Table 13.8.

According to macrokinetic data (Table 13.8), in the first stage, the system is characterized by greater strength and orientation of intermolecular bonds. The activation energy of the dissolution of paraffin deposits in the next stage is slightly reduced, which is explained by the decrease in the structural strength of saturated paraffinic hydrocarbons, mainly n-alkanes. The decrease in the activation energy of the dissolution of paraffin deposits is due to the transition of the system from a discontinuous state to a free-dispersed state, and the interaction of the paraffin components of the sample becomes weak.

The effect of additives on the solubility of distilled crude oil fractions under the duration process of 2 h at a temperature of 35 °C is presented in Fig. 13.11. Analysis of the obtained data showed that the use of distilled crude oil fractions for the dissolution of deposits with high paraffin content does not provide the required efficiency. The maximum degree of dissolution when using fractions of 110–150 °C and 150–200 °C was 0.55 parts.

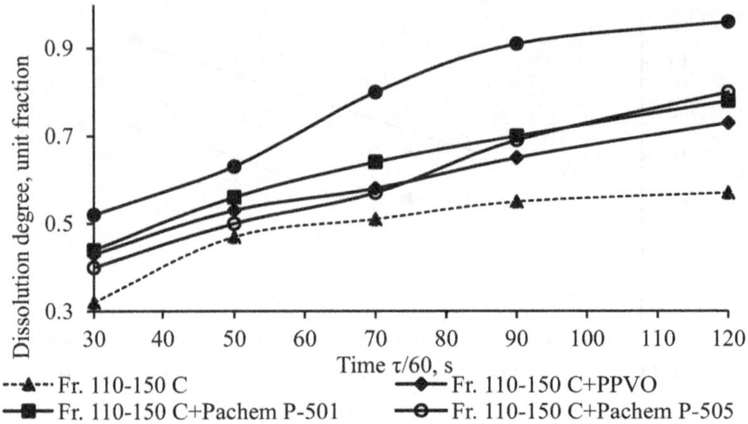

Fig. 13.11 The dissolution degree of ARPDp in pure distilled crude oil fraction of 110–150 °C and in distilled crude oil fraction of 110–150 °C with 5% weight of the additives: PPVO; Pachem P-505; Pachem P-501; a mixture of additives – 1.5% weight of PPVO, 1.5% weight of Pachem P-505 and 2% weight of Pachem P-501

Additives are not solvents, but their introduction has increased not only the "true" solubility of the components but also the ability to disperse the ARPD. This is evidenced by the increase in the degree of solubility from 0.57 to 0.73 (PPVO), 0.78 (Pachem P -501), 0.8 (Pachem P -505), and 0.96 (composite mixture of three additives).

On the paraffin-type ARPD, when a composite mixture of additives is introduced into the solvent, a positive synergistic effect is observed at an almost equal ratio of the additive components. At the same time, their ability to dissolve is almost 2 times higher in comparison with the base solvent. The synergistic effect of the action of three-component additives can significantly increase the efficiency of the distilled crude oil fraction to the level of industrial reagents-inhibitors.

ARPDp are capable to form a stable bond with high-molecular-weight hydrocarbons. In this case, more dispersed paraffin particles pass into the solution and to prevent their re-aggregation requires a high content of additives (up to 5% wt.).

With increasing concentration of additives above 5% weight there is an increase in the surface layers of associates with additive molecules, which reduce the surface activity due to the weak affinity with the core of deposit aggregates and, accordingly, a smaller effect of interaction with paraffin-asphaltene components. In addition, with an excessive concentration of additives more than 5% weight they are adsorbed on the surface of the ARPD, and the resulting polymolecular layer prevents the penetration of the active components of the additives to the surface of the ARPD.

The effective action of the composite mixtures of additives on ARPD with a high content of paraffins is primarily due to the increased solubility of resins and asphaltenes present in the deposits, which are cementing agents of crystals of hard paraffins.

The obtained results confirm the feasibility of using multifunctional additives and their compositions for the destruction of ARPD, taking into account the synergistic effect and the peculiarities of the joint course of dissolution and dispersion processes.

13.4 Conclusions

Mechanical disposal of oil deposits does not solve the problem of their rational utilization, environmental and economic aspects of the problems associated with their use as a potential and valuable source of additional hydrocarbons.

It is established that differences in the composition of deposits determine the peculiarities of their dissolution in reagents of different chemical structure. A set of experimental studies showed that distilled crude oil fractions of 110–150 °C and 150–200 °C were more effective for dissolving paraffinic ARPD. For deposits with a high content of asphaltene components, it is advisable to use a composition of aromatic-aliphatic solvents based on benzene and hexane in a ratio of 1:1 at a maximum allowable temperature of 35 °C.

It is proved that the use of additives-inhibitors increased the solubility of the components of the ARPD due to the dispersing ability from 0.57 to 0.73 when using an additive of vegetable origin, to 0.78 (Pachem P-501), to 0.8 (Pachem P-505), and up to 0.96 when using a composition of three additives. The results confirmed the feasibility of using multifunctional additives and their compositions, taking into account the synergistic effect and the peculiarities of the joint processes of dissolution and dispersion of ARPD.

It is substantiated that the Erofeev-Kolmogorov equation can be used to calculate the kinetic parameters of ARPD dissolution. The dynamics of change in the degree of dissolution of ARPD is exponential and includes two stages. The maximum dissolution rate was observed in the first minutes of the process. After 40 min, the dissolution rate gradually decreases. Processing of experimental data allowed to establish the rate constants and conditional activation energy of each stage.

Data on the influence of the chemical nature of ARPD solvents, as well as the studied patterns and features of dissolution kinetics in different hydrocarbon systems, are the scientific basis for evaluating the effectiveness of solvents used to remove ARPD of different nature and can be used for technological decisions.

References

Ahmetov AF, Gerasimova EV, Nuriazdanova VF (2008) Analiz laboratornyh metodik opredeleniya effektivnosti rastvoritelej asfal'to-smolo-parafinovyh otlozhenij. Bashkirskij himicheskij zhurnal 1:65–67

Delmon B (1969) Introduction a la cinetique heterogene. Editions Technip, Paris. Russian edition Del'mon B. (1972) Kinetika geterogennyh reakcij (trans: Bazhina NM, Boldyrev VV). Mir, Moskva

Gorshkov AM, Chekanceva LV, Shishmina LV (2012) Issledovanie fazovogo povedeniya asfal'tenov v model'noj sisteme i v nefti. CHernye metally, spec. Vypusk 4:76–78

Gurvich LM, Sherstnev NM (1994) Mnogofunkcional'nye kompozicii PAV v tekhnologicheskih operaciyah neftedobychi. VNIIOENG, Moskva

Kolpakova NA, Romanenko SV, Kolpakov VA (2008) Sbornik zadach po himicheskoj kinetike. Izd-vo Tomskogo politekhnicheskogo universiteta, Tomsk

Kononov OV, Mastobaev BN (2010) Analiz i klassifikaciya sushchestvuyushchih sposobov bor'by s otlozheniyami v neftyanyh emkostyah. Istoriya nauki i tekhniki 6:60–68

Tertyshna OV, Roienko KV, Martynenko VO, Polishchuk VV, Horpynko YuH, Pushak AP (2018a) Hrafitove mastylo z napovniuvachem vidkhodiv naftopererobky. Patent Ukrainy na vynakhid 116077

Tertyshna OV, Roienko KV, Martynenko VO, Polishchuk VV, Horpynko YuH, Pushak AP (2018b) Kaltsiieve mastylo z napovniuvachem vidkhodiv naftopererobky. Patent Ukrainy na vynakhid 117417

Tertyshna OV, Roienko KV, Martynenko VO, Snizhko LO (2015) Vykorystannia asfaltosmoloparafinovykh vidkladen yak dodatkiv do naftovykh dorozhnikh bitumiv. Naftohazova haluz Ukrainy 3:33–35

Tertyshnay EV, Martynenko VO, Girenko AA et al (2018). The influence of an additive of vegetable origin on the aggregative stability of petroleum. SOCAR Proceedings. №1 052–058

Chapter 14
Cavitation Treatment of Gas Condensate Gasoline, Modified with Monohydric Alcohols

Sergii Kudryavtsev, Oleksii Tselishchev, Sergii Boichenko, and Marina Loriia

Nomenclature

CO_2	Carbon dioxide
IPA	Isopropanol
MON	The octane number by the motor method
RON	Research method

14.1 Introduction

The development of new chemical technologies for the production of motor fuels with improved technical, economical, and environmental characteristics is an important task for science and industry. In the future, reducing the cost of production of gasoline in chemical processes can be achieved by changing the method of supplying energy to the reaction zone.

Today, the main method of supplying energy to the reaction zone is the thermal method – heating the raw material and (or) catalyst to the desired temperature. But the selectivity of this method of energy supply is very low. Therefore, energy consumption is the second largest in the calculation of the cost of motor fuels. The first place remains in the cost of raw materials. Moreover, high-octane oxygen-containing additives to gasoline (alcohols, ethers, etc.) have a much higher cost than low-octane hydrocarbon fraction.

S. Kudryavtsev (✉) · O. Tselishchev · M. Loriia
Volodymyr Dahl East Ukrainian National University, Severodonetsk, Ukraine
e-mail: kudriavcev@snu.edu.ua; celischev@snu.edu.ua; loria@snu.edu.ua

S. Boichenko
National Aviation University, Kyiv, Ukraine

S. Boichenko et al. (eds.), *Chemmotological Aspects of Sustainable Development of Transport*, Sustainable Aviation, https://doi.org/10.1007/978-3-031-06577-4_14

Although not always, traditional technologies of homogeneous and heterogeneous catalysis can be effectively used for the production of high-octane gasolines. The use of traditional catalysis is complicated by the presence of mechanical impurities in the raw material, the content of catalytic poisons in the raw material, low concentrations of reagents, and many other factors. Promising alternative technological processes can be processes that use physical methods of impact on raw materials and catalyst. The main advantage of physical methods of energy supply over thermal methods is the possibility of selective energy supply to reaction centers. In the long run, this will help reduce the total energy consumption for the production of motor fuels in industrial processes.

There are different ways of physical activation of raw materials and catalysts: acoustic methods (sound oscillations of different frequencies and shock waves), electromagnetic methods (constant or electric and (or) magnetic field), optical and radiation methods (different types of radiation), mechanical and hydromechanical methods (grinding, emulsification, mixing, pseudo- and vibration liquefaction, filtration, etc.), and mass transfer (crystallization, drying, etc.).

Processing of raw materials (mixtures of hydrocarbon fractions with high-octane additives) in a cavitation field is a promising mechanical method of energy supply for industrial implementation. This method uses the effect of hydrodynamic cavitation – the generation of a significant number of bubbles in the liquid raw material during the passage of this raw material through a special nozzle reactor and the subsequent significant increase in temperature and pressure when the bubbles collapse. Thus, each bubble formed during hydrodynamic cavitation treatment is an individual mini-reactor in which chemical transformations are carried out at elevated temperatures and pressures (Boichenko et al. 2017; Milotskyi and Hanzha 2016). A number of experiments on the modification of gasolines with monohydric alcohols were performed under cavitation treatment. The obtained results show a significant increase in the octane number of such gasoline mixtures in comparison with simple mechanical mixing of alcohols and gasolines. The composition of motor gasolines and their compliance with EURO-3 (4,5) standards are set by DSTU 7687: 2015. According to this standard, gasolines can contain up to 10% bioethanol, up to 12% isopropanol, and up to 15% isobutanol and other alcohols and ethers. Thus, the combination in one technological process of stages of mixing gasoline with alcohol and cavitation treatment of the mixture will allow to obtain an additional increase in octane number and reduce the cost of raw materials and energy for the production of high-octane gasoline.

14.2 Theoretical Substantiation of Chemical Transformations That Cause an Increase of the Octane Number of Gasoline-Alcohol Mixtures During Their Processing in a Cavitation Field

Cavitation treatment technology actively influences the chemical raw material, which is in the liquid phase. Therefore, it is most expedient to carry out processes with chemical transformations in the liquid phase or at the interface of the phases by the method of cavitation treatment. According to the theoretical foundations of hydrodynamic cavitation, in cavitation cavities, which are formed during the passage of raw materials through the cavitation nozzle-reactor, and then collapse, there are zones of high temperatures and pressures. The local increase in temperature can reach thousands of degrees, and the pressure can increase up to a hundred times. Under such conditions, cavitation caverns are mini-reactors in which chemical reactions take place with high efficiency. In recent years, numerous studies have been conducted on the impact of cavitation treatment on petroleum products. It is proved that in the conditions of cavitation processing, the reactions of cracking of hydrocarbons intensify. In some modes of cavitation, the reactions of compaction, polymerization, and coking can intensify. During the cavitation treatment of gasoline in the presence of hydrogen peroxide, the course of reactions leading to the formation of methanol was proved (Milotskyi and Hanzha 2016). Theoretically, the formation of higher-molecular-weight alcohols is possible. Own studies of the composition of alcohol-gasoline mixtures after cavitation treatment showed an increase in the content of ethers. The mechanism of chemical transformations (Tselishchev et al. 2020a, b) was proposed, which explains the increase in the OC of alcohol-gasoline mixtures under the action of cavitation.

The first chemical reactions that take place in an alcohol-gasoline mixture in cavitation caverns are reactions of formation of radicals from alkanes and hydrocarbons of other classes. Given the high temperature and pressure at the time of collapse of the cavitation cavity, the most likely way to form radicals is the thermal dissociation of molecules (14.1) and (14.2):

$$C_{(n+m)}H_{2(n+m)+2} + kav \rightarrow \dot{C}_{(n+m)}H_{2(n+m)+1} + \dot{H} \tag{14.1}$$

$$C_{(n+m)}H_{2(n+m)+2} + kav \rightarrow \dot{C}_nH_{2n+1} + \dot{C}_mH_{2m+1} \tag{14.2}$$

The bond energy between the carbon atoms in the alkane molecule is almost 20% less than the bond energy between the carbon and hydrogen atoms. Therefore, the probability of reaction (14.2) under cavitation treatment is higher than the probability of reaction (3.1). Reactions (14.1) and (14.2) and other similar transformations are chain nucleation reactions in a radical-chain reaction.

All subsequent transformations of alkyl radicals can occur in several directions. All these transformations will be reactions of chain continuation in the radical-chain scheme of the mechanism of this process:

Conversion of radicals into olefins (14.3) and cycloalkanes (14.4) with six carbon atoms in the cycle:

$$\dot{C}_nH_{2n+1} \rightarrow \dot{C}H_3 + C_{(n-1)}H_{(2n-2)} \tag{14.3}$$

$$\dot{C}_nH_{2n+1} \rightarrow C_6H_{12} + \dot{C}_{(n-6)}H_{(2n-11)} \tag{14.4}$$

Conversion of radicals into aromatic compounds (on the example of the reaction that leads to the formation of benzene) (14.5):

$$\dot{C}_nH_{2n+1} \rightarrow C_6H_5 - R + \dot{C}_XH_Y \tag{14.5}$$

Interaction of alkyl radicals with monohydric alcohols with the formation of ethers and the generation of free radicals (14.6):

$$\dot{C}_nH_{2n+1} + R - O - H \rightarrow C_nH_{2n+1} - O - R + \dot{H} \tag{14.6}$$

The disappearance of alkyl radicals (chain break reactions) can occur by the following reactions:

Interaction with the reactor material (in the real experiment there is a slight corrosion of the equipment):

$$\dot{C}_nH_{2n+1} + Me \rightarrow \text{Corrosion products} \tag{14.7}$$

Recombination of radicals:

$$\dot{C}_nH_{2n+1} + \dot{C}_mH_{2m+1} \rightarrow C_{(n+m)}H_{2(n+m)+2} \tag{14.8}$$

other ways.

Analysis of potentially possible ways of chemical transformations of hydrocarbons and alcohols in the cavitation field allows us to conclude that the products of cavitation processing are high-octane components of motor fuels. The reactions of formation of ethers from alcohols will allow to obtain chemicals that are stabilizers of emulsions. This allows you to get alcohol-gasoline fuel, resistant to delamination, even in the presence of small amounts of water in the mixture. This opens a new way to modify low-octane motor gasolines with bioethanol and other mixtures of alcohols of biochemical origin, which contain water impurities.

14.3 Substantiation of Expediency and Prospects of Use of Monohydric Alcohols in Productions of High-Octane Gasolines

Bioethanol This is an ordinary ethyl alcohol produced from vegetable raw materials. Bioethanol can be used as a stand-alone motor fuel, or as a high-octane additive to gasoline. The use of bioethanol may be economically feasible. This alcohol has a high-octane number – 107–108 units by research method. The presence of an oxygen atom in the molecule reduces the formation of soot during the combustion of bioethanol and bioethanol-gasoline mixtures in car engines. This extends engine life and reduces emissions of toxicants in the exhaust gases.

Bioethanol is a renewable resource. The use of bioethanol as a fuel does not lead to the formation of greenhouse gases. During the production and subsequent combustion of bioethanol, as much CO_2 is released as was removed from the atmosphere by plants, which were then processed into bioethanol.

The use of bioethanol reduces exhaust toxicity by almost 21%. Oxygen present in ethanol allows you to burn more hydrocarbons. Impurities of 10% ethanol in gasoline can reduce emissions of particulate matter by 50% and carbon monoxide by 30%. Bioethanol is biodegradable and does not irreversibly pollute the environment.

Some studies show that bioethanol has a negative energy balance. Bioethanol production consumes more energy than can be obtained from the combustion of bioethanol in engines. But the energy balance can vary greatly in the production of bioethanol from different plants. In addition, the production of high-octane hydrocarbon gasolines by traditional thermocatalytic methods also requires more energy than is released during the combustion of these gasolines. That is, bioethanol is a real competitor to hydrocarbons in the production of high-octane gasoline.

Bioethanol has its drawbacks. In bioethanol, the heat of combustion is lower than in hydrocarbon gasolines – 25 MJ/kg in bioethanol against 42 MJ/kg in gasoline. Also, the presence of oxygen in the molecule makes bioethanol less energy-intensive – 1 l of ethanol contains 37% less energy than 1 l of hydrocarbon gasoline. This leads to a higher specific consumption of alcohol fuel compared to hydrocarbon fuel. Bioethanol has a low saturated vapor pressure and a high evaporation temperature. This makes it difficult to start the engine at low temperatures. Fuel containing bioethanol with water impurities will have high corrosion activity. In some engine modes, incomplete combustion of bioethanol may occur. This will lead to the formation of toxic aldehydes in the exhaust gases. Due to the presence of these disadvantages, bioethanol is usually used not in pure form but as a mixture with hydrocarbon fuels. This improves the performance of gasoline and saves hydrocarbon resources.

Isopropanol (IPA) is another alcohol that can be used to increase the octane number of gasolines. IPA is obtained by combining propylene with water through the hydration process or through the dehydrogenation of acetone. There are two ways of the hydration process, and both of these processes include a distillation process to separate the resulting isopropyl alcohol from water. IPA is also a by-product of bioethanol production.

The urgency of the use of IPA is associated with an increase in the production of propylene from catalytic cracking plants. Isopropyl alcohol will increase in quantity in the coming years and will be available as a raw material, as it is mainly made of propylene, and propylene will be reduced in use due to the global trend to reduce consumption of plastics, including polypropylene.

Isobutanol This alcohol can become a real competitor to methanol and bioethanol as a high-octane additive to gasoline due to its relatively low cost. Also, isobutanol together with IPS and isoamyl alcohol form the basis of fusel oils, which are a waste in the production of bioethanol and can be effectively used to increase the octane number of gasoline. Such use will increase the environmental performance of gasoline production processes and reduce the amount of waste from bioethanol production. In general, the use of alcohols of biochemical origin for the modification of gasoline is very promising. Such technologies will reduce the consumption of oil and gas condensate, which in turn will reduce greenhouse gas emissions.

14.4 Laboratory Unit and Method of Research of Processes of Modification of Gasolines by Alcohols in a Cavitation Field

The main task for the laboratory unit to study the effect of cavitation treatment on the octane number of alcohol-gasoline mixtures is the closest possible approximation to the conditions of future industrial technology. Therefore, the laboratory unit reproduces all the necessary technological stages (six) of mixing alcohol with hydrocarbon gasoline fractions, cavitation treatment, recycling of raw materials, separation of alcohol-gasoline mixture from gaseous products, and analysis of liquid and gas phases.

In this laboratory unit, the principle of hydrodynamic cavitation is implemented on a nozzle with an outlet pressure of 9.0 MPa. This is enough pressure to carry out the necessary chemical transformations with minimal energy costs. Changing the number of cycles of cavitation treatment of raw materials leads to a change in the intensity of mechanochemical activation of the alcohol-gasoline mixture and affects the composition of products and the octane number. By varying the intensity of mechanochemical activation through cavitation, it is possible to achieve the required octane number for alcohol-gasoline mixtures and obtain brands of gasoline A-95 and A-98 with the minimum possible consumption of alcohol, which in future industrial technology will reduce production costs by reducing raw material costs. Experimental studies were performed on the original laboratory unit, which is schematically presented in Fig. 14.1.

The experiments were performed as follows. Gasoline is mixed with alcohol in the right proportions and enters the tank, from which the raw material is dosed by the regulator using a high-pressure pump on the nozzle-reactor. The nozzle reactor allows you to create a pressure at the outlet of the nozzle at the level of 9.0 MPa. Pressure measurement is carried out by the RI manometer, and change of pressure is possible by replacement of a nozzle. At the outlet of the nozzle, nozzles are formed cavitation cavities, which are mini-reactors in which chemical transformations take place.

Cavitation cavities are also formed when droplets of raw material collide from the nozzle at a speed of more than 140 m/s, with the reactor walls and with a cone-shaped bumper, which is located in the separator C directly in front of the nozzle.

In Fig. 14.2 shows a general view of the laboratory unit (a), which was developed by the authors, and the appearance of the cavitation reactor (b). Productivity of unit makes more than 5 l of the raw materials, which are subject to cavitation processing, for a minute.

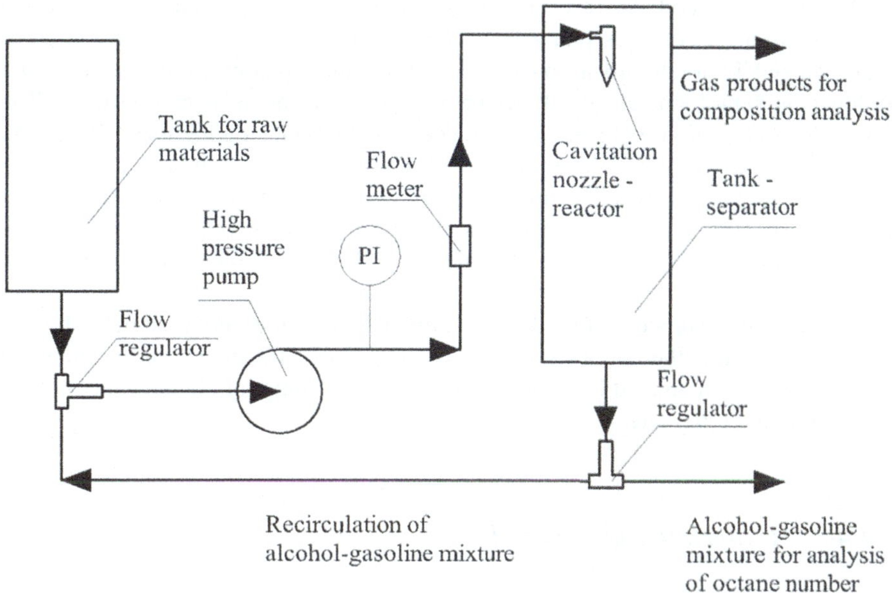

Fig. 14.1 Laboratory unit for studying the effect of cavitation treatment on the octane number and composition of gasolines modified with alcohols (Tselishchev et al. 2020a, b)

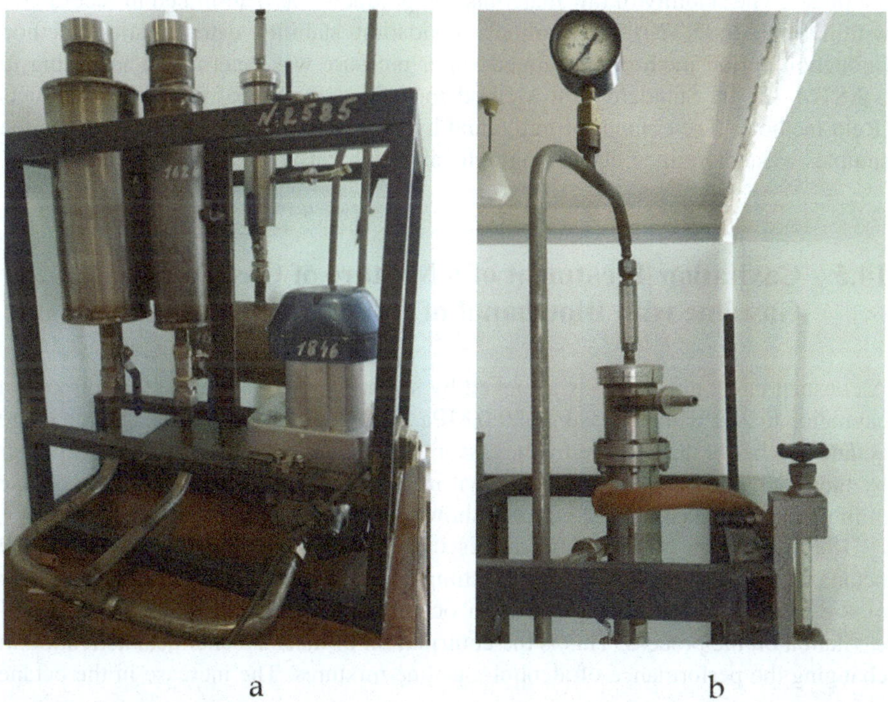

Fig. 14.2 Laboratory author's unit for studying the effect of cavitation on the octane number and fractional composition of alcohol-gasoline mixtures: (**a**) general view; (**b**) cavitation reactor (Tselishchev et al. 2020a, b)

The separator collects raw materials and degas them from air and possible gaseous products. The gas phase from the top of the separator is analyzed chromatographically for hydrocarbon content. The liquid phase, which is the target product, is sent for analysis of octane number and chemical composition. The intensity of cavitation treatment is regulated both by replacing the nozzle (pressure change) and by varying the number of cycles of cavitation treatment. For this purpose, the possibility of recirculation of raw materials on laboratory unit is provided. If stratification of the liquid alcohol-gasoline mixture occurred during the experiment, this can be seen only after unloading the reaction products from the separator tank.

Analysis of the process of modification of gasoline in the cavitation field allowed us to conclude that the initial parameters of the process (fractional composition and evaporation) are regulated by changes in mixture consumption, bioethanol content, and nozzle inlet pressure.

14.4.1 Experimental Methods

The fractional composition of raw materials and products was determined according to ISO 3405-88. The octane number was determined by the research method using a portable octanometer SHATOX SX-150 with a measurement error of not more than 0.2 points. The stability of raw materials and products was determined in accordance with DSTU 7685: 2015 "Gasoline". Oxidation stability determination method (induction period method). Saturated vapor pressure was determined according to 6 ASTM 323-08 Standard Test Method for vapor pressure of petroleum products (Reid method). The content of individual hydrocarbons, alcohols, and ethers in the samples was determined chromatographically on a chromatograph Crystal-2000.

14.5 Cavitation Treatment of a Mixture of Gas Condensate Gasoline with Bioethanol of Different Concentrations

A constant cavitation effect is achieved by several passes of the mixture through a cavitation nozzle with a pressure of 9.0 MPa. Experimental data on the change in the octane number of the mixture by the research method (RON) and the octane number by the motor method (MON) are shown in Table 14.1, and the increase in octane number due to the cavitation effect is shown in Table 14.2.

The increase in octane number was calculated as the difference between the octane number of the mixture after reaching a constant value and the octane number of the initial mixture. The increase in octane number characterizes the effect of cavitation on the process. This is the contribution of mechanochemical activation in changing the performance of alcohol-gasoline mixtures. The increase in the octane

Table 14.1 Change (RON) and (MON) from the number of cavitation cycles for gas condensate with the addition of bioethanol

	The number of cycle cavitation treatment										
Octane number	0	1	2	3	4	5	6	7	8	9	10
Gas condensate without the addition of bioethanol											
RON	78.6	78.3	78.7	79.0	79.1	79.2	79.2	79.3	79.3	79.5	79.6
MON	76.0	75.8	76.2	76.3	76.4	76.5	76.5	76.6	76.5	76.7	76.7
Gas condensate with the addition of 0.5% vol. bioethanol											
RON	87.2	87.9	87.9	88.1	89.1	88.6	88.9	89.3	89.3	89.8	89.4
MON	80.6	81.0	81.0	81.2	81.6	81.4	81.5	81.7	81.7	81.9	81.7
Gas condensate with the addition of 1% vol. bioethanol											
RON	81.1	82.5	83.1	83.6	84.6	84.8	85.5	85.8	86.0	85.5	86.2
MON	77.6	78.4	78.7	78.5	79.6	79.6	79.9	80.0	80.1	79.9	80.2
Gas condensate with the addition of 1.5% vol. bioethanol											
RON	91.5	91.9	92.1	92.1	92.1	92.2	92.4	92.4	92.4	92.3	92.3
MON	82.8	83.0	83.2	83.1	83.1	83.2	83.4	83.5	83.4	83.3	83.4
Gas condensate with the addition of 2% vol. bioethanol											
RON	90.8	91.1	91.3	91.3	91.7	91.6	91.4	91.4	91.8	91.6	91.7
MON	82.4	82.5	82.7	82.6	82.8	82.8	82.8	82.7	82.9	82.8	82.9
Gas condensate with the addition of 3% vol. bioethanol											
RON	93.9	94.4	94.1	94.4	94.4	94.5	94.5	94.6	94.6	94.7	94.6
MON	84.1	84.4	84.1	84.4	84.4	84.5	84.5	84.6	84.6	84.7	84.6
Gas condensate with the addition of 4% vol. bioethanol											
RON	94.9	95.0	95.1	95.1	95.1	95.3	95.7	95.4	95.4	95.5	95.7
MON	84.9	85.0	85.1	85.1	85.1	85.3	85.7	85.4	85.4	85.5	85.7
Gas condensate with the addition of 5% vol. bioethanol											
RON	96.3	96.5	96.6	96.9	97.0	97.0	97.0	97.0	96.9	96.8	97.2
MON	86.3	86.7	87.0	87.2	87.2	87.2	87.2	87.2	87.2	87.2	87.2

Table 14.2 Change in octane number by the research method (RON) and by the motor method (MON) before and after cavitation treatment depending on the content of bioethanol in the gas condensate

		RON		MON		Increment of octane number, points	
№	The content of bioethanol in the gas condensate, % vol.	Before	After	Before	After	RON	MON
1	0	78.6	79.6	76.0	76.7	1.0	0.7
2	0.5	87.2	89.4	80.6	81.7	2.2	1.1
3	1.0	81.1	86.2	77.6	80.2	5.1	2.6
4	1.5	91.5	92.3	82.8	83.4	0.8	0.6
5	2.0	90.8	91.7	82.4	82.9	0.9	0.5
6	3.0	93.9	94.6	84.1	84.6	0.7	0.5
7	4.0	94.9	95.7	84.9	85.7	0.6	0.8
8	5.0	96.3	97.2	86.3	87.2	0.9	0.9

number is achieved due to the passage in the cavitation-activated liquid of chemical reactions and due to the effective emulsification of alcohol-gasoline mixtures with small impurities of water.

In the absence of bioethanol additives, the octane number of gas condensate increased by almost 1 point by the research method and by 0.7 points by the motor method during cavitation treatment. That is, the hydrocarbon mixture is subjected to cavitation treatment and chemical transformations occur, which lead to an increase in the gas condensate of high-octane components.

A mixture of 99% gas condensate and 1% bioethanol proved to be the most effective. Simple mechanical mixing of condensate and bioethanol increased the RON by 2.5 points and the MON increased by 1.7 points. The total increase in RON during cavitation was 5.1 points, and MON, 2.6 points. This does not delaminate the mixture.

A further increase in the amount of bioethanol in the mixture reduces the effect of cavitation treatment, although the initial RON and MON are much higher. For example, for a mixture of 97% condensate and 3% bioethanol, RON increased by almost 15.3 points compared to pure gas condensate, and the increase in MON was 8.1 points. During cavitation treatment, the increase in RON and MON was 0.7 and 0.5 points, respectively. That is, there is some optimal concentration of bioethanol in the mixture with gas condensate, at which cavitation treatment gives the largest increase in octane number. To date, this is an experimentally established concentration of 1.0% by volume of bioethanol. Mixtures with a concentration of 2%, 3%, 4%, and 5% of bioethanol show approximately the same increase in octane number due to cavitation treatment. Starting at a concentration of 2%, a small amount (approximately 0.1–0.2% by volume of the mixture) of the heavy phase peels off. Probably, it is water with admixtures of alcohols and ethers.

Figure 14.3 shows the nature of the change in octane number by the research method from the number of cycles of cavitation treatment. It can be noted that three to five cycles of cavitation treatment were sufficient to achieve a constant value of RON. Although a mixture with an impurity of 1.0% bioethanol required a slightly higher intensity of cavitation treatment – a constant value of the octane number was reached after seven cycles.

Mixtures of gas condensate gasoline with 0.5% and 1.0% bioethanol showed a slightly different nature of the change in octane number compared to other studied mixtures. For low concentrations of bioethanol cavitation, treatment of alcohol-gasoline mixture showed the greatest effect. The number of cycles of cavitation treatment, after which a constant value of the octane number was established, was greater than for all other studied mixtures. This may be due to the distribution of water droplets in bioethanol in the hydrocarbon gasoline phase. This forms a stable emulsion, the properties of which change with the formation of esters with each new cycle of cavitation treatment of alcohol-gasoline mixture. For higher concentrations of alcohol in gasoline, this effect is not so noticeable.

Increasing the number of cycles of cavitation treatment to 30 did not give an effect higher than that achieved for five to seven cycles. Although, it is possible that by increasing the pressure on the nozzle above 9.0 MPa will be able to achieve

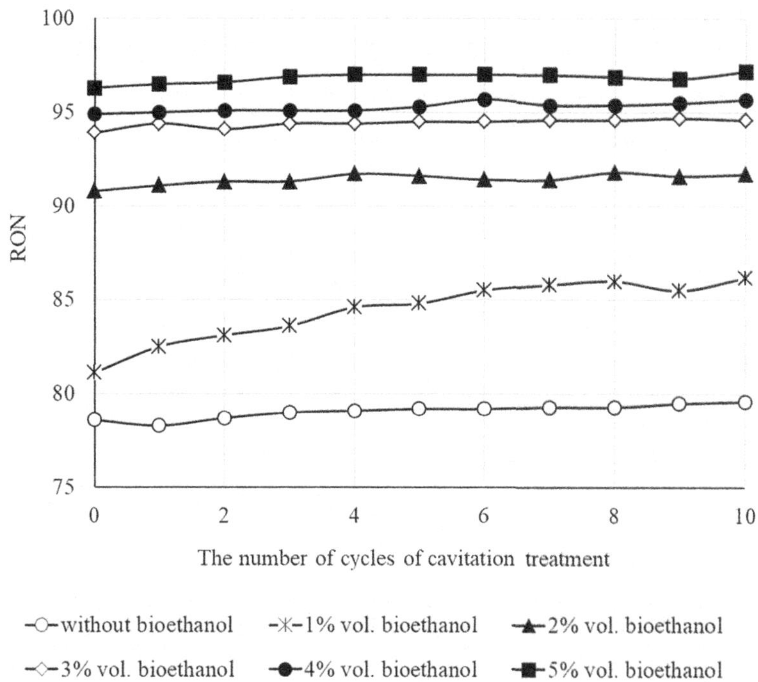

Fig. 14.3 The effect of the number of cycles of cavitation treatment on the octane number of gas condensate gasoline modified with bioethanol (Boichenko et al. 2020)

higher increases in octane number. The study of higher cavitation pressures is planned in the next stages of research.

14.6 Cavitation Treatment of a Mixture of Gas Condensate Gasoline with Isopropanol of Different Concentrations

The results of studies to study the change in RON and MON from the number of cycles of cavitation treatment for different concentrations of IPA in gas condensate gasoline are given in Table 14.3.

Studies have focused on finding IPA concentrations and cavitation regimes that are suitable for raising the RON to 95 and 98 points. These RON values correspond to the indicators of A-95 and A-98 gasoline brands.

Comparison of the achieved growth (RON) and (MON) with the initial indicators are given in Table 14.4. As final indicators, (RON) and (MON) values of OC which do not change after achievement of a constant value are accepted.

Listed in Table 14.3, results show that the addition of IPA leads to an increase (RON) and (MON) in the entire range of studies. As the number of cavitation cycles

Table 14.3 Change of octane number by research method (RON) and by motor method (MON) from the number of cycles of cavitation treatment for gas condensate with the addition of isopropanol

	The number of cycles cavitation treatment										
Octane number	0	1	2	3	4	5	6	7	8	9	10
Gas condensate without the addition of isopropanol											
RON	92.3	92.4	92.4	92.6	92.6	92.8	92.9	92.9	92.5	93.0	93.1
MON	83.3	83.5	83.5	83.7	83.7	83.7	83.8	83.8	83.8	83.8	83.8
Gas condensate with the addition of 0.5% vol. isopropanol											
RON	93.0	93.3	93.0	93.4	93.4	93.4	93.5	93.4	93.5	93.5	93.7
MON	83.8	83.8	83.8	83.8	83.8	83.8	83.9	83.8	83.9	83.9	83.9
Gas condensate with the addition of 1% vol. isopropanol											
RON	93.2	93.8	93.8	93.8	93.7	93.9	93.6	94.0	94.0	94.2	94.1
MON	83.8	84.0	84.0	84.0	83.9	84.0	83.9	84.0	84.1	84.2	84.1
Gas condensate with the addition of 1.5% vol. isopropanol											
RON	94.0	93.9	94.1	94.2	94.2	94.4	94.4	94.5	94.2	94.5	94.6
MON	84.1	84.0	84.1	84.2	84.2	84.4	84.4	84.5	84.2	84.5	84.6
Gas condensate with the addition of 2% vol. isopropanol											
RON	94.6	94.6	94.7	94.7	94.7	94.7	94.8	94.9	94.9	94.9	95.0
MON	84.6	84.6	84.7	84.7	84.7	84.7	84.8	84.9	84.9	84.9	85.0
Gas condensate with the addition of 2.5% vol. isopropanol											
RON	94.9	94.9	94.8	95.0	95.0	95.0	95.0	95.1	95.5	95.7	95.8
MON	84.9	84.9	84.8	85.0	85.0	85.0	85.0	85.1	85.5	85.7	85.8
Gas condensate with the addition of 3% vol. isopropanol											
RON	95.0	95.0	95.1	95.1	95.4	95.1	95.4	95.6	95.7	95.8	95.9
MON	85.0	85.0	85.1	85.1	85.4	85.1	85.4	85.6	85.7	85.8	85.9
Gas condensate with the addition of 3.5% vol. isopropanol											
RON	94.9	95.3	95.5	95.7	95.7	95.8	95.8	95.8	95.8	95.8	95.8
MON	84.9	85.3	85.5	85.7	85.7	85.8	85.8	85.8	85.8	85.8	85.8

Gas condensate with the addition of 4.0% vol. isopropanol											
RON	95.8	96.0	96.1	96.1	96.2	96.2	96.2	96.2	96.3	96.4	96.3
MON	85.8	86.0	86.1	86.1	86.2	86.2	86.2	86.2	86.4	86.5	86.3
Gas condensate with the addition of 5.0% vol. isopropanol											
RON	96.5	96.5	96.7	96.7	96.8	97.0	97.1	97.0	97.1	97.0	97.1
MON	86.7	86.8	87.1	87.1	87.2	87.2	87.2	87.2	87.2	87.2	87.2
Gas condensate with the addition of 7.0% vol. isopropanol											
RON	97.6	97.9	97.9	98.0	98.0	98.1	98.1	98.1	98.2	98.3	98.5
MON	87.6	87.9	87.9	88.0	88.0	88.1	88.1	88.1	88.3	88.4	88.7
Gas condensate with the addition of 10.0% vol. isopropanol											
RON	101.5	101.7	101.7	101.8	102.0	102.0	102.0	102.1	102.2	102.2	102.1
MON	91.2	91.4	91.4	91.5	91.7	91.7	91.7	91.7	91.9	91.9	91.7
1	2	3	4	5	6	7	8	9	10	11	12
Gas condensate with the addition of 12.0% vol. isopropanol											
RON	104.3	104.2	104.3	104.3	104.4	104.5	104.5	104.6	104.6	104.6	104.5
MON	93.8	93.7	93.8	93.8	93.9	94.0	94.0	94.1	94.1	94.1	94.0

Table 14.4 Change of octane number by the research method (RON) and by the motor method (MON) before and after cavitation treatment depending on the content of isopropanol in the gas condensate

№	The content of isopropanol in gas condensate, % vol.	RON		MON		Increment of octane number, points	
		Before	After	Before	Before	After	Before
1	0	92.3	92.88	82.3	82.8	0.58	0.50
2	0.5	93.0	93.53	83.8	83.88	0.53	0.08
3	1.0	93.2	94.08	83.8	84.1	0.88	0.3
4	1.5	94.0	94.45	84.1	84.45	0.45	0.35
5	2.0	94.6	94.93	84.6	84.43	0.33	0.33
6	2.5	94.9	95.53	84.9	85.53	0.63	0.63
7	3.0	95.0	95.75	85.0	85.75	0.75	0.75
8	3.5	94.9	95.8	84.9	85.8	0.9	0.9
9	4.0	95.8	96.3	85.8	86.35	0.5	0.55
10	5.0	96.5	97.05	86.7	87.2	0.55	0.5
11	7.0	97.6	98.28	87.6	88.38	0.68	0.78
12	10.0	101.5	102.15	91.2	91.8	0.65	0.6
13	12.0	104.3	104.58	93.8	94.08	0.27	0.27

increases, the values (RON) and (MON) first increase, reach a certain value, and remain constant during subsequent cavitation cycles. That is, there is a certain amount of energy introduced into the system, at which the equilibrium state of the system is achieved. Most likely, equilibrium concentrations of substances are established.

After seven to eight cycles of cavitation treatment (Table 14.3), a constant value (RON) and (MON) is achieved, which does not increase during 10 treatment cycles or more. In Fig. 14.4, you can see that 3% vol. IPA is enough to create an indicator (RON) of 95 points for mechanical mixing of gas condensate gasoline and IPS, which is sufficient for the production of gasoline brand A-95 according to DSTU 7687: 2015. The use of cavitation treatment (more than seven cycles at a pressure of 9.0 MPa) allows you to reduce the number of IPA, sufficient to achieve an indicator (RON) of 95 points, up to 2.5% vol.

The situation is almost similar, with the achievement (RON) of 98 points. Mechanical mixing requires the addition of IPA in the amount of 8.1% vol. And the use of cavitation in the amount of four to seven cycles can reduce the amount of IPA to 7.0% vol. Extrapolating the results to the scale of an industrial plant shows that the use of cavitation will reduce the cost of raw materials – IPS – by 17%. The required amount of IPA in the mixture will decrease from 3.0% vol. up to 2.5% vol. in the production of gasoline brand A-95. For the production of A-98 gasoline, the reduction will be 14–8.1% vol. up to 7.0% vol.

A constant value of the octane number is achieved in six to eight cycles of cavitation treatment (Fig. 14.4), which is slightly more than the required number of cycles of cavitation in the modification of gas condensate gasoline with

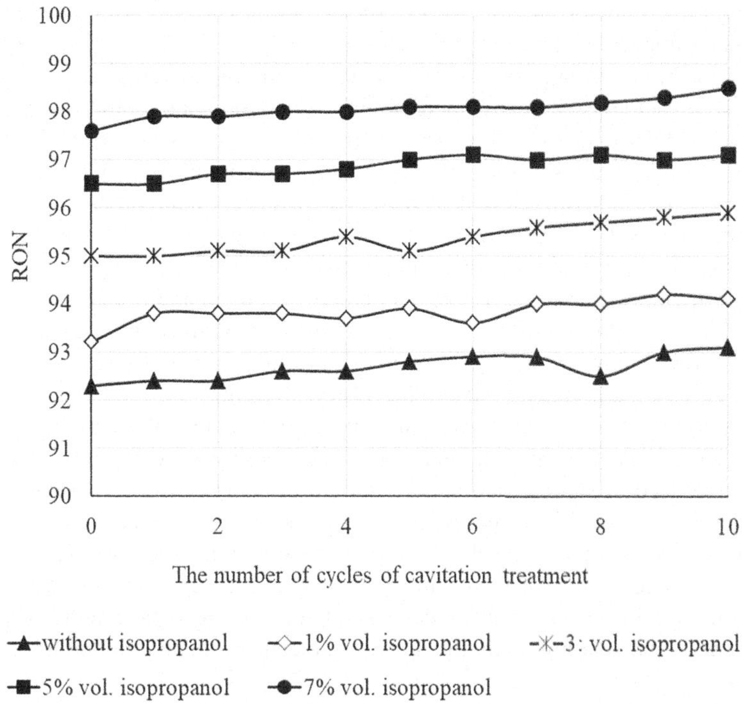

Fig. 14.4 The effect of the number of cycles of cavitation treatment on the octane number of gas condensate gasoline modified with isopropanol (Leonenko et al. 2020)

bioethanol. That is, the use of isopropanol requires the supply of more energy to the reaction zone.

For high concentrations of IPA, there is a tendency to further a small increase in octane number with increasing intensity of cavitation treatment. But carrying out 30 cycles of cavitation treatment did not show an increase in octane number higher than that achieved in eight cycles.

14.7 Cavitation Treatment of a Mixture of Gas Condensate Gasoline with Isobutanol of Different Concentrations

Isobutanol can become a real competitor to methanol and bioethanol as a high-octane additive due to its relatively low cost. The use of biobutanol, which is one of the main components of fusel oils, will further reduce the cost of production of gasoline A-95 and A-98. Biobutanol is sparingly soluble in hydrocarbons, but its use will stabilize the alcohol-gasoline mixture under conditions of water in this mixture. Biobutanol has a lower specific oxygen content in the molecule, so biobutanol-based fuels will have a higher specific energy consumption than fuels with bioethanol and isopropanol impurities. Experimental data are shown in Tables 14.5 and 14.6.

It should be noted that the use of isobutanol requires more cycles of cavitation treatment to achieve a steady increase in octane number than bioethanol and isopropanol. For industry, this will mean higher energy consumption for the production of similar brands of fuel. But the prospect of isobutanol is that higher energy costs for cavitation treatment will be offset by lower raw material costs.

During this series of experiments, a significant increase in the temperature of the mixture of gas condensate gasoline with alcohol was observed after 15–25 cycles of cavitation treatment. The increase in temperature led to an additional increase in the octane number of the mixture, so visually there is a significant increase in octane number immediately after cavitation treatment and a gradual decrease in octane number by 0.6–1.2 points due to cooling of the mixture.

In Fig. 14.5, it is clear that a constant value of the octane number is achieved after eight to ten cycles of cavitation treatment. Among the studied alcohols, isobutanol showed the need to supply the greatest amount of energy through cavitation treatment to achieve a lasting effect of increasing the octane number. If we compare bioethanol, isopropanol, and isobutanol, it can be noted that the lower the molecular weight of alcohol, the less energy must be supplied to achieve a lasting effect of cavitation. But in absolute terms, the largest increase in octane number showed bioethanol – the alcohol with the lowest molecular weight among the studied alcohols.

From Fig. 14.5, it is clear that the main effect is achieved in the range of six to ten cycles of cavitation treatment. That is, there is some optimal amount of energy supplied to the reaction zone by mechanical means, in which the system undergoes irreversible chemical transformations and changes in the octane number.

Further research will be aimed at studying the effect of the intensity of cavitation treatment on the octane number of gas condensate gasolines modified with mixtures of alcohols. It is also planned to study the modification of gas condensate gasoline with fusel oils and other by-products of bioethanol production. Of interest is the involvement in the production of high-octane gasolines of any alcohols of biochemical origin.

In the future, this will allow the production of A-95 and A-98 gasolines on an industrial scale with the addition of the least possible amount of alcohol and reducing the cost of alcohol.

Table 14.5 Change of octane number (ON) by research (RON) and motor (MON) methods from the number of cycles of cavitation treatment for gas condensate with the addition of isobutanol

	The number of cycles cavitation treatment													
ON	0	1	2	3	4	5	6	7	8	9	10	15	20	25
1	2	3	4	5	6	7	8	9	10	11	12	13	14	15
Gas condensate without the addition of isobutanol														
T	25.3	24	24	24.5	24.7	25.1	25.3	26	25.9	26.4	27	27.4	27.9	28.6
RON	93.4	94	94	94.1	94.3	94.4	94.3	95	94.7	94.7	94.8	94.9	95	95
MON	83.8	84	84	84.1	84.3	84.4	84.4	85	84.7	84.7	84.8	84.9	85	85
AKI	88.8	89	89	89.1	89.3	89.4	89.6	90	89.8	89.8	90	90.3	90.7	90.7
Gas condensate with the addition of 0.5% isobutanol														
T	26.5	26	27	27.2	28.1	28.8	29.3	30	30.5	30.8	30.8	30.8	30.6	30.2
RON	94.7	95	95	94.8	94.9	94.9	95.1	95	95.7	95.8	95.8	95.9	95.9	95.9
MON	84.7	85	85	84.8	84.9	84.9	85.1	85	85.7	85.7	85.8	85.9	85.9	85.9
AKI	89.7	90	90	89.8	89.9	90	90.1	90	90.7	90.8	90.8	90.9	90.9	90.9
Gas condensate with the addition of 1.0% isobutanol														
T	25.8	26	26	26.8	27.4	28.4	29.4	30	30.2	30.8	30.9	31.1	31.6	31.7
RON	94.8	95	95	94.9	95	95.1	95.6	96	95.8	95.9	95.9	96.2	96.2	96.2
MON	84.8	85	85	84.9	85	85.1	85.1	86	85.7	85.8	85.9	85.9	86.1	86.2
AKI	89.8	90	90	89.9	90	90.1	90.6	91	90.8	90.9	90.9	91.1	91.2	91.2
Gas condensate with the addition of 3.0% isobutanol														
T	29.5	29	30	30.1	30.4	31	31.6	32	32.5	32.8	33.5	33.5	33.5	33.6
RON	96.4	96	96	96.5	96.9	96.9	97.3	97	97.4	97.5	97.6	97.6	97.7	97.7
MON	86.5	87	87	86.8	87.1	87.2	87.3	87	87.4	87.5	87.6	87.6	87.7	87.7
AKI	91.6	92	92	91.8	92	92.2	92.3	92	92.4	92.5	92.6	92.6	92.7	92.7
Gas condensate with the addition of 5.0% isobutanol														
T	31.5	30	30	30.5	30.9	31.6	32.4	33	33.3	33.6	34	34.1	34.4	34.1

(continued)

Table 14.5 (continued)

	The number of cycles cavitation treatment													
ON	0	1	2	3	4	5	6	7	8	9	10	15	20	25
1	2	3	4	5	6	7	8	9	10	11	12	13	14	15
RON	97.8	98	98	97.8	97.9	98	98.3	98	98.7	99	99.1	99.1	99.2	99.2
MON	87.8	88	88	87.8	87.9	88	88.4	89	88.9	89.1	89.1	89.2	89.2	89.2
AKI	92.8	93	93	92.8	92.9	93	93.4	94	93.8	94	94.1	94.1	94.2	94.2
Gas condensate with the addition of 10.0% isobutanol														
T	25.6	26	26	26.3	26.7	27.8	28.4	29	29.6	30	30	28.1	28.7	29.8
RON	101.3	101.3	101.3	101.5	101.7	101.8	102	102.2	102.4	102.4	102.6	102.6	102.6	102.6
MON	91	91	91	91.4	91.5	91.7	91.8	92	92	92.1	92.2	92.2	92.2	92.2
AKI	96.2	96	97	96.7	96.8	97	97.1	97	97.3	97.3	97.4	97.4	97.4	97.4
Gas condensate with the addition of 15% isobutanol														
T	28.9	29	28	28.6	28.9	30.1	30.1	30	30.6	30.5	30.6	30.8	31	31
RON	105.6	105.6	105.7	105.8	105.9	106	106.1	106.2	106.3	106.4	106.5	106.6	106.5	106.5
MON	95.5	96	96	95.8	95.9	96.1	96.1	96	96.3	96.5	96.5	96.6	96.6	96.6
AKI	100.5	100.6	100.7	100.8	100.9	101	101.1	101.2	101.3	101.4	101.4	101.4	101.4	101.4

Table 14.6 Change of octane number by the research method (RON) and by the motor method (MON) before and after cavitation treatment, depending on the content of isobutanol in the gas condensate

№	The content of isobutanol in gas condensate, % vol.	RON		MON		Increment of octane number, points	
		Before	After	Before	After	Before	After
1	0	93.4	94.97	83.8	84.97	1.57	1.17
2	0.5	94.7	95.90	84.7	85.90	1.2	1.2
3	1.0	94.8	96.20	84.8	86.07	1.40	1.27
4	3.0	96.4	97.67	86.5	87.67	1.27	1.17
5	5.0	97.8	99.17	87.8	89.20	1.37	1.40
6	10.0	101.3	102.60	91	92.20	1.3	1.2
7	15.0	105.6	106.53	95.5	96.60	0.93	1.1

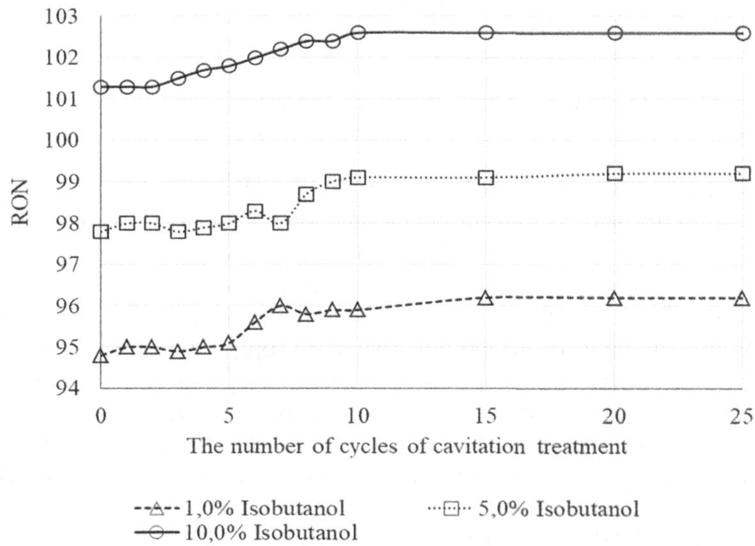

Fig. 14.5 The effect of the number of cycles of cavitation treatment on the octane number of gas condensate gasoline modified with isobutanol (Tselishchev et al. 2020a, b)

14.8 Comparison of the Effectiveness of the Effect of Cavitation Treatment on the Increase in the Octane Number of Gas Condensate Gasoline with the Addition of Bioethanol and Isopropanol

Graphical dependences of the change in octane number by the research method for gasoline with different amounts of added bioethanol and isopropanol are shown in Fig. 14.6. When using bioethanol, the effect of cavitation is much more pronounced

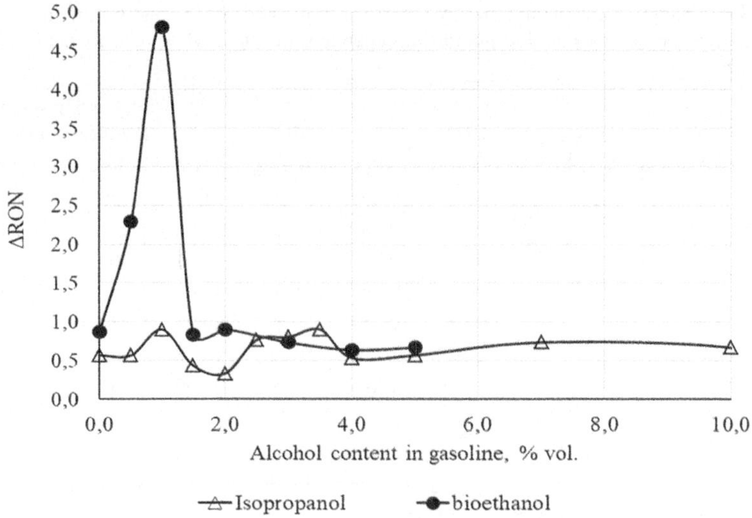

Fig. 14.6 Change of octane number by the research method during cavitation treatment of gas condensate gasoline at different concentrations of alcohols (Tselishchev et al. 2020a, b)

at low concentrations of alcohol – up to 1% vol. This can be due to both the intensification of chemical transformations and the homogenization of the emulsion formed when water from bioethanol enters the gasoline.

It should be noted that the phenomenon of resonant influence of the intensity of mechanical activation, which includes cavitation, was observed for other technologies that use mechanical methods of selective supply of energy to the reaction centers. For example, in a method known as "aerosol nanocatalysis" (Leonenko 2017), the catalyst is activated by changing the oscillation frequency of the vibrorector. Studies of various hydrocarbon processing processes by this method have shown the presence of anomalous maximum dependences of the effective constant of the rate of cracking reactions at certain oscillations, as well as changes in the yield of individual products.

For cavitation treatment, the resonance phenomenon can be explained by creating the optimal concentration of activated particles. The rate of consumption of activated particles in the target reactions exceeds the rate of consumption of side reactions. As the concentration increases, nontarget reactions, such as n-alkane production reactions and compaction reactions, accelerate.

Figure 14.6 illustrates the possibility of establishing the optimal concentrations of alcohol, at which the technology of cavitation processing of raw materials gives the maximum effect. The initial octane number of the gasoline fraction can be changed by combining it with high-octane fractions. Further addition of IPA or other alcohols will allow to organize the process of obtaining gasoline A-95 and A-98 with minimal energy consumption and consumption of raw materials.

14.9 Recommendations for the Construction of an Experimental Industrial Plant for the Modification of Gasoline with Alcohols in the Cavitation Field

The research and industrial plant must ensure the mixing of the gasoline fraction with alcohol in the required proportions, homogenization of the mixture, treatment in a cavitation field, and separation (if necessary) of the gas and water phases from commercial gasoline. It is planned that the installation (Fig. 14.7) will work in periodic mode. At the installation, it will be possible to replace the nozzles to change the intensity of cavitation treatment. The plant will be able to use as raw materials any alcohols or mixtures thereof. Hydrocarbon raw materials can be both gas condensate gasoline and gasoline fractions from refineries.

One period of operation of the installation will include the following stages: (1) dosing the required amount of gasoline fraction from T1 and alcohol from T2 in the tank-mixer T3; (2) mixing the mixture with a mechanical stirrer M; (3) heating (in winter) to a temperature of 308–323 K in the heat exchanger T; (4) cavitation treatment of the mixture in the cavitator Cav, the required pressure on the nozzle is created by a high pressure pump P; and (5) degassing and separation of the aqueous phase in separator S. To achieve the desired values of octane number, the alcohol-gasoline mixture may be subject to several cycles of cavitation treatment – for this recycling of raw materials provided by the circuit "separator S – pump P – cavitator Cav – separator S". On the scale of the research and industrial installation, it will be

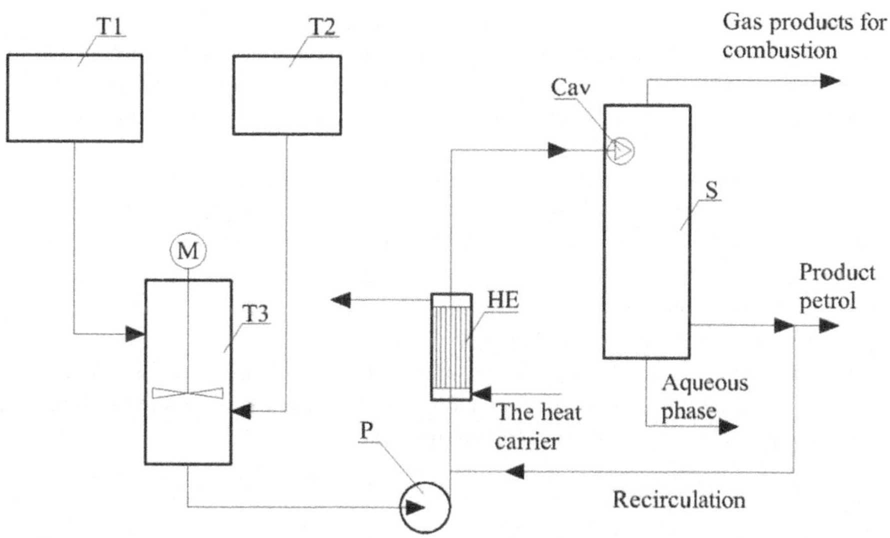

Fig. 14.7 Schematic diagram of a pilot plant for the process of modification of gasoline with alcohols with simultaneous cavitation treatment of the mixture (Tselishchev et al. 2017). T1, tank with gasoline fraction; T2, container with alcohol; T3, tank-mixer; M, mechanical drive of the stirrer; P, high pressure pump; HE, heat exchanger; Cav, cavitation nozzle; S, separator

possible to optimize the technological regime by varying the intensity of cavitation treatment and the composition of the alcohol-gasoline mixture to achieve maximum economic effect when market prices for raw materials and electricity.

14.10 Conclusions

In this paper, the relevance of the introduction into gasoline production of such chemical technologies that use cavitation processing of raw materials and selective energy supply for the reaction zone was substantiated. The expediency of production of high-octane gasolines on the basis of a combination of processes of mechanical mixing of hydrocarbon gasolines with alcohols and processes of cavitation processing of alcohol-gasoline mixes is also substantiated.

The influence of the intensity of cavitation treatment on the increase of the octane number is studied, and it is proved that there is some optimal intensity at which a constant value of the octane number of the mixture is achieved. As the bioethanol content of the mixture increases, the number of cavitation cycles (intensity) required to achieve a constant octane value decreases from eight cycles of gas condensate without bioethanol to four cycles with a bioethanol content of 3% and above. To achieve the octane values of the mixture corresponding to gasolines of grades A-92 and A-95, it is necessary to add 2% and 5% of bioethanol, respectively.

The most effective for the action of cavitation was a mixture of 99% gas condensate and 1% bioethanol, when the total increase in RON was 5 points and MON was 2.6 points. A further increase in the amount of bioethanol in the mixture reduces the effect of cavitation treatment, although the initial RON and MON are much higher. The higher the molecular weight of the alcohol, the more cycles of cavitation must be performed to achieve a steady increase in octane number.

The effect of IPA concentration on the octane number of modified gasoline under cavitation treatment is nonlinear in nature with several maxima at certain concentrations – 1.0% vol. and 3.5% vol. and 6.5% vol. IPA for the indicator (RON) and 3.5% vol. and 6.5% vol. IPA for indicator (MON). By varying the initial concentration of IPA and the octane number of the gasoline fraction, it is possible to optimize the technological mode of production of gasoline A-95 and A-98 in terms of raw material consumption and energy consumption.

All studied alcohols showed the possibility of modification of gas condensate gasoline in the cavitation field with an additional increase in octane number by 1.2–2 points compared with simple mechanical mixing.

References

Boichenko SV, Lanetskyi VH, Cherniak LM, Radomska MM, Kondakova OH (2017) Doslidzhennia vplyvu kavitatsiinoi obrobky na oktanove chyslo avtomobilnoho benzynu (research of cavitation influence on automobile gasoline octane number). Enerhetyka: ekonomika, tekhnolohii, ekolohiia 2:107–114

Kudryavtsev S, Tselishchev A, Leonenko S, Boichenko S, Loria M (2020) Determining the influence of cavitation treatment on the octane number of gas-condensate gasoline modified with isopropanol. East-Eur J Enterp Technol 6(6–108):116–123

Leonenko SV (2017) Study of catalytic cracking process of fuel oil to obtain components of motor fuels using aerosol nanocatalysis technology S.V. Leonenko, S.A. Kudryavtsev, I.M. Glikina. Adsorpt Sci Technol 35:878–883

Milotskyi VV, Hanzha SN (2016) Sposib pidvyshchennia oktanovoho chysla hazokondensatnykh i naftovykh priamohonnykh benzyniv (method of increasing octane number of gas condensate and oil straight-run gasolines). Visnik of the Volodymyr Dahl East Ukrainian National University 14(203):85–88

Tselishchev OB, Loriia MG, Kudriavtsev SO (2020a) Sposoby peretvorennia n-alkaniv u vysokooktanovi komponenty motornykh palyv: monohrafiia (methods of conversion of n-alkanes into high-octane components of motor fuels: monography). Publishing center of V. Dahl East Ukrainian National University, Sievierodonetsk

Tselischev OB, Kudryavtsev SO, Loriya MG et al (2020b) Modification of motor gasoline with bioethanol in the cavitation field. Voprosy Khimii i Khimicheskoi Tekhnologii 6:171–178

Tselishchev OB, Loriia MH, Boichenko SV et al (2017) Transformation of n-alkanes of gasoline into components of motor fuels in cavitation field. TEKA Commission of Motorization and Energetics in Agriculture 17(2):25–30

Index

A

Additive, 3, 20–22, 24, 25, 28, 37, 38, 40–42, 44, 129, 150, 159, 213, 214, 218, 222, 224, 227, 228, 232, 243, 251, 263–265, 267, 268, 270, 271, 276, 281
Adenosine triphosphate, 130
Agriculture, 87–89, 128, 156
Alcohol, 10, 11, 16, 20, 50–79, 102, 135, 187–206, 215–217, 267–288
Alfalfa, 154, 155
Alkaline transesterification, 51, 52, 64, 78, 79
Alternative biofuel, 190
Alternative fuel, 2, 4, 149, 209, 222, 243
Ammonium nitrogen, 96, 97
Asphalt-resin-paraffin deposits, 247
Aviation fuel, 2, 4–11, 13, 14, 17, 127, 128, 232–236, 240–244

B

Bacteria, 83, 136, 154–156, 210–212, 216, 217, 219, 221–224, 232, 237, 242
Biocidal additive, 213–215, 218, 222, 224, 226–228
Biodiesel, 50, 52, 53, 57, 65, 67, 79, 80, 87, 90–92, 94, 97, 128, 154
Biodiesel purification, 78
Bioethanol, 50, 54, 60, 127, 128, 268, 270–272, 274–277, 280–282, 285, 286, 288
Biofertilizer, 88–89, 97
Biofuel, 2, 3, 12, 13, 17, 50, 81–97, 127–151, 212, 241, 243
Biogenic element, 93, 94, 96, 97
Biopollution, 244
Boundary lubricant film, 35–37, 40
Butoxide, 52, 55–57, 59, 74, 79
Butyl esters, 50, 52, 75, 76, 78, 79

C

Camelina, 4, 6, 7, 127–151
Carbon dioxide, 2, 91–93, 95–97, 101–103, 109, 121, 124, 128, 135
Catalytic hydrothermolysis, 14
Cavitation treatment, 267–288
Cavitation treatment intensity, 274, 276, 281, 287, 288
Certification, 3, 4, 12, 14, 16, 17
Coal, 5, 102, 149, 167–185
Combustion dynamics, 176

D

Decarbonization, 2, 17
Dependence, 20, 33, 35, 46, 106, 115, 119, 161, 169–173, 181, 195, 196, 201–204, 250, 252, 253, 259–261, 263, 285, 286
Deposits, 55, 210, 211, 239, 247–265
Detection, 214, 241, 242
Diesel engine, 187–206
Diesel fuel, 87, 101, 149, 200–202, 209, 212, 216, 218, 219, 221–223, 225, 228, 233, 236
Dissolution, 26, 59, 216, 247–265

S. Boichenko et al. (eds.), *Chemmotological Aspects of Sustainable Development of Transport*, Sustainable Aviation, https://doi.org/10.1007/978-3-031-06577-4

E
Ecological and economic estimation, 185
Economic efficiency, 20, 90–92, 156, 157
Emission, 2, 3, 6–9, 17, 93, 95–97, 101–103, 121, 128, 181–183, 190, 191, 270–272
Energy efficiency, 130, 140, 148, 150
Environmental safety, v, vii
Ethoxide, 56, 57, 60, 61, 64, 78, 79
Ethyl esters, 55, 56, 61, 63, 65–67, 70–74, 78, 79, 87, 88

F
Fatty acid methyl, 50, 86, 209
Feedstocks, 2, 4–9, 11–16, 51, 53, 54
Fermentation, 11, 88, 89, 92, 129–132, 134–143, 148, 150, 151
Ferroalloys, 168, 176–181, 185
Fischer-Tropsch (FT) process, 4
Free fatty acids, 51, 60, 78, 79
Friction, 31–33, 35–46, 142, 145, 211
Fuels and lubricants, 210, 231
Fuel system, 5, 10, 210–213, 221, 224, 231–244
Functional bread, 127–151
Fungi, 83, 156, 210–214, 216–218, 220–222, 224, 225, 232–234, 237, 242

G
Gas condensate gasoline, 267–288
Glycerol, 51–53, 55, 56, 58, 59, 61–65, 67–80

H
Heat utilization, 198, 204, 205
Heavy oils, 20, 21, 26, 154
Hydrocarbons, 3, 5–9, 11–16, 26, 42, 43, 87, 101–121, 153–155, 167, 191, 193, 197, 203, 204, 211, 212, 214, 231–234, 236–238, 243, 248, 249, 251–255, 259–261, 263–265, 267–272, 274, 276, 281, 286–288
Hydrotreated esters and fatty acids (HEFA), 4, 6–8, 14

I
Isobutanol, 268, 271, 281–285
Isopropanol, 268, 271, 278–282, 285

J
Jet fuel, 3–7, 9–16, 231–244

K
Kinematic viscosity, 22, 26, 38, 39, 50, 54, 60, 66, 68, 105, 210
Kinetics, 31, 32, 38, 39, 43, 44, 46, 108, 109, 111–113, 115, 117, 121, 170–173, 205, 247–265

L
Lactic acid bacteria, 128–133, 135–140, 151
Lactulose, 132, 133, 135–139, 151
Lifting force, 134, 135, 139
Local temperature, 37–41

M
Methanol conversion, 102, 114–118, 121, 197
Microalgae, 81–97
Microbiological pollution, 209–224, 240, 242
Microbiological stability, 242, 244
Microorganisms, 8, 11, 88, 127, 132, 142, 150, 155, 210–218, 222, 224, 231–238, 241–243
Microstructure, 44
Mineral fibers, 108

N
Nitrates, 81, 96, 97, 164
Nitrogen oxides, 8, 181, 191, 202, 203, 205

O
Octane, 190, 191, 268, 270–278, 280–283, 285–288
Oil, 3, 4, 6, 7, 19–22, 26, 28, 36, 38, 40, 42–44, 46, 50–64, 66–70, 72, 73, 75–79, 87, 92, 101, 102, 149, 153–165, 188, 191, 209, 211, 219, 220, 222, 223, 225, 227, 228, 231, 232, 234, 247–265, 272
Oil contamination, 154, 155, 234
Oil fraction, 249, 253, 254, 260–265
Oil transportation, 247

P
Petroleum products, 154, 155, 232, 234, 235, 248, 269, 274
Phosphates, 81, 93, 96, 97, 135
Physicochemical properties of oil, 22
Polymers, 15, 20–22, 24–26, 28, 66, 72, 82, 85
Power, 102, 120, 132, 141–145, 147, 148, 154, 155, 180, 181, 211, 237
Prevention, 129, 175, 214, 242

Properties, 2, 5–10, 12–16, 20, 22, 26, 28, 31, 32, 37, 40–42, 44, 50, 56, 65, 66, 88, 114, 128, 131, 155–157, 161–165, 167, 172, 177, 178, 206, 209–224, 232, 236, 238, 239, 243, 248, 249, 254, 276

R
Rapeseed oil, 52–54, 73, 77, 79, 219, 221–223, 225, 227, 228
Rate constants, 109, 112, 117, 121, 250, 258–263, 265
Redox potential, 132, 133
Reduction of oil viscosity, 21
Rheological properties, 20–25, 130, 131, 137

S
Sensors, 158, 160–164, 200, 240
Shear rate gradients, 38–40, 42, 44
Shear stress, 20–24, 28
Sliding, 36–40, 42–46, 142
Smokeless solid fuel, 167–185
Soil, 88, 153–165, 210, 232, 234
Soil management, 155
Soil parameters, 161, 162, 164
Solvent, 27, 52, 55, 248–256, 259–262, 264, 265
Soy, 6
Specific fuel consumption, 198, 202, 205
Sustainable aviation fuel, 2–4
Synthesized isoparaffins, 4, 8
Synthesized kerosene with aromatics, 12
Synthesized paraffinic kerosene, 10

T
Temperature of destruction, 38, 42
Thermocyclic ionic nitriding, 33, 38, 39, 43–46
Tribological characteristic, 35, 36

V
Vacuum distillation, 52, 56, 65–70, 72, 73, 78, 79

W
Wasted frying oils, 53, 60, 66, 67, 73–75, 77, 79
Wear, 31–35, 38, 42–47, 177, 191, 211

Y
Yeast, 9, 130, 132, 135, 136, 215, 218, 242

The manufacturer's authorised representative in the EU is Springer Nature Customer Service Centre GmbH, Europaplatz 3, 69115 Heidelberg, Germany. If you have any concerns regarding our products, please contact ProductSafety@springernature.com

Printed and bound by CPI Group (UK) Ltd, Croydon, CR0 4YY
28/09/2026
02224093-0005